Smart Grids and Big Data Analytics for Smart Cities

Chun Sing Lai • Loi Lei Lai • Qi Hong Lai

Smart Grids and Big Data Analytics for Smart Cities

 Springer

Chun Sing Lai
Department of Electrical Engineering
Guangdong University of Technology
Guangzhou, China

Department of Electronic and Computer
Engineering
Brunel University London
London, UK

Qi Hong Lai
Sir William Dunn School of Pathology
University of Oxford
Oxford, UK

Loi Lei Lai
Department of Electrical Engineering
Guangdong University of Technology
Guangzhou, China

ISBN 978-3-030-52154-7   ISBN 978-3-030-52155-4   (eBook)
https://doi.org/10.1007/978-3-030-52155-4

This Springer imprint is published by the registered company Springer Nature Switzerland AG
The registered company address is: Gewerbestrasse 11, 6330 Cham, Switzerland

Preface

This book was written in response to the increasing interest in smart city technology and its deployment worldwide. There is a strong belief that smart city technology will produce an all-win solution with regard to environmental, social, and economic impact.

Major environmental, economic, and technological challenges such as: climate change; economic restructuring; pressure on public finances; digitalization of the retail and entertainment industries, and the growth of urban and ageing populations has generated huge interest for cities to be run differently and smartly. The term "smart city" was coined to describe such cities, and they promise a significant improvement in the quality of life of its citizens through the combination of information and communication technology (ICT), new services and improved city infrastructure. The evolutionary process in the development of a smart city is mainly driven by an innovative, user-centric vision—specifically by tackling urban issues from the perspective of citizens and taking into account their need to engage with city management and planning. The approach is based on emerging technology, whereby the solution obtained through integration of human and social capital allows their significant interaction as it is adopted to a city. The application of the Internet of Things (IoT) to city operation is of special interest to support the aim of efficiently transforming cities to acquire substantial and sustainable development as well as high quality of life.

The mission of building smart cities is based on achieving improved utilization of renewable energy, safeguarding of the environment, and waste reduction. At the same time, fostering cohesion between citizens to obtain shared benefits derived from the eco-sustainability vision which is headed by effective industrial and urban development to allow pressing needs to be met without compromising the imminent generations' capacity. When considering an eco-sustainable method, practicality is essential in the various facets and at different layers of the development process such as environment, social services, and mobility. A smart city employs various kinds of electronic IoT sensors to amass data and such data is used to control resources and assets efficiently. The data is often sourced from devices, assets, and citizens and is processed and studied to then understand and improve crime detection, transportation and traffic systems, water supply networks, hospitals, information systems, waste management, power plants, and additional community services.

Smart city is now a popular term; however, its definition and specifics remain unclear. This has led to different interpretations of a smart city. Most commonly, a smart city can be described by six important pillars, namely smart people (social and human capital), smart living (quality of life), smart economy (competitiveness), smart mobility (transport and ICT), smart governance (participation), and smart environment (natural resources). Smart city programs and technologies have now been developed in many cities worldwide including London, New York, Hong Kong, Singapore, Paris, Tokyo, Amsterdam, Barcelona, Dubai, Stockholm, and Copenhagen—some of which will be discussed in more detail in case studies.

This book focuses on delivering comprehensive and detailed analysis of the following areas of smart cities: smart energy, smart mobility, smart health, and smart water. The purpose is to inform the reader firstly through more general but comprehensive coverage of the concept of smart cities before diving into more specific areas without excessive specialization as to avoid merely not only presenting qualitative data and numerical techniques, but also providing, where feasible, practical case studies and project discussions.

Chapter 1 discusses what a smart city should be. In this chapter, characteristics, functionality, and domain of smart city will be explained. Different elements of a smart city, such as smart energy, smart water, smart health, smart infrastructure, and big data analytics will be examined. Case studies will be used to demonstrate the work done to help to establish a smart city deployment and some benefits derived from the effort spent. Some examples of smart cities worldwide will be reported. Challenges and opportunities derived from future smart city projects will be discussed.

Due to the need to use a large number of renewables in the near future and the requirement to have a stable energy system, Chapter 2 covers data analytics for solar energy in promoting smart cities. In this chapter, a comprehensive review of high penetration of photovoltaic (PV) and an overview of electrical energy storage (EES) for PV systems is presented. Solar power forecasting techniques for operation and planning of PV and EES are included. A deterministic approach for sizing PV and ESS with anaerobic digestion (AD) biogas power plant is developed to achieve a minimal levelized cost of energy (LCOE). The aim is to minimize energy imbalance between generation and demand due to AD generator constraint and high penetration of PV. For data analytics, the chapter presents the issues in correlation analysis due to imbalanced data and data uncertainty in real-life solar data. A robust correlation framework was proposed and tested on real-life solar irradiance and weather condition data. For solar data cluster analysis, a novel method with Fuzzy C-Means with dynamic time warping distance was proposed to determine patterns in daily clearness index (CI) profiles. CI profiles could be varied significantly in different seasons.

Based on high security, transparency, tamper-proof, and decentralization, blockchain is suitable for microgrids with high renewable penetration and advanced supervisory control and data acquisition (SCADA) sensors as there is a need for a new market approach to facilitate the power generation and load balance and make the optimal use of low carbon energy sources. Chapter 3 presents blockchain applications in microgrid clusters. Microgrids with blockchain can give a more resilient, cost-efficient, low-transmission-loss, and environment-friendly grid. Smart contract-based hybrid peer-to-peer (P2P) energy trading model with cryptocurrency named localized renewable energy certificate (LO-REC) will be discussed. The advantages and challenges of combining blockchain with microgrids are identified. This chapter serves as a guide for future research on blockchain applications in microgrids.

Water management is a critical task and impacts on the environment and economics. Chapter 4 deals with a time-synchronized ZigBee building network for smart water management. It is essential for the development of a flexible, reliable, and scalable sensor network to install and replace water sensors in buildings. Wireless communication will be of utmost importance. Nevertheless, incorrect network time synchronization will create packet loss and long latency which reduces the network performance. In this chapter, time-synchronized ZigBee building network is proposed for water management according to the node-to-node time synchronization. The simulation result shows that the mean synchronization error and variance are reduced. Also, an interference-mitigated ZigBee-based advanced metering infrastructure solution was created for high-traffic smart metering.

Without energy, any city cannot be in proper operation. Also, for the convenience of the citizens, electrical vehicles would be needed. As a result, there will be many batteries within a city. However, this could give risk to human and it is essential to minimize the damage. Chapter 5 reports a narrowband internet of things (NB-IoT)-based temperature prediction for valve-regulated lead-acid battery (VRLA). Due to its huge market, VRLA gained a significant part in industries. However, VRLA safety has been a wide concern since it is prone to self-heating problems which generate extra cost or even cause accidents when the internal temperature (IT) of VRLA is exceeded. To prevent potential hazards, effective internal VRLA temperature monitoring methods are required. In the method, the internal temperature is estimated by ambient temperature (AT) and input current (IC) through a pre-trained prediction model. The measured temperature data will be sent to the backend server using NB-IoT. A kind of recurrent neural network, namely nonlinear autoregressive exogenous is applied to determine the potential relationship between the input AT, IC, and the output IT.

It is learnt that over 60% of adult drivers experienced sleepiness during driving and over 40% of traffic accidents are created by intoxicated drivers. Chapter 6 reports a health detection scheme for drunk drivers. The integration of the wearable sensors facilitates the real-time monitoring of human conditions under different scenarios including patient tracking and human safety. In this chapter, an electrocardiogram (ECG)-based status of human detection (ECG-HSD) scheme was proposed to sense both drowsy and intoxicated conditions. In ECG-HSD scheme, resemblances of ECG signals during ordinary, drowsy, and intoxicated conditions are collected and the equivalent feature vector was constructed. The essential data points on ECG samples are weighted to improve detection accuracy. With multiple criteria decision-making approach, the results showed that ECG-HSD scheme could achieve acceptable accuracy and rapid testing time.

This book addresses the most up-to-date problems of a smart city and their solutions in a cohesive manner. It is the product of contributions from world-class experts, educators, and students so to cover all levels of understanding to optimize its delivery. Therefore, we are confident that it will provide invaluable insight for decision-makers, engineers, doctors, educators, system operators, managers, planners, and researchers across all levels of career and academic progression.

London, UK   Chun Sing Lai
Guangzhou, China   Loi Lei Lai
Oxford, UK   Qi Hong Lai
25 May 2020

Acknowledgments

The authors wish to thank Mr. Michael McCabe of Springer Nature and his team in supporting this project. Special help from Mr. Menas Donald Kiran, Ms. Mohanarangan Gomathi, and Ms. Cynthya Pushparaj in producing the book is very much appreciated.

The authors wish to thank friends, colleagues, and students, without their support this book could not have been completed. In particular, the authors thank Dr. Kim Fung Tsang, Professor Ruiwen He, Professor Zhao Xu, Professor Wing W. Y. Ng, Dr. Youwei Jia, Dr Haoliang Yuan, Dr. Zhuoli Zhao, Dr. Fang Yuan Xu, Dr. Yifei Wang, Ms. Liping Huang, Ms. Yingshan Tao, Ms. Xin Cun, Ms. Mengxuan Yan, Mr. Zhiheng Huang, Mr. Xiaoqing Zhong, and Professor Mohammad Shahidehpour. The permission to reproduce copyright materials by the IEEE and Elsevier for a number of papers mentioned in some of the chapters is most helpful.

Last but not least, the authors appreciate the extraordinary support given from their family during the preparation of the book. In particular, to Ms. Qi Ling Lai in designing some of the art works and Ms. Li Rong Li in providing a workable environment under a pandemic.

Contents

About the Authors

Chun Sing Lai received the BEng (1st Hons.) in Electrical and Electronic Engineering from Brunel University London, UK, in 2013 and DPhil in Engineering Science from the University of Oxford, UK, in 2019. He is currently a Lecturer at the Department of Electronic and Computer Engineering, Brunel University London, UK, and also a Visiting Academic with the Department of Electrical Engineering, Guangdong University of Technology, China. He is a member of Brunel Institute of Power Systems. From 2018 to 2020, Dr. Lai was an Engineering and Physical Sciences Research Council (EPSRC) Research Fellow with the Faculty of Engineering and Physical Sciences, University of Leeds. He is Working Group Chair for the IEEE Standards Association P2814 standard, Publication Co-chair of the 6th IEEE International Smart Cities Conference (ISC2 2020), Secretary of the IEEE Smart Cities Publications Committee, and Acting Editor-in-Chief for IEEE Smart Cities Newsletters. He organized the workshop on Smart Grid and Smart City in the 2017 IEEE International Conference on Systems, Man, and Cybernetics (SMC2017) in Canada and workshop on Blockchain for Smart Grid in the 2018 IEEE International Conference on Systems, Man, and Cybernetics (SMC2018) in Japan. His current research interests are in power system optimization, energy system modelling, data analytics, and energy economics for low carbon energy networks and energy storage systems. Dr. Lai is a member of IET and Senior member of IEEE.

Loi Lei Lai received the BSc (1st Hons.), PhD, and DSc from the University of Aston, UK, and City, University of London, UK, respectively, in Electrical and Electronic Engineering. Presently, he is University Distinguished Professor at the Guangdong University of Technology, Guangzhou, China. He is a member of the IEEE Smart Cities Steering Committee and Chair of the IEEE Systems, Man and Cybernetics Society (IEEE/SMCS) Standards Committee. He was a member of the IEEE Smart Grid Steering Committee; Director of Research and Development Centre, State Grid Energy Research Institute, China; Pao Yue Kong Chair Professor at Zhejiang University, China; Vice President with IEEE/SMCS; Professor and Chair in Electrical Engineering at the City, University of London, UK; and a Fellow

Committee Evaluator for IEEE Industrial Electronics Society. He was awarded an IEEE Third Millennium Medal, IEEE Power and Energy Society (IEEE/PES) UKRI Power Chapter Outstanding Engineer Award in 2000, IEEE/PES Energy Development and Power Generation Committee Prize Paper in 2006 and 2009, IEEE/SMCS Outstanding Contribution Award in 2013 and 2014, and the Most Active Technical Committee Award in 2016. His research team has just received a best paper award from the 6th IEEE International Smart Cities Conference. His current research areas are in smart cities and smart grid. He is Fellow of IET and IEEE.

Qi Hong Lai attended Harrow International School Beijing where she was awarded funding under the IEEE Systems, Man, and Cybernetics (SMC) Society Pre-College Activities initiative to set up a program on Brain-Computer Interface. She received the BSc (1st Hons.) in Biomedical Science from King's College London, UK, in 2019. At present, she is working towards her DPhil in Molecular Cell Biology in Health and Disease at the Sir William Dunn School of Pathology, University of Oxford. Her current research interests are in transcriptional gene regulation, bioinformatics, and smart health.

Chapter 1
Smart City

1.1 Introduction

Major technological, economical, and environmental changes have generated interest in smart cities, including climate change, economic restructuring, the move to online retail and entertainment, aging populations, urban population growth, and pressures on public finances. A smart city is considered as an idealistic city, where the quality of life for citizens is significantly improved by combining Information and Communication Technology (ICT), new services, and new urban infrastructures [1]. The main innovation in the smart city evolutionary process includes a user-centric vision and accounting urban issues from the perspective of the need of the citizens with their engagement in the city management and operation. That is, smart city concept may be defined as an integrated solution in which human and social capital heavily interact, using emerging technology. The application of the Internet of Things (IoT) paradigm to urban scenarios is of special interest to support the smart city vision that aims to efficiently achieve sustainable and resilient development and a high quality of life on the basis of a multi-stakeholder, municipality-based partnership.

The mission is to accelerate city transformation processes to obtain a better use of renewable resources, reducing wastes and safeguarding the environment, while at the same time promoting the cohesion between citizens that are to be joined to obtain shared benefit in terms of quality of life.

Turning to an eco-sustainable vision, it consists of promoting a respectful urban and industrial development, able to address current needs without compromising the capacity of future generations. The eco-sustainable approach has to be applied in several aspects and at several layers of the evolutionary process, such as mobility, environment, and social services. A smart city uses different types of electronic Internet of things (IoT) sensors to collect data and then use these data to manage assets and resources efficiently. This includes data collected from citizens, devices, and assets that are processed and analyzed to monitor and manage traffic and

C. S. Lai et al., *Smart Grids and Big Data Analytics for Smart Cities*,
https://doi.org/10.1007/978-3-030-52155-4_1

transportation systems, power plants, water supply networks, waste management, crime detection, information systems, schools, libraries, hospitals, and other community services.

In brief, a smart city integrates information and communication technology (ICT), and various physical devices are connected to the IoT network to optimize the efficiency of city operations and services and connect to citizens. Smart city technology allows city officials to interact directly with both community and city infrastructure and to monitor what is happening in the city and how the city is evolving. ICT is used to enhance quality, performance, and interactivity of urban services, to reduce costs and resource consumption, and to increase contact between citizens and government. Smart city applications are developed to manage urban flows and allow for real-time responses. A smart city may therefore be more prepared to respond to challenges than one with a simple transactional relationship with its citizens. However, the term itself remains unclear to its specifics and, therefore, open to many interpretations. Examples of smart city technologies and programs have been implemented in Singapore, Dubai, Milton Keynes, Southampton, Amsterdam, Barcelona, Madrid, Stockholm, Copenhagen, China, and New York.

A smart city may be described by six fundamental pillars, namely, smart economy (competitiveness), smart people (social and human capital), smart governance (participation), smart mobility (transport and ICT), smart environment (natural resources), and smart living (quality of life), as shown in Table 1.1.

As detailed in Table 1.2, each characteristic is defined by a number of factors, which are described by a number of indicators.

Based on the above six characteristics, the following could be derived:

- A smart city uses innovative connectivity model and high technology-based infrastructure are used to enhance its economic efficiency, promoting social, urban, and cultural development.

Table 1.1 Characteristics of a smart city

Smart economy	Productivity; entrepreneurship; innovation attitude; international dimension; ability to transform; labor market flexibility
Smart people	Openminded; participation in public life; creativity; adaptability; lifelong learning; social and ethnic pluralism
Smart governance	Transparent governance; public and social services; participation in decision-making; political strategies and viewpoint
Smart mobility	Efficient, innovative, sustainable and safe transportation; accessibility
Smart environment	Low carbon economy; pollution; sustainable resources management and planning; environmental protection; natural resources exploration
Smart living	Safety; health conditions; housing quality; education facilities; social cohesion; cultural; cultural integration; tourist attractions

Table 1.2 Categories and aims for smart city initiatives

Category	Aim
Understanding smart cities: research and evaluation	Improve the knowledge base for and provide lessons for European policy
Designing smart city initiatives and strategies	Design of initiatives and city-level action plans
Smart city governance	Provide governance guideline and facilitate learning
Supporting the development of smart cities	Measure other than direct support that can be used to stimulate smart city development
Promoting smart cities: replication, scaling and ecosystem seeding	Create conditions to the expansion and extension of the most promising smart city approaches

- A smart city must be attractive and friendly toward the new business realities that intend to promote urban progress.
- A smart city must promote social inclusion, allowing a homogeneous development along the entire city.
- A smart city must invest in high technology-based instruments and in the educational process finalized to create high-skilled people.
- The people, in the meanwhile of this evolutionary process, must be encouraged to use modern technologies.
- A smart city must consider the social and environment sustainability as the most important strategy to pursue.

New community and technology leaders, managers, and solutions providers are needed to develop the smart city ecosystem. They need to operate in the intersection of technology, innovation, business, operations, strategy, and people. Smart cities are comprised of a "system of systems." These can include smart lighting systems, building automation systems, emergency management systems, security and access control systems, smart grids, renewables, water treatment and supply, transportation, and more.

Data are the lifeblood of the smart city. Open data, generated by municipal organizations, are only one source of data. When supplemented with data created by businesses and private citizens, it yields richer insights and better outcomes. Smart city ecosystem architects utilize the full extent of the ecosystem to create city data. They plan and build data marketplaces, robust data sharing and privacy policies, data analytics skills, and monetization models that facilitate the sourcing and usage of city data. The age of the internet of things (IoT) has brought with it an increasingly broad range of sensors and IoT platforms. Many of these have made their way into the smart cities sector. In the industrial internet of things architecture, smart city platforms perform many functions, including analytics, data management, remote asset monitoring, performance management, decision support, cybersecurity, device management, and visualization. IoT technology holds the best promise for providing unification and context to the huge array of data generated by smart cities and turns this data into actionable, contextualized information that can be used to reduce energy consumption, lifecycle cost, and operational costs while improving the safety and quality of life of citizens. For example, LED lighting systems provide a good value proposition because of overall reduced energy consumption and dramatically reduced maintenance schedule. With the combination of IoT-enabled smart lighting systems, this could further deliver even greater value.

The technology infrastructure must be in place. Information collected must be protected and used in accordance with the wishes of the owners. Effective architects unite the needs of policymakers, technologists, and innovators to create sensible policies that create the right outcomes. That is, policies, legislation, and technology must be continuously aligned to maintain the right balance of protection, privacy and transparency. The infrastructure must be robust, resilient, and reliable. Cybersecurity and technology policies, processes, and systems must be revised to be smart city focused. Digital skills, from data analytics, machine learning to software engineering, must be the new competencies of the smart city.

In a data-driven society, the large volume of data is accelerating fast. The reliance on human–machine collaboration to be successful will require the velocity, veracity, value, speed, security, and the universal interoperability of data. The explosion in hardware vendors, the number of communication protocols, and the lack of standardization of metadata and labeling among system integrators have created an environment in which data brokering between devices may be lost in translation. The desired flow of data back and forth between databases, levels of the technology stack, applications, industries, regions, countries, and freely throughout the global economy does not yet exist. Situational awareness is one of the future huge challenges. The need for interoperability is essential and important.

1.2 Functional Domains

1.2.1 Sensors and Intelligent Electronic Devices

Sensors can provide data required by smart applications to improve system efficiency. Intelligent Electronic Devices (IEDs) have been deployed extensively in power automation systems recently due to the integration and interoperability features of the IEDs. IED handles additional features like self and external circuit monitoring, real-time synchronization of the event monitoring, local and substation data access, programmable logic controller functionality, and an entire range of software tools for commissioning, testing, event reporting, and fault analysis.

1.2.2 Communication Networks and Cyber Security

Communication network technologies are constantly evolving and underpin almost everything. The next generation of network and security must be able to support and enhance the world economy—whether that be through social developments, medical systems, and low carbon development. One of the major challenges is cyber security which is the application of technologies, processes and controls to protect systems, networks, programs, devices, and data from cyberattacks. It aims to reduce the risk of cyberattacks and protect against the unauthorized exploitation of systems, networks, and technologies.

1.2.3 Systems Integration

System integration is defined in engineering as the process of bringing together the component subsystems into one system. With an aggregation of cooperating subsystems cooperating so that the system is able to deliver the required functionality to ensure the sub-systems function together as a system, and in information technology as the process of linking together different computing systems and software applications physically or functionally.

1.2.4 Intelligence and Data Analytics

Data have intrinsic value which is essential to extract that value and convey the information in the data through presentable visualizations. Organizations and governments want to exploit data to predict behaviors and extract valuable real-world insights. Billions of devices and social media conversations are accelerating the rate at which data are produced. There is an urgent need to understand data and make systems, policies, and governance models more efficient and effective.

1.2.5 Management and Control Platforms

Data management and control platform are used to monitor and control smart objects in the internet of things (IoT). By combining IoT-specific features and protocols such as HTTP, the platform allows anomaly detection in IoT devices and real-time error reporting mechanisms.

1.2.5.1 Smart City Domains and Sub-domains

Table 1.3 shows the relationship between smart city domains and sub-domains with five different smart cities functionalities. "X" means that there is a close relationship between the domains/sub-domains and functionalities.

Table 1.3 Smart city domains, sub-domains, and functionalities

	Sensors and intelligent electronic devices	Communication networks and cyber security	Systems integration	Intelligence and data analytics	Management and control platforms
Energy					
Advanced demand response				X	
Microgrid/ nanogrid			X		X
Smart and energy-efficient buildings	X		X	X	X
Distributed energy resources integration	X				
Energy analytics and visibility	X			X	
Energy services				X	X
Smart and energy-efficient lighting		X		X	
Transportation					
Smart traffic and congestion management		X		X	
Fleet management		X			X
Smart public transit system		X	X		X
Shared mobility solutions				X	
EV charging station network					X
Connected vehicles and transport		X	X		
Vehicle to grid		X			
Smart parking	X				
Pedestrian management				X	
Health and safety					
Smart crowd management		X		X	
Smart security system	X	X			

(continued)

Table 1.3 (continued)

	Sensors and intelligent electronic devices	Communication networks and cyber security	Systems integration	Intelligence and data analytics	Management and control platforms
Disaster management and emergency services		X			X
Environmental monitoring and response systems	X		X	X	
Wellness services	X				
Food and agriculture					
Open data and urban info systems					X
Smart retail solutions	X		X		
Connected community		X			
Virtual learning					X
Economic development			X	X	
Command center					X
Infrastructure planning				X	X
Digital city work management					X
Public service management		X			
Digital citizen self-service		X			X
Water					
Water reclamation	X		X		
Water AMI	X	X			
Water services					X
Smart agriculture	X			X	
Connected water monitoring and response systems	X		X		
Waste					
Smart waste collection	X				
Waste to energy					X
Smart recycling	X				

(continued)

Table 1.3 (continued)

	Sensors and intelligent electronic devices	Communication networks and cyber security	Systems integration	Intelligence and data analytics	Management and control platforms
Battery second life/recycling			X	X	
Waste diversion (lifestyle extension)				X	
Total count	14	13	10	15	15

1.3 Elements of a Smart City

It is forecasted that the total population living in cities will be increased by 75% by 2050, as a result, there is an increased demand for smart, sustainable environments that offer citizens a high quality of life. This leads to the evolution to smart cities. A smart city will bring together technology, government and society to enhance elements, namely smart energy, smart economy, smart mobility, smart environment, smart economy, smart living and smart governance.

1.3.1 Smart Energy

In this sub-section, few areas associated with smart energy will be presented.

1.3.1.1 Hourly Unit Commitment with Resilience-Constrained

In this part, hourly unit commitment with resilience-constrained electricity grids will be presented.

Nomenclature

Variables and Functions	
C_k	Total cost for scenario k
$\overline{C}$	Average total cost for overall sampled scenarios
$F(\cdot)$	Probability function for accumulated outage
$F_{ci}(\cdot)$	Fuel consumption function for unit i
$h_0(\cdot)$	Baseline function for PHM
$h(\cdot)$	Proportional hazard model function
$H(\cdot)$	Power flow entropy function

$I_{i,t}$	Commitment state for unit i at time t
$LS_{d,t}$	Load shedding amount for load d at time t
$n_{m,t}$	Total number of lines with line loading rate falls into mth interval $[(m-1)*u, m*u)$ at time t
$P_{i,t}$	Generation for unit i at time t
$PL_{l,t}$	Real power flow for line l at time t
PL_l^{max}	Thermal limit for line t
$P_{b,t}^{inj}$	Net real power injection for bus b at time t
$r_{l,t}$	Loading rate for line l at time t
T	Outage time
$Z_2(t)$	Line loading rate at time t
u_m, v_m	Auxiliary variables used to linearize $|r_{j,t} + r_{k,t}|$ for the objective function (1.11)
u_n', v_n'	Auxiliary variables used to linearize $|r_{j,t} - r_{k,t}|$ for the objective function (1.11)
u_j'', v_j''	Auxiliary variables used to linearize $|r_{j,t}|$ for the objective function (1.11)
ω_1, ω_2	Penalty term coefficients
γ_1, γ_2	Weight coefficients for the $Z_1(t)$ and $Z_2(t)$
$\psi(\cdot)$	Link function of PHM
$Z_1(t)$	Weather condition for each line at time t
σ	Average hourly power flow entropy
D	Average hourly line power flow variance
E_a	Average loading rate for all lines
E_s	Average loading rate for affected lines
Constants and Sets	
a, b	Scale and shape parameters for $h_0(\cdot)$
$D_{d,t}$	Real power demand for load d
ns	Scenarios Number
$SF_{l,b}$	Shift factor for line l to the bus b
M	Successive intervals number used to calculate $H(\cdot)$
NL	Total transmission lines number
NAL	Lines number affected by extreme weather
NCL	Number of pair combinations of transmission lines
NT	Number of times
NG	Generation units number
ND	Load number
NB	Bus number
$SU_i(SD_i)$	Start-up (shutdown) cost function for unit i
S_L	All system transmission lines set
S_{AL}	Transmission lines affected by weather set
S_m	Auxiliary variables um, vm set
$S_{m'}$	Auxiliary variables $u_{m'}, v_{m'}$ set
S_n	Auxiliary variables u_n, v_n set
VOLL	Value for lost load
ε	Convergence threshold for Monte Carlo process

Introduction

A large number of cascading outages and blackouts has exhibited the vulnerability of power systems and the shortcoming to sustain major outages under catastrophic circumstances. Generally, the power system infrastructure is designed to be highly reliable under normal conditions and often not highly resilient under extreme conditions [2]. Therefore, resilience-based operation strategies will provide more specific and cost-effective approaches in critical conditions and play a decisive role in initiating blackout preventions and resilience enhancements.

Resilience has different definitions [3–6] according to the context of extreme events. In general, the definitions can be divided into two types. One is expressed as the ability for a power system to bounce back after certain disturbances have occurred, i.e. restoration. The other one is represented as the power system capability to change its state to respond to some unexpected events, i.e. adaptation. Figure 1.1 shows the resilience index curve in which an extreme event starts at time t_0 and forced outages occur at time t_1. Power system resilience will start corrective or preventive strategy. Corrective strategies mainly focus on power grid islanding [7] and reconfiguration [8, 9]. However, preventive strategies, as illustrated in Fig. 1.1, can boost the resilience curve more effectively. This section presents a preventive option instituting a resilience-constrained unit commitment (RCUC) strategy is used for power system stability enhancement under extreme conditions.

Previously, research mainly focused on the coordination of preventive SCUC operation strategies with $N - k$ contingencies [10–13], maintenance scheduling [14], reliability indices [15–17], and risk indices [18, 19], etc. Benders decomposition was proposed in [10] for coordinated SCUC with maintenance scheduling and N-1 contingencies. The stochastic SCUC, which was modeled based on mixed-integer programming (MIP) with robust optimization [11] to guarantee a balanced power under any $N - k$ contingencies. Reference [12] proposed a two-stage robust optimization for $N - k$ contingency. The unit commitment and transmission switching were simultaneously optimized in [13] with $N - 1$ contingency. Reference [14] proposed and integrated the SCUC framework with maintenance scheduling considering severe weather effects. Stochastic forced outages and reliability indices were considered in [15]. Reference [16] sought the trade-off between cost and reliability in which the cost of maintaining a certain reliability level was quantified.

Fig. 1.1 Power system resilience curves

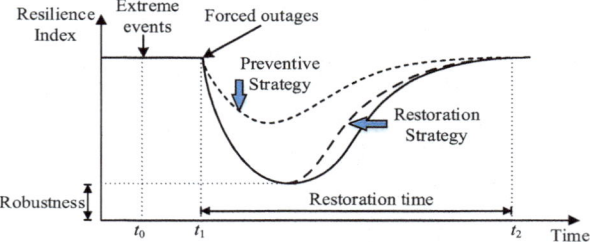

Reliability constraints were modeled, linearized, and applied using an MIP method in [17].

The reliability-based SCUC was widely discussed in the literature. However, there are limited studies on SCUC considering extreme events and cascading outages. A proactive SCUC model under hurricane events was proposed in [20] in which the operation cost and expected recourse cost were optimized in a two-stage stochastic MIP framework. An integration of preventive and emergency strategies was studies in [21] to coordinate the preventive action and emergence response. A proactive SCUC framework presented in [22] introduced a Markov process to model power system state transitions in extreme events. SCUC was sequentially solved within each system state. Researches have discussed the impact of microgrid on power system operation resilience [23, 24]. Reference [25] proposed a proactive microgrid management for enhancing the power system resiliency in which a two-stage adaptive and robust pre-disturbance scheduling takes into account several uncertainties for reducing the damaging consequences of islanding events.

Although these works have studied preventive strategies for SCUC under extreme conditions, SCUC strategies would require additional work to consider more uncertain energy resources, cascading outages, the coordination of electricity and other large infrastructures, potential malfunctions in protection systems as we add more phasor measurement units, and enhance the automation in electric power systems. In essence, it is imperative to model SCUC for enhancing the resilience of power systems and optimize its solution for reducing the probability of component outages in blackout types of incidents.

Secure and effective preventive strategies against extreme events could reduce the probability of cascading outages and boost the power system resilience. Large-scale power transfers resulting from the outages of heavily loaded lines are the main causes of cascading failures. Moreover, heavy transmission loading rate tends to increase the probability of relay malfunctions and transformer outages. Approximately half of the recorded blackouts are triggered by weather-related events. Therefore, managing the power system operations in extreme weather conditions and reducing the transmission loading rate often signify effective preventive strategies in power system operations.

The self-organized criticality (SOC) [26] is a critical state in a large system where a minor event can lead to a catastrophe. Reference [27] demonstrated that the SOC is an essential characteristic of blackouts in large power systems. Reference [28] illustrated that power system loading that is close to the system operating limit is the key condition leading to cascading outages. According to the system structure and operating state, Ref. [29] proposed an entropy-based metric to evaluate the robustness of power grid with respect to cascading failures. References [30, 31] studied the network entropy in terms of its topology and structure. Reference [32] showed the correlations between SOC and the heterogeneity of power flow distribution by introducing the power flow entropy index. Accordingly, the larger the power flow entropy, the more routinely a power system state can lead to SOC. Therefore, lower power flow entropy can prevent power systems from evolving into SOC.

This part focuses on developing an RCUC solution towards cascading outage preventions, power flow entropy reductions, and resilience promotions. The main contributions of this study are summarized as follows:

- Develop a proactive and sequential RCUC framework that considers interactions among power system operation states and random component outages, in which power flow entropies, component forced outages, and system operation costs are simultaneously addressed.
- Develop two penalty terms, modeled by the absolute value function, for improving the homogeneity of power flow distribution and regulating power line loading rate affected by extreme weather.
- Develop a general linearization method for representing the absolute value function in MIP model.
- Introduce the proportional hazard model (PHM) to quantify the effect of weather conditions and line-loading rate on component-forced outage rates. In addition, present a recursive sampling process in order to meet sequential simulation framework requirements.

The rest of the sub-section is organized as follows. Section "Description of the Proposed RCUC Framework" describes the proposed framework. Section "Proposed RCUC Model" presents the RCUC model. Section "RCUC Solution Methodology" introduces the solution method. Section "Case Studies" illustrates case studies. Section "Conclusions" draws conclusions.

Description of the Proposed RCUC Framework

The most effective operation strategy under extreme weather events is to adjust RCUC and power flow solutions proactively. Before we proceed with the development of the proposed framework, we present the forced outage rate and the sampling method as follows:

Forced Outage Rate of Transmission Lines

Transmission lines are subject to random outages which are affected by weather conditions and real-time loading. The proportional hazard model (PHM) [33] is introduced to represent the forced outage rate in (1.1) below:

$$
\begin{aligned}
h(t,Z(t)) &= h_0(t) \cdot \psi(Z(t)) \\
&= a^{-b} \cdot b \cdot t^{b-1} \cdot \exp\left(\gamma_1 \cdot Z_1(t) + \gamma_2 \cdot Z_2(t)^2\right)
\end{aligned}
\tag{1.1}
$$

where $h_0(t)$ is the baseline function for the basic degradation process, the Weibull hazard rate function is adopted here, in which a is the scale parameter, and b is the shape parameter of the distribution. $\psi(Z(t))$ is the link function to quantify the impact of influencing factors $Z(t)$ (i.e., covariates). Here two covariates, weather condition $Z_1(t)$ and line loading rate $Z_2(t)$, are considered, γ_1 and γ_2 are the weight coefficients of the two covariates. For $Z_1(t) \in \{0, 1, 2\}$ and $Z_2(t) \in [0, 1]$, in weather condition function $Z_1(t)$, 0 means normal weather, 1 means severe weather, and 2 means major storm disaster. To address the nonlinear relationship, a quadratic function is added to the load rate $Z_2(t)$ which poses a minute failure impact if the line is lightly loaded.

In this section, weather conditions are assumed to be known in advance through meteorological services and line loading rate is dynamically obtained by the RCUC solution. The main contributions of this study are not specific to certain weather events and the corresponding SCUC solutions. Instead, the proposed model is concerned with the overall system performance in terms of SOC which provides an effective preventive solution toward resilience enhancement. Reference [34] normalizes the weather intensity into three categories which are used widely in the power system reliability evaluation. Without the loss of generality, we follow this classification instead of specifying individual weather events.

Sequential Sampling Method

Given a component survived until t_0, the reliability function at time t is calculated as

$$R(t|t_0) = P(T > t | T > t_0) = \exp\left\{-\int_{t_0}^{t} h(x, Z(x)) dx\right\} \tag{1.2}$$

The inverse transform sampling is adopted in the proposed sequential sampling process which includes two steps.

1. Sample u from the uniform distribution unif(0, 1) interpreted as probability.
2. Return the maximum t such that $F(t) \leq u$.

In the second step, we accumulate the hourly outage probability $F(t)$ as

$$F(t) = 1 - R(t) = 1 - R(t|t-1) R(t-1) = 1 - R(t|t-1)(1 - F(t-1))$$
$$= 1 - \exp\left\{-\int_{t=1}^{t} h(x, Z(x)) dx\right\} + F(t-1)\exp\left\{-\int_{t=1}^{t} h(x, Z(x)) dx\right\} \tag{1.3}$$

This is a recursive process in which the outage probability $F(t)$ is determined by $F(t-1)$ and the covariates conditions are stated at hour t. For example, if we find $F(t) \geq u$ at hour t while $F(t-1) < u$, then one sampled random outage is obtained. For simplicity, the outage is assumed to begin at the end of hour t, i.e. the line is tripped at hour $t+1$ upon any outage, and a repair time is sampled according to an exponential distribution. We consider minimal maintenance in which forced outage rates before and after maintenance remain the same.

Random outages have an impact on RCUC and power flow solutions which can alter the outage process of remaining lines. Furthermore, the RCUC solution and the sampling process interact with each other. RCUC and power flow solutions affect the accumulated outage probability at hour t. In turn, the sampled outages affect the next-hour RCUC solution. Therefore, the RCUC solution and sampling process are executed alternatively and sequentially.

Proposed RCUC Framework

The proposed RCUC framework is sequential and Monte Carlo-based. By sequential, we mean an initial RCUC is calculated for the first period and the scenario within the period is dynamically generated according to the weather state and line loading rate, which is used for processing the initial status of the next period. The sequential process is shown in the dashed box in Fig. 1.2. By applying a Monte Carlo-based method, we assert that if the generated samples cannot represent the global feature, the sequential simulation process will be repeated until the convergence condition is met.

The RCUC framework in Fig. 1.2 includes three steps which are discussed as follows:

1. The area weather conditions and the information on line loading and network availability are identified at the initialization step. The initial status of all components is assumed to be in normal operation.
2. The line status sampling and RCUC are executed sequentially as follows:

    (a) Calculate RCUC and line loadings at period t.
    (b) Evaluate the accumulated outage probability of normal lines by the proposed sequential sampling method based on the weather conditions and line-loading rates.
    (c) Determine whether any random outage of normal lines has occurred. If so, sample repair times; if not, record the accumulated outage probability for the next sequential sampling.
    (d) Update the remaining repair time for the lines that are under repair. In this way, the status of all lines is generated which will be used as the initial status for the next period.
    (e) Repeat steps to (d) until the optimization horizon is concluded in which an RCUC solution is obtained for one generated scenario.

3. The coefficient of variation (CV) of costs over multiple scenarios is used as convergence condition. If the condition is not met, go back to Step 1 for engaging additional Monte Carlo iterations.

The CV is used as a measure of relative uncertainty in the Monte Carlo simulation [35] and judge whether the sampled results can represent the global features for the RCUC solution. The CV of total costs is adopted as convergence condition,

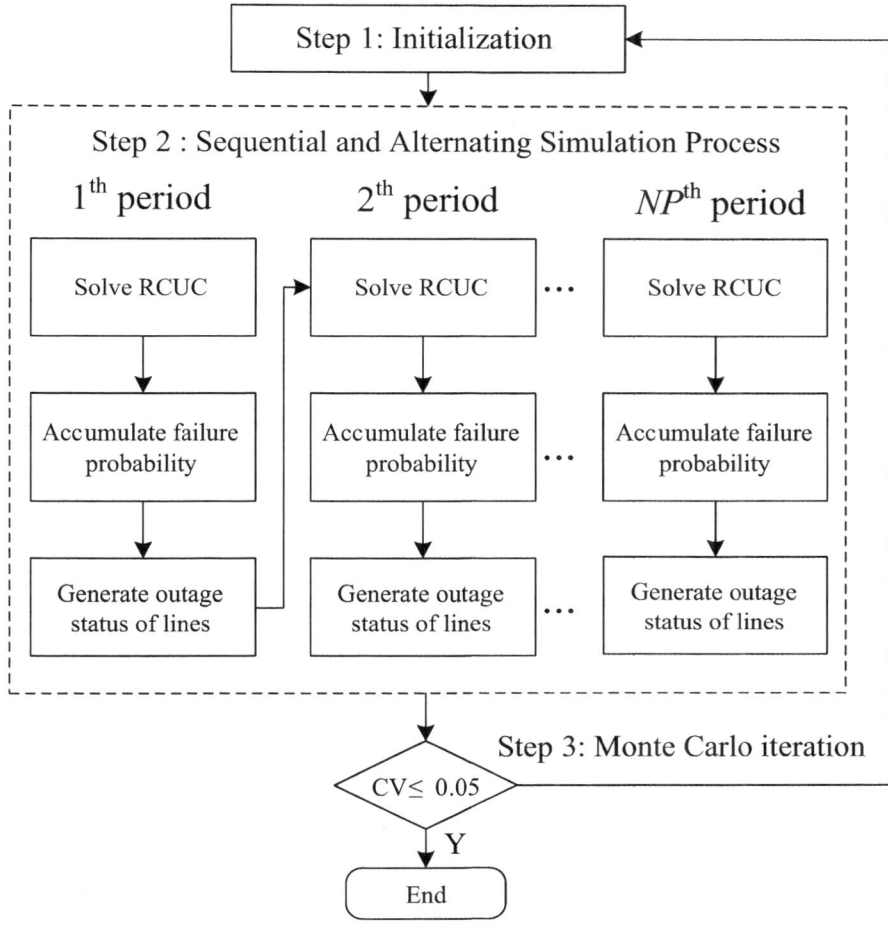

Fig. 1.2 Sequential and Monte Carlo-based RCUC framework

$$\varepsilon = \sqrt{\frac{1}{ns(ns-1)} \sum_{k=1}^{ns} (C_k - \overline{C})^2} / \overline{C} \qquad (1.4)$$

If ε is larger than 0.05, a new scenario should be generated.

Proposed RCUC Model

In this section, power flow entropy is introduced and the proposed RCUC model is solved in each Monte Carlo iteration using power flow measurements. The following formulations are applied to individual iteration. Therefore, we omit the iteration index in presenting the following model.

Power Flow Entropy and Proposed Penalty Terms

Power flow entropy is proposed as a measurement of the global heterogeneity for the power flow distribution in an electricity grid [32]. It is defined as

$$H(t) = -\sum_{m=1}^{M} \frac{n_{m,t}}{\text{NL}} \log_2 \frac{n_{m,t}}{\text{NL}} \tag{1.5}$$

where M is the total number of successive intervals stated as $[0, u)$, $[u, 2 * u)$, ..., $[(M - 1) * u, M * u)$. $n_{m,t}$ is the total number of lines with a loading rate absolute value $r_{l,t} = \left| \text{PL}_{l,t} / \text{PL}_l^{\max} \right|$ that falls into the mth interval $[(m - 1) * u, m * u)$ at time t. NL is the total number of transmission lines.

The power flow entropy $H(t)$ provides a measure of power flow distribution uniformity. Accordingly, $H_{\min}(t) = 0$ when all transmission lines loading rates are within the same interval. In this case, the grid load distribution is homogeneous and all lines carry loads within their rated capacities. The maximum entropy $H_{\max}(t) = \log_2 M$ occurs when $n_{m,t}/\text{NL} = 1/M$, i.e., number of lines in arbitrary interval is identical. Therefore, higher power flow entropy means greater heterogeneity in power flow distribution.

When entropy is high, a few transmission lines could be carrying heavy loads while others are lightly loaded. The heavily loaded lines could fail more easily and the mass transfer of power flow on such lines could trigger cascading failures. Reference [32] showed that the power flow entropy has a close relation with both the dynamic propagation course and the static black size of cascading failures. The power flow entropy can represent an index for the short-term operation defense against large-scale blackouts.

In would be difficult to append (1.5) to the SCUC formulation and optimize the power flow entropy directly. Therefore, we consider a penalty term that would reflect the heterogeneity of power flow distribution as the sum of the absolute value of the difference between any two absolute transmission loading rates, states as

$$\text{pn}_1 = \sum_{j,k \in S_L, j \neq k} \left(\left\| r_{j,t} \right| - \left| r_{k,t} \right\| \right) \tag{1.6}$$

The minimum value of (1.6) could reduce $H(t)$ and improve the power flow distribution. Since (1.6) is non-convex, it would require convexification. Using $\|r_{j,t}| - |r_{k,t}\| \leq |r_{j,t} + r_{k,t}|$ and $\|r_{j,t}| - |r_{k,t}\| \leq |r_{j,t} - r_{k,t}|$, we have

$$\left\| r_{j,t} \right| - \left| r_{k,t} \right\| \leq \frac{1}{2} \left(\left| r_{j,t} + r_{k,t} \right| + \left| r_{j,t} - r_{k,t} \right| \right) \tag{1.7}$$

In (1.7), a lower value of $\frac{1}{2} \left(\left| r_{j,t} + r_{k,t} \right| + \left| r_{j,t} - r_{k,t} \right| \right)$ could reduce (1.6). So, (1.8) is the first penalty term adopted in the objective function where S_L is the set of all transmission lines.

$$\text{pn}_1 = \sum_{j,k \in S_L, j \neq k} \left| r_{j,t} + r_{k,t} \right| + \left| r_{j,t} - r_{k,t} \right| \tag{1.8}$$

In addition to the effect of the global heterogeneity of power flow distribution, outages of heavily loaded lines which could result in large-scale power transfers are the one of the critical causes of cascading failures. Since weather-related events could often trigger blackouts, it is imperative to reduce the loading rate of transmission lines in areas affected by extreme weather. Therefore, the second penalty term is established as

$$\text{pn}_2 = \sum_{l \in S_{AL}} \left| r_{l,t} \right| \tag{1.9}$$

where S_{AL} is the set of weather-affected transmission lines.

The two penalty terms fulfill different tasks in power system operation states. However, additional compromise and coordination would be required between the two penalty terms as reductions in certain line flows could increase the flow in other lines. In essence, we would reduce power flows through weather affected areas in order to lower the power flow entropy effectively.

Proposed RCUC Objective Function

In the DC power flow model, the loading rate r_l is presented as a linear function of generator outputs $P_{i,t}$ using the power flow tracing method [36] as

$$r_{l,t} = \frac{PL_{l,t}}{PL_l^{\max}} = \frac{\sum_{b=1}^{NB} SF_{l,b} \cdot P_{b,t}^{\text{inj}}}{PL_l^{\max}} \tag{1.10}$$

where $SF_{l,b}$ is shifter factor which is the sensitivity of line flows to changes in bus injections. Shift factor reflects how a line power flow changes with a change in bus injection power. Shift factor matrix only depends on the grid topology, grid parameters, and the location of reference bus.

The objective function for NT operation periods is shown as

$$\begin{aligned}
\min. \quad & \sum_{t=1}^{NT} \sum_{i=1}^{NG} \left[F_{ci}(P_{i,t}) \cdot I_{i,t} + SU_{i,t} + SD_{i,t} \right] + VOLL \cdot \sum_{t=1}^{NT} \sum_{d=1}^{ND} LS_{d,t} \\
& + \omega_1 \cdot \sum_{t=1}^{NT} \sum_{j,k \in S_L, j \neq k} \left| r_{j,t} + r_{k,t} \right| + \left| r_{j,t} - r_{k,t} \right| + \omega_2 \cdot \sum_{t=1}^{NT} \sum_{l \in S_{AL}} \left| r_{l,t} \right|
\end{aligned} \tag{1.11}$$

where $F_{ci}(P_{i,t}) = a_i + b_i \cdot P_{i,t} + c_i \cdot P_{i,t}^2, \forall i, \forall t$ is the cost of unit i at hour t. To construct a proactive defense strategy, the two penalty terms are introduced into the RCUC objective function. The first penalty term indicates the difference in transmission loading rates which can improve the power flow homogeneity and reduce the power flow entropy. The second penalty term represents local power flows in areas affected by extreme weather conditions. If we substitute (1.10) into (1.11), the penalty terms are the absolute value functions of the linear combination of generator outputs.

Using (1.11), we simultaneously consider operation cost, power flow homogeneity, and weather-affected power flows. One can adjust ω_1 and ω_2 to find an appropriate trade-off between economics and resilience. Higher ω_1 and ω_2 means a larger power flow homogeneity and a more resilient operation state. The downside in this case is that more expensive generators could be dispatched. In extreme weather conditions, ω_1 and ω_2 will increase to enhance the system resilience.

RCUC Constraints

The RCUC model may include the following constraints, with the extended list of constraints made available in [14].

- Generation capacity constraints
- Unit startup/shutdown cost
- Unit ramping up/down capability
- Unit minimum ON/OFF time constraints
- System power balance
- Transmission flow constraints

The proposed RCUC model is deterministic and its solution varies in each Monte Carlo iteration. However, the proposed penalty terms and operation strategy, which can improve the power flow distribution and reduce the blackout risk, pertain to the overall system performance rather than individual Monte Carlo solutions.

RCUC Solution Methodology

This section proposes a solution methodology to smooth the model (1.11) and solve it more efficiently. A more general form of the programming model shown as model (P) is used,

$$(\text{P}) \quad \min. \quad F(x,y) + \sum_{k=1}^{K}\left|\sum_{i=1}^{n} d_{k,i} x_i\right| \tag{1.12}$$

$$\text{s.t.} \quad \text{model constraints}$$

Alternatively, we consider the following programming model (Q) where we prove that its optimal solution is the same as that of model (P).

$$\text{min.}\quad F(x,y)+\sum_{k=1}^{K}\left(u_k+v_k\right)$$

$$(\text{Q})\quad\text{s.t.}\quad \sum_{i=1}^{n}d_{k,i}x_i+u_k-v_k=0,\quad k=1,2,\ldots,K$$

$$u_i\ge 0,\, v_i\ge 0,\quad \forall i$$

$$\text{and model constraints}$$

(1.13)

Lemma If (x^*,y^*,u^*,v^*) is the optimal solution of model (Q), then $u_i^*\cdot v_i^*=0,\forall i$ and the optimal objective value is

$$q^*=F\left(x^*,y^*\right)+\sum_{k=1}^{K}\left|\sum_{i=1}^{n}d_{k,i}x_i\right|$$

(1.14)

Proof (Proof of contradiction) If $u_i^*\cdot v_i^*\ne 0$, then $u_j^*>0,v_j^*>0$. When $u_j^*\le v_j^*$, construct the following feasible solution

$$\begin{cases}x=x^*\\y=y^*\end{cases};\quad u_k=\begin{cases}0,&\text{for } k=j\\u_k^*,&\text{for } k\ne j\end{cases}\quad v_k=\begin{cases}v_j^*-u_j^*,&\text{for } k=j\\v_k^*,&\text{for } k\ne j\end{cases}$$

It is easy to verify that the constructed solution is feasible. Since $q^*=F\left(x^*,y^*\right)+\sum_{k=1}^{K}\left(u_k^*+v_k^*\right)$ is the optimal value of objective function, the difference between its value and that of q for the constructed solution is

$$q-q^*=F\left(x,y\right)+\sum_{k=1}^{K}\left(u_k+v_k\right)-F\left(x^*,y^*\right)-\sum_{k=1}^{K}\left(u_k^*+v_k^*\right)$$

$$=\left(0+v_j^*-u_j^*\right)-\left(u_j^*+v_j^*\right)=-2u_j^*<0$$

which contradicts the hypothesis that (x^*,y^*,u^*,v^*) is the optimal solution. The same reasoning applies to the situation where $u_j^*\ge v_j^*$. Hence $u_k^*\cdot v_k^*=0$ for arbitrary k.

According to the equality constraint in (1.13), if $u_k^*=0$, then $\sum_{i=1}^{n}d_{k,i}x_k^*=v_k^*\ge 0$;

if $v_k^*=0$, then $\sum_{i=1}^{n}d_{k,i}x_k^*=-u_k^*\le 0$. In any case, $u_k^*+v_k^*=\left|\sum_{i=1}^{n}d_{k,i}x_i\right|$. Substitute this

equation into q^* and the lemma is proved.   ∎

Theorem If (x^*,y^*,u^*,v^*) is the optimal solution of model (Q), then (x^*,y^*) is necessarily the optimal solution of model (P).

Proof: If (x^*, y^*, u^*, v^*) is the optimal solution of model (Q), (x^*, y^*) is necessarily a feasible solution of model (P) since (x^*, y^*) satisfies all the constraints of model (P). Let $(\hat{x}, \hat{y})$ be the optimal solution of model (P), then

$$\hat{p} = F(\hat{x}, \hat{y}) + \sum_{k=1}^{K} \left| \sum_{i}^{n} d_{k,i} \hat{x}_i \right| \leq p^* = F(x^*, y^*) + \sum_{k=1}^{K} \left| \sum_{i}^{n} d_{k,i} x_i^* \right| \tag{1.15}$$

Construct a feasible solution of model (Q) based on the presumed optimal solution $(\hat{x}, \hat{y})$ of model (P) as

$$\begin{cases} \hat{x}_k, \hat{y}_k, \hat{u}_k = 0, \quad \hat{v}_k = \sum_{i=1}^{n} d_{k,i} \hat{x}_i & \text{if } \sum_{i=1}^{n} d_{k,i} \hat{x}_i \geq 0 \\ \hat{x}_k, \hat{y}_k, \hat{u}_k = -\sum_{i=1}^{n} d_{k,i} \hat{x}_i, \hat{v}_k = 0 & \text{if } \sum_{i=1}^{n} d_{k,i} \hat{x}_i < 0 \end{cases}$$

It is easy to verify that (x^*, y^*, u^*, v^*) is a feasible solution of model (Q) since $(\hat{x}, \hat{y}, \hat{u}, \hat{v})$ satisfies the model (Q) constraints. Comparing the objective function value $\hat{q}$ of $(\hat{x}, \hat{y}, \hat{u}, \hat{v})$ with the optimal objective function value (1.11), we have

$$\begin{aligned} \hat{q} &= F(\hat{x}, \hat{y}) + \sum_{k=1}^{K} \left| \sum_{i=1}^{n} d_{k,i} \hat{x}_i \right| \\ &\geq F(x^*, y^*) + \sum_{k=1}^{K} \left| \sum_{i=1}^{n} d_{k,i} x_i^* \right| \end{aligned} \tag{1.16}$$

Since q^* is the optimal (minimum) objective value of model (Q). By comparing (1.15) and (1.16), we conclude that $\hat{p} = \hat{q} = p^* = q^*$. So, (x^*, y^*) is the optimal solution of model (P). ∎

So far, (1.6) which measures the power flow entropy is changed to a convex formulation (1.7). Further, penalty terms in (1.11) are changed to linear smooth formulations. Accordingly, the penalty terms of (1.11) are stated as

$$\begin{aligned} \text{min.} \quad & \sum_{t=1}^{NT} \sum_{i=1}^{NG} \left[F_{ci}(P_{i,t}) + SU_{i,t} + SD_{i,t} \right] \\ & + \omega_1 \cdot \sum_{t=1}^{NT} \sum_{m=1}^{NCL} (u_{m,t} + v_{m,t}) + \omega_1 \cdot \sum_{t=1}^{NT} \sum_{n=1}^{NCL} (u'_{n,t} + v'_{n,t}) \\ & + \omega_2 \cdot \sum_{t=1}^{NT} \sum_{j=1}^{NAL} (u''_{j,t} + v''_{j,t}) \end{aligned} \tag{1.17}$$

s.t.
$$r_{j,t} + r_{k,t} + u_{m,t} - v_{m,t} = 0, \quad j,k \in S_L, \forall t$$
$$r_{j,t} - r_{k,t} + u'_{n,t} - v'_{n,t} = 0, \quad j,k \in S_L, \forall t$$
$$r_{j,t} + u''_{j,t} - v''_{j,t} = 0, \quad j \in S_{AL}, \forall t$$
$$u_{m,t} \geq 0, \quad v_{m,t} \geq 0, \quad u'_{n,t} \geq 0, \quad v'_{n,t} \geq 0, \quad u''_{j,t} \geq 0, \quad v''_{j,t} \geq 0$$

and other model constraints

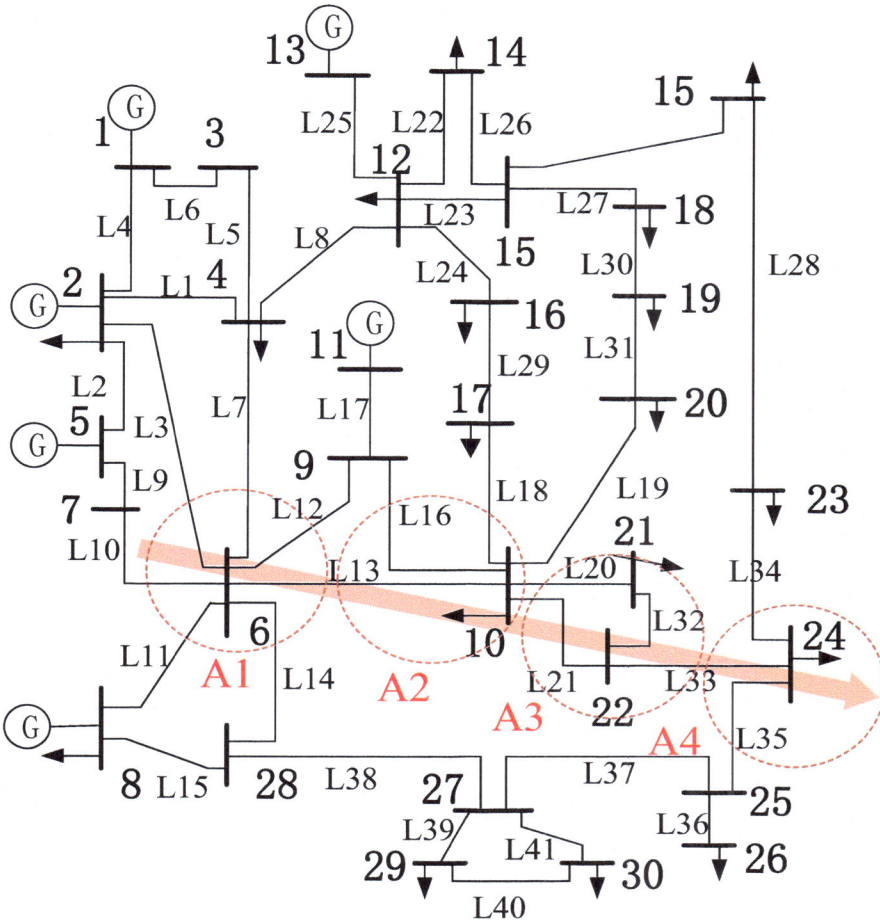

Fig. 1.3 IEEE 30-bus system and weather moving trajectory

Table 1.4 Regions, transmission lines and weather intensity

Affected area	Impacted transmission line	Weather intensity
A1	L3, L7, L11, L12, L13, L14	$Z_1(t) = 1$
A2	L13, L16, L18, L19, L20, L21	$Z_1(t) = 2$
A3	L20, L21, L22, L33	$Z_1(t) = 2$
A4	L33, L34, L35	$Z_1(t) = 1$

where $NCL = C_{NL}^2$ is the number of pair combinations of all transmission lines and NAL is the number of weather-affected lines. To smooth one absolute value function term, two auxiliary variables and one auxiliary constraint are introduced. In total, 4NCL + 2NAL auxiliary variables and 2NCL + NAL auxiliary constraints are added for one period. Note that our proposed method is not specific to any form of the original objective function $F(x,y)$ or model constraints. Therefore, the linear and smooth transformation approach applies to any optimization model.

Case Studies

In this section, we perform case studies on the modified IEEE 30-bus system. The cases are tested with the MATLAB R2014b and the Gurobi solver on a desktop computer with a 3.20 GHz i5 processor and 8 GB RAM.

The IEEE 30-bus system is composed of six generators, 21 load buses, and 41 transmission lines. The topology of the system and the moving trajectory of the tempest are shown in Fig. 1.3. Table 1.4 shows the affected region, transmission lines, and different intensities in the moving path.

IEEE 30-Bus System

The following three cases are discussed to verify the value of RCUC model and solution methodology.

Case 1: Test RCUC versus conventional SCUC without considering random outages (Monte Carlo framework). Case 1 is studied to verify the feasibility of the linearization method for absolute value functions and the effectiveness of penalty terms in the proposed RCUC model.
Case 2: Add random outages to Case 1. The interactions between forced outages and system operation states are addressed.
Case 3: Add varying weather intensity and moving trajectory to Case 2. The validity of the sequential and alternating process between RCUC and outage sampling is demonstrated.

These cases are presented as follows:

Case 1

In this case, no weather effect is considered, i.e. all transmissions lines are considered to be available in the 24 h horizon. However, we still assign a specific area to reduce the local power flow within the area. The area A1 in Fig. 1.3 is selected and the affected lines are shown in Table 1.4. The Case 1 results are shown in Fig. 1.4 and Tables 1.5, 1.6, 1.7.

Figure 1.4 consists of four subfigures. For the traditional model, there is no penalty term which means $\omega_1 = \omega_2 = 0$. Figure 1.4(a) shows the 24-h power flow entropy

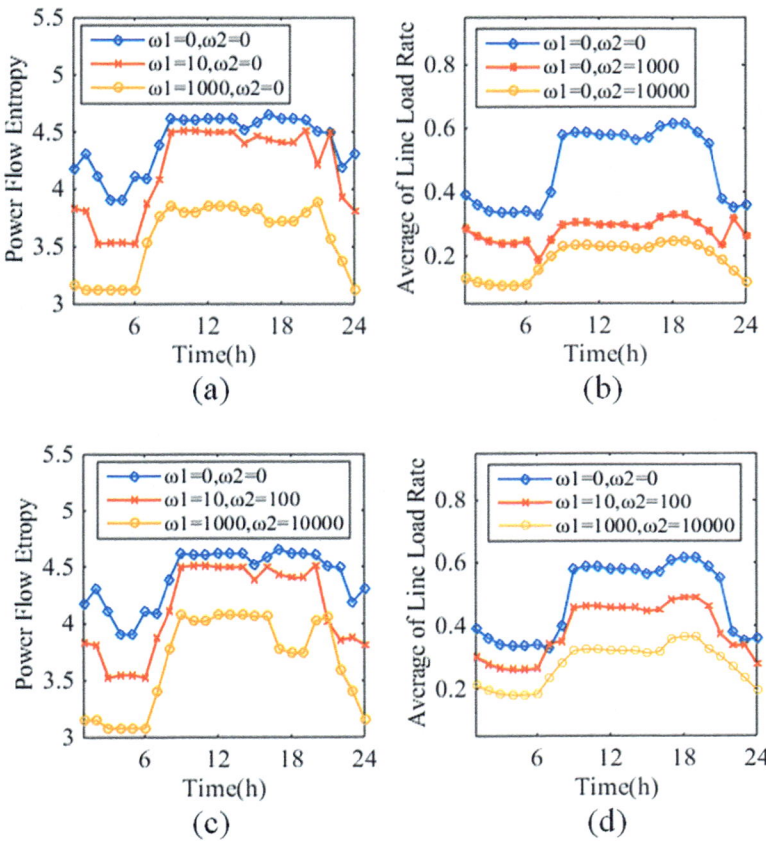

Fig. 1.4 Case 1 results in the 24-h horizon. (**a**) Power flow entropy. (**b**) Average loading rate for A1. (**c**) Power flow entropy. (**d**) Average loading rate for A1

Table 1.5 Results of Case 1 when ω_1 varies and $\omega_2 = 0$

ω_1	Generation cost ($\$$)	Power flow entropy σ	Power flow variance D	Average $r_{l,t}$ of all lines E_a	Maximum $r_{l,t}$
1	550,709.86	4.396	0.035	0.36	1.00
10	558,898.06	4.152	0.027	0.31	0.83
100	603,284.47	3.607	0.017	0.28	0.60
100	605,339.59	3.561	0.016	0.28	0.60
1000	608,620.29	3.507	0.015	0.29	0.60

when only the first penalty term is considered, i.e. ω_1 varies and $\omega_2 = 0$. Here, the power flow entropy drops as ω_1 increases, which demonstrates the effect of improving the uniformity of power flow distribution. Figure 1.4(b) illustrates that the power flowing through one specific area can be adjusted by the second penalty term.

Table 1.6 Results of Case 1 when ω_2 varies and $\omega_1 = 0$

ω_2	Generation cost ($)	Average $r_{l,t}$ of selected area E_s	Maximum $r_{l,t}$ of selected area
1	550,695.04	0.48	0.78
10	550,695.11	0.48	0.78
100	550,699.56	0.48	0.79
1000	564,342.52	0.28	0.62
10,000	593,952.91	0.19	0.62

Table 1.7 Results of Case 1 when ω_1 and ω_2 are considered together

ω_1 and ω_2	Generation cost ($)	Power flow entropy σ	Power flow variance D	Average $r_{l,t}$ of all lines E_a	Average $r_{l,t}$ of selected area E_s	Maximum $r_{l,t}$
$\omega_1 = \omega_2 = 0$	550,824.17	4.4	0.035	0.365	0.479	0.995
$\omega_1 = 10$ $\omega_2 = 100$	560,503.63 (+1.76%)	4.118 (−6.41%)	0.026 (−27.50%)	0.310 (−14.96%)	0.383 (−20.10%)	0.678
$\omega_1 = 100$ $\omega_2 = 1000$	602,575.51 (+9.40%)	3.706 (−15.76%)	0.017 (−50.69%)	0.280 (−23.22%)	0.268 (−44.08%)	0.597
$\omega_1 = 1000$ $\omega_2 = 10,$ 000	604,652.35 (+9.79%)	3.651 (−17.03%)	0.016 (−54.3%)	0.282 (−22.73%)	0.274 (−42.79%)	0.597

Figures 1.4(c), (d) demonstrate the power flow entropy and the power flow rate in the selected area when two penalty terms are considered. The proposed model will find a trade-off between the power flow uniformity of total lines and the local power flow in the selected area.

Table 1.5 shows the detailed results of generation cost and the uniformity of power flow distribution. Three indices are adopted, including the average hourly power flow entropy σ, average hourly line flow variance D, and average loading rate of all lines E_a, to measure the power flow uniformity in 24 h. In Table 1.5, the generation cost increases while σ and D decrease as ω_1 increases. Moreover, E_a and maximum loading rate become smaller because the proportion of the first penalty term in the objective function becomes larger as ω_1 increases. The optimal solution tends to reduce the first penalty term to minimize the objective function. Similarly, the average loading rate in a selected area E_s and the maximum loading rate of affected transmission lines become smaller in Table 1.6 when ω_2 is increased, which is due to the higher proportion of the second penalty term.

There are a compromise and coordination between the two penalty terms as demonstrated in Table 1.7. As ω_1 and ω_2 increase, the flow homogeneity of all lines and line flows through specific areas can be reduced simultaneously. The larger the ω_1 and ω_2, the higher the generation cost and the better the power flow distribution will be. However, if ω_1 and ω_2 exceed 1000 and 10,000, respectively, the power flows cannot be improved any further. That is, the resilience enhancement margin has its own limitations considering a preventive operation strategy. The limitation depends on capacities of generators and lines, network topology, system load, etc. In such

Table 1.8 Results for different load in Case 2

System load	Base case load		Increase base case load by 50%	
Model	Traditional model	Proposed model	Traditional model	Proposed model
Number of outages	0.91	0.35 (−61.54%)	2.48	1.14 (−54.03%)
Outage duration (h)	6.69	2.23	16.27	7.18
LS (MW)	143.248	31.549 (−77.98%)	1382.094	615.411 (−55.47%)
Generation cost ($)	549,884.83	601,251.49 (+9.34%)	778,727.40	833,649.51 (+7.05%)
Total cost ($)	836,381.16	664,349.70 (−20.57%)	3,542,915.4	2,064,471.7 (−41.73%)

cases, additional power system dispatch and unit commitment will not improve the resilience. However, considering new generators and transmission lines, enlarging line capacities, and developing enhanced demand response programs can support the resilience enhancement.

In Table 1.7, for the traditional SCUC ($\omega_1 = \omega_2 = 0$), the generation cost is 550,824.17 and $\sigma = 4.4$, $D = 0.035$, $E_a = 0.365$ and $E_s = 0.479$. For the proposed model when $\omega_1 = 1000$ and $\omega_2 = 10,000$, the uniformity indices σ, D, and E_a decrease by 17.03%, 54.3%, and 22.73%, respectively, the average loading rate of selected area decreases by 42.79%, whereas the generation cost increases by merely 9.79%. This is because higher ω_1 and ω_2 corresponds to additional expenses for scheduling more costly generators to enhance the power flow homogeneity and a reduction in specific line flows. Therefore, in the proposed RCUC, ω_1 and ω_2 represent a trade-off between operation cost and system resilience. Cases 2 and 3 will show the benefits when power systems are under extreme events.

Case 2

In this case, the proposed sequential and proactive framework of RCUC is studied when the system is subjected to extreme events. To obtain better system resilience, we choose $\omega_1 = 1000$ and $\omega_2 = 10,000$ in Case 2. In PHM, we set $\gamma_1 = 2$, $\gamma_2 = 5$, $a = 10,950$, $b = 1$. The parameters are adopted as [37]. If exhaustive historical data are considered a, b, γ_1 and γ_2 can be obtained using the maximum likelihood estimation method [33]. The weather is assumed major storm disaster, i.e., $Z_1(t) = 2$ in the link function for the 24-h simulation horizon. The weather-affected area is A1. The repair rate is 2 and the mean repair time is 1/2 day, i.e., 12 h.

In Table 1.8, the expected line outage time and duration, the expected generation cost, and the expected load shedding over all Monte Carlo iterations are demonstrated. Table 1.8 shows the results for two load profiles. In the fourth row, the outage duration means expected total repair time in one scenario which is calculated as the average of the total repair time in one scenario over all sampled scenarios. There is an evident decline in the expected outage times, outage duration, and load shedding when the proposed model is applied as compared with those of the traditional SCUC model. This is because line flows are much more homogeneous and line loadings are in proportion to line capacities when the proposed approach is applied. Thus, the traditional situation is avoided when some lines undertake heavy loading while others undertake lighter loading. In extreme events, the probability of relay malfunction and cascading outages will increase. For the traditional SCUC model, after heavily loaded lines are tripped, the power flow will be shifted to remaining lines rather than becoming more uniform, which gives rise to more forced outages and load shedding.

In Table 1.8, the generation cost in the proposed RCUC approach is only 9.34% higher than that of traditional SCUC model. However, RCUC introduces a 77.98% reduction in load shedding cost and lowers the total cost by 20.57%. When the base case system load is increased by 50%, the proposed model is still effective as indicated in Table 1.6. In this case, the load shedding increase is distinctly nonlinear which is because there are more transmission lines with high loading rates as the load demand increases. In the traditional SCUC model, at hour 18 with the highest loading, there are 5-line flow rates that exceed 0.8 and 13 lines over 0.7. A higher system load and line loading rate could lead to additional relay malfunctions and cascading outages.

In extreme conditions, the proposed RCUC model gains a more efficient power system operation strategy which is due to the effectiveness of penalty. However, the effectiveness of the proposed model will decrease when comparing the two loading cases. The load shedding decreases by 77.98% for the base case load and 55.47% for the higher load. The same trend is followed by other system performance indices. There is a limit on the enhancement of power system resilience which depends on the characteristics of generation resources, power network, and individual loads. In general, a heavier loading will result in a smaller available margin for generation and line flow capacity and a smaller chance for attaining higher resilience.

Case 3

In this case, the moving trajectory and varying intensity of the tempest are shown in Fig. 1.3 and the impacted transmission line in the path of tempest at different time periods is considered in Table 1.4. The simulation horizon is expanded to 48 h. At the same time, base hourly loads in 24 h used in Cases 1 and 2 are extended to 48 h in this case. Assume each area is impacted by weather events for 12 h and the weather intensity in areas A1, A2, A3, A4 is 1, 2, 2, 1, respectively.

Table 1.9 demonstrates the results in Case 3 where the proposed sequentially proactive RCUC framework can consider the effect of line loading rate and time-varying extreme weather events. Compared with the total cost decrease in Case 2, the total cost decline in Case 3 is a bit smaller. Two reasons may be stated for this phenomenon. First, the weather-affected area is moving, which precludes certain lines from being exposed to severe weather. The weather impact is dispersed to some extent. Second, weather conditions are set to vary from severe to extreme in the 48 h horizon while the weather in Case 2 was set to be extreme in the horizon.

Table 1.9 Obtained results in Case 3 with different models

Model	Number of outages	Outage duration (h)	LS (MW)	Generation cost ($)	Total cost ($)
SCUC model	1.262	15.396	317.763	1,105,296.76	1,740,822.74
RCUC model	0.713	6.624	225.807	1,193,989.76 (+8.02%)	1,645,602.87 (−5.47%)

Fig. 1.5 Convergence trend of CV

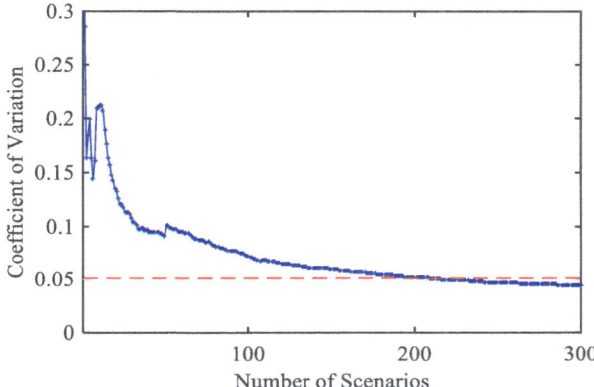

Therefore, Case 3 not only proves the effectiveness of the proposed model for considering the real-time status of the system, but also indicates that the proposed model can offer more resilient in response to extreme events. A significant advantage in resilience is obtained by increasing the generation cost which is no more than 10% in all three studied cases.

Convergence Performance of RCUC

Figure 1.5 shows the convergence trend of the proposed RCUC framework. The Monte Carlo's convergence condition, i.e., the CV of total cost to be less than 0.05, is satisfied in about 150–200 iterations. For each iteration, the running time for the 30-bus system is about 40 s which is slightly longer than that of the traditional SCUC since pair combinations of all lines are considered by adding $C_{41}^2 = 820$ absolute value functions. One can select certain transmission lines, including a few heavily loaded lines to construct the penalty term for obtaining a higher power flow homogeneity and acceptable computation burden. The Monte Carlo iterations are independent which could be subject to parallel computation and a lower CPU time. In general, the proposed RCUC framework would provide a reliable solution and gain a better insight on power system operations within a reasonable CPU time.

Conclusions

In this section, a Monte Carlo-based proactive and sequential RCUC framework is introduced. A sequential sampling method is proposed to generate random outages base on PHM. To address the non-convexity caused by the penalty terms in the RCUC model, a convex approximation approach is presented and its validity is discussed. The major findings are as follows:

1. The sequential optimization considering the real-time system status is an effective means for dealing with extreme weather and enhancing the system resilience.
2. The power flows are concentrated excessively in certain lines in order to improve the uniformity of the power flow distribution. This approach can reduce the risk of relay malfunctions and cascading outages consequently reduce the required load shedding during extreme weather conditions. Particularly, generator outputs are adjusted to reduce flows in those transmission lines that are affected by extreme weather.
3. The relationship between load shedding and system load demand, under extreme weather conditions, is not a linear function. In essence, the required load shedding for resilience would be much higher when demand increases.
4. There is an adjustment limit for the power flow distribution homogeneity which depends on the network topology and parameters, generator capacities, and system load demand. Generally, higher load levels correspond to smaller generation margins for resilience improvement.

1.3.1.2 Managing Transactive Energy in a Multi-microgrid System

In this section, a reconfigurable distribution network for managing transactive energy in a multi-microgrid system for smart energy is introduced.

Nomenclature

Variables and Functions	
$C(\cdot)$	Microturbine generation cost function
$C_{ij}^{\text{del}}(\cdot)$	Transfer cost function
ΔE_i	The trading adjustment amount at bus i
$P_{\text{pur},iD}^t$	Power purchased by MGi from DSO at time interval t (kW)
$P_{\text{pur},ij}^t$	Power purchased by MGi from MGj at time interval t (kW)
$P_{\text{sel},i}^t$	Power sold by MGi at time interval t (kW)
$L(\cdot)$	Lagrangian function
K	Iteration number in bi-level programming model
$P_{\text{MT},i}^t$	Microturbine (MT)i output at time interval t

Pd_i^t	Equivalent load of MGi at time interval t
P_m^{inj}	Power injection at bus m, equal to 0 if $m \in NN_s$
P_{mn}	The active power flow from bus m and n
q_{mn}	The reactive power flow from bus m and n
r_l	Auxiliary variable, equal to $V_m V_n \cos(\theta_m - \theta_n)$ where V_m, V_n are voltage amplitude and θ_m, θ_n are voltage angles of the two terminal buses of line l
t_l	Auxiliary variables, equal to $V_m V_n \sin(\theta_m - \theta_n)$
u_m^l	Auxiliary variables, equal to u_m if line l is connected, and 0 otherwise.
u_m	Voltage amplitude at bus m
α_l	Binary variable for network configuration; 1 if the line l is connected, and 0 otherwise
β_{mn}	Binary variable; equal to 1 if bus n is the parent of bus m, and 0 otherwise.
λ_i	Lagrangian multiplier for subproblem i, indicate the sale price of MGi

Constants and Sets

a, b, c	Coefficients of MT generation cost function
d_{ij}^t	Equivalent electrical distance between MGi and MGj at time interval t
p, q	Lagrangian multipliers update parameters
A_l	$g_l^2 + \left(b_l + b_l^{sh}/2\right)^2$
B_l	$g_l^2 + b_l^2$
b_l^{sh}	Series conductance and susceptance of a line l
C_l	$g_l^2 + b_l\left(b_l + b_l^{sh}/2\right)$
D_l	$g_l b_l^{sh}/2$
g_l, b_l	Series conductance and susceptance of line l
I_{max}	Maximal current flow allowed through a line
L	Set of branches
N	Set of buses
$N(m)$	Set of buses that connected to bus m
N_s	Set of substation buses
N_{switch}	The maximum times of the change of switch status between every two adjacent scheduling time
NT	Length of a scheduling interval
NMG	Number of MGs
$P_{MT}^{min}, P_{MT,i}^{max}$	Minimum and maximum MT output of MGi
PD, QD	Forecasted load (MW, MVAr)
φ_m	Reactive and real power ratio at bus m
V_{max}, V_{min}	Maximal and minimal voltage amplitude
η_{LMP}^t	LMP at time t

Introduction

In traditional power systems, power is generated centrally by large power plants and flown unidirectionally to load centers through transmission and distribution systems. With the increasing penetration of distributed energy resources and MGs, transactive energy emerges in a new power market which enables end-to-end energy trading in a coordinated and distributed system operation [38–40].

In recent years, researchers are pushing forward the coordination between microgrids (MGs) and power grid. Some literatures focus on using MGs to support the main grid's operation. Reference [41] employed the corrective control to relieve post-contingency overflows to support main grid's security control with MGs. References [42–46] enhanced the system resilience by using MGs. In [42], controllable and islandable MGs were used to enhance the resiliency of power grid. Four resilience indices were introduced to measure the impact of extreme conditions from different aspects. Reference [43] quantified and enabled the resiliency of a power distribution system with multiple MGs by using analytical hierarchical process and percolation theory. Reference [44] proposed a resilience-oriented service restoration method using MGs to restore critical load after natural disasters. Reference [45] proposed a two-level hierarchical outage management scheme for resilient operation of multi-MGs while the autonomy of MGs was guaranteed. In [46], dynamically forming MGs was proposed to continue supplying critical loads after natural disasters to enhance the resilience of distribution system.

However, these aforementioned studies only focused on the system operation optimization with MGs. With the development of power market, MGs could participate in power market for transactive energy trading, which was also attracting researchers' concerns. Reference [47] modeled wholesale and local markets by considering the MG electricity auction in energy communities. Reference [48] defined a virtual energy sharing coordinator among prosumers. Not using dual prices, a pricing model was presented based on the feed-in tariff, grid electricity tariff, and the supply and demand ratio. However, the trading decision of MGs in [47, 48] was made in a centralized manner. The autonomy of MGs was ignored.

To guarantee the MGs' autonomy, Ref. [49] presented a multi-MGs energy management strategy where the dual variable of total power balance constraint was used as the distribution marginal cost. But the model does not provide the trading price for each MG. Reference [50] adopted Lagrangian relaxation to describe the transactive energy trading process among multi-MGs where Lagrangian multipliers were interpreted as clearing prices for each MG. Reference [51] expanded the model in [50] with augmented Lagrangian relaxation and proposed an energy management model for multi-MGs. Reference [52] proposed a two-stage energy exchange strategy for multi-MGs which makes use of electric vehicles for curbing peak power exchanges. The power exchange constraints between MGs and the power distribution grid are modeled whose dual variables are used as price signal. Not treating the energy as a homogeneous product, Reference [53] classified the energy demand into several classes according to their preference for source/load. Augmented Lagrangian relaxation was introduced to obtain a distributed structure where Lagrangian multipliers were used as trading prices. Reference [54] designed an incentive mechanism using the Nash bargaining theory to encourage proactive energy trading and fair benefit sharing in multi-MGs. Reference [55] adopted the game theory to design a seller level game for MGs where trading strategies and pricing mechanisms were discussed. However, these studies only considered the transactive energy trading among multi-MGs. The system operation is ignored. In practice, the energy trading among multi-MGs will have an impact on the system operation, which may affect the energy trading among multi-MGs conversely.

Fewer works considered distribution system operation in the transactive energy market among MGs [56–61]. In the limited works, two issues are mainly concerned: transactive energy management [56–59] and transactive energy trading and pricing [60, 61]. Reference [56] proposed a bi-level transactive energy model where the upper and lower levels are the operation of distribution and multi-MG, respectively. Distflow equations were used to model the power flow in the upper level. The power losses were ignored, which may lead to serious error in distribution systems. Reference [57] considered energy interactions among MGs and between MGs and the distribution system. The former was described by a bi-level model, the latter was modeled by the game theory. The genetic algorithm was used to solve the bi-level problem. Reference [58] developed the coordinated operation of multi-MG in a power distribution system. Both grid-connected and islanded modes were considered in MGs. However, distribution power losses were ignored in order to apply convex power flow equations. Reference [59] decomposed the distribution system operation into power distribution and multi-MG subproblems with an alternating direction for multipliers. References [56–59] addressed the coordinated energy management of networked MGs and the distribution system. However, the end-to-end energy sharing and trading among MGs were not taken into account. Reference [60] developed an inter-MG auction in electricity market to manage the excess supply or residual demand where market imbalances were assigned to the utility grid. In this case, although a distribution level market was devised, the distribution power flow was ignored. In [61], a distribution system operator (DSO) acted as an intermediary for MG energy exchanges. DSO calculated a reference trading level for MGs based on which a penalty term was added to market objectives to minimize the trading mismatch. In this case, all MGs traded with DSO by considering the reference trading level. However, in [60, 61], MG trading models were not peer-to-peer, and the MG autonomy was limited.

In practice, energy trading among MGs needs the support from the grid facility. Moreover, energy trading among MGs may account for line congestions. Thus, it is reasonable and fair that MGs should make a payment to DSO for their energy trading. In [50], the cost of trading energy between MGs was considered. The trading cost was assumed to be common for each MG. However, the impact of network topology on the energy trading among MGs was ignored. MGs locate at different nodes in practical distribution system. In this study, we consider that MGs' trading cost is related to the trading amount and the distance between the traders. Different network topology will lead to different distances among MGs, and MGs may make different trading decisions.

Table 1.10 Contribution comparisons

References	Realizing decentralized trading decision for MGs	Considering distribution system operation	Considering network reconfiguration	Considering energy trading cost
[47, 48]	×	×	×	×
[49–55]	√	×	×	×
[56–61]	×	√	×	×
[50]	√	×	×	√
This work	√	√	√	√

Fig. 1.6 Proposed bi-level framework

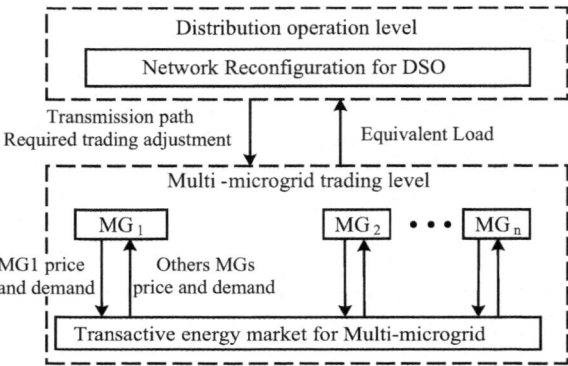

The contribution made is shown in Table 1.10 by comparing with the reviewed literature. The main contributions of this study are summarized below:

- The distribution reconfiguration is considered in the transactive energy market for multi-MGs. The distribution reconfiguration manages the trading among MGs by calculating the proper path dynamically.
- A distributed end-to-end multi-MG trading model is proposed considering MGs positions in the distribution network. The trading level is updated dynamically in bi-level iterations according to the equivalent distance between MGs.
- The impact of network topology on the energy trading among MGs is considered. We consider that MGs' trading cost is related to the trading amount and the distance between the traders to make the trading model more practical.
- The proposed transactive energy market framework would ensure the individual MG autonomy while maintaining the cooperation among MGs and the power distribution system.

Proposed Framework for Transactive Energy

The proposed framework is depicted in Fig. 1.6 where the DSO and multi-MG are independent entities representing their individual objectives. MGs consider trading with their peers at the lower level and with the DSO at the upper level. A distributed end-to-end energy trading and pricing model is established using Lagrangian relaxation and decomposition methods, respectively. MGs make their trading decisions independently and submit their equivalent load to DSO. The DSO reconfigures the network and determines the flow path between MGs. If network constraints cannot be satisfied at the upper level, the trading adjustment will be considered iteratively at the lower level. The iterations at the lower level and between lower and upper levels will continue until the convergence criterion is satisfied.

The proposed multi-MG transactive energy trading model considered at the lower level is described as follows:

- Each participating MG is represented by an equivalent dispatchable microturbine (MT) in the distribution system.
- MGs trade energy in an end-to-end process (without the DSO's trading intervention). Each MG can also buy energy from the DSO.
- Transactive energy purchaser will bear the energy delivery cost considering network and trading constraints.
- Each MG offers transactive energy quantity and price and updates its strategy using other MGs offers.

The DSO's network reconfiguration considered at the upper level of Fig. 1.7 affects the transactive energy delivery cost and correspondingly changes the multi-MG behavior as well as the transactive energy trading quantity and price. The use of AC power flow equations ensures the accuracy of distribution but introduces non-convexity. We consider a convex problem by applying a feasible convex superset in the proposed optimal distribution reconfiguration. Different supersets could be considered for the relaxation method including the second-order cone programming (SOCP) and the semi-definite programming. When the distribution network is radial, SOCP is the tightest relaxation which gains the fastest solution [62, 63]. Accordingly, we apply the mixed-integer SOCP in this study to optimize the distribution network reconfiguration.

Multi-Microgrid Transactive Formulation

The multi-MG trading at the lower level is modeled by Lagrangian relaxation in which each MG forms a decomposed subproblem, while the network reconfiguration at the upper level is modeled by SOCP. The proposed bi-level transactive energy problem for multi-MGs is formulated as follows:

$$
\begin{aligned}
&\min && (1.31) \\
&\text{s.t.} && (1.32)-(1.51) \\
& && Pd_i^t \in \arg\min (1.19) \\
&\text{s.t.} && (1.21)-(1.27), (1.52)
\end{aligned}
\tag{1.18}
$$

where the upper level in (1.18) is to minimize the operation cost and trading adjustment requirement. The lower level is to minimize the total trading cost of all MGs. Next, we will present the detailed objectives and constraints for the transactive energy trading process.

Lower Level: Lagrangian Relaxation Solution for Multi-microgrid

At the lower level, the primal objective function of multi-MG trading is

$$\min \sum_{t=1}^{NT} \sum_{i=1}^{NMG} \left(C\left(P_{MT,i}^t\right) + \eta_{LMP}^t \cdot P_{pur,iD}^t + \sum_{j=1}^{NMG} C_{ij}^{del}\left(P_{pur,ij}^t\right) \right) \tag{1.19}$$

where the first term is the MTi generation cost stated as $C_i(x) = a_i x^2 + b_i x + c_i$. The second term is the power purchase cost from DSO. The third term is the delivery cost of MGi power purchase. This item considers power losses which have a quadratic relationship with power transfer. Accordingly, C_{ij}^{del} is defined as

$$C_{ij}^{del} = \sigma_{ij} \cdot \frac{P_{pur,ij}^{t\ 2} + Q_{pur,ij}^{t\ 2}}{\hat{V}_i^2} d_{ij} = \sigma_{ij}\left(\varphi^2 + 1\right) P_{pur,ij}^{t\ 2} d_{ij} \tag{1.20}$$

where V_j is the voltage of node j, assumed to be 1 in the delivery cost estimate. In addition, we assume that when a MG buys active power, it also imports reactive power $Q_{pur,\ ij} = \varphi P_{pur,\ ij}$ in proportion to its power factor. We use a coefficient σ_{ij} to estimate the delivery cost. The unit of σ_{ij} is \$/(kW$^2\cdot\Omega$) and it affects the marginal delivery cost if MG i buys power from j and is predetermined for exchanging transactive energy. d_{ij}^t is the equivalent distance between MGs i and j which represents the transmission path resistance between the two MG nodes. The transmission path between node i and node j is dynamically calculated based on the latest topology. The proposed transmission path searching method is as follows:

Initialization
  • Input the latest network topology and the node-branch incidence matrix, the initial node m and the target node n.
  • Define the set N_f denoting the nodes that have been found before. The starting $N_f = \{m\}$.
  • Define the set N_{nf} as the nodes that have not been found before. We have $N_f \cup N_{nf} = N$, $N_f \cap N_{nf} = \varnothing$.
  • Let the set N_k denotes the neighbor nodes found in the k-th search.
  • Initialize the route record matrix $R \in NL \times NB$.
Repeat the iteration while $\{$target node $n \notin N_f\}$
    for each node $i \in N_{k-1}$
        find the node $j \in N_{nf}$ that next to node i;

Add the branch *Lij* to the transmission path from initial node m to node i;

Record the transmission path in the j-th column of R;

add node j to the set N_k;

end

Update $N_f = N_f \cup N_k$, $N_{nf} = NB/N_f$;

$k = k + 1$;

End

Output: The n-th column of matrix R which represents the transmission path between initial node m and target node n.

The constraints for (1.18) are shown as

$$P_{\text{MT},i}^t + \sum_{j \neq i, j \in \text{MG}} P_{\text{pur},ij}^t + P_{\text{pur},iD}^t - P_{\text{sel},i}^t - Pd_i^t = 0, \quad i \in \text{MG} \tag{1.21}$$

$$Q_{\text{MT},i}^t + \sum_{j \neq i, j \in \text{MG}} Q_{\text{pur},ij}^t + Q_{\text{pur},iD}^t - Qd_i^t = 0, \quad i \in \text{MG} \tag{1.22}$$

$$P_{\text{pur},ij}^t \geqslant 0, \quad P_{\text{pur},ii}^t = 0, \quad Q_{\text{pur},ij}^t \geqslant 0, \quad Q_{\text{pur},ii}^t = 0, \quad i, j \in \text{MG} \tag{1.23}$$

$$P_{\text{pur},iD}^t \geqslant 0, \quad Q_{\text{pur},iD}^t \geqslant 0, \quad i \in \text{MG} \tag{1.24}$$

$$P_{\text{sel},i}^t \geqslant 0, \quad Q_{\text{sel},i}^t \geqslant 0, \quad i \in \text{MG} \tag{1.25}$$

$$P_{\text{MT},i}^{\min} \leqslant P_{\text{MT},i}^t \leqslant P_{\text{MT},i}^{\max}, \quad i \in \text{MG} \tag{1.26}$$

$$P_{\text{sel},i}^t = \sum_{j=1}^{\text{NMG}} P_{\text{pur},ji}^t, \quad i = 1, 2, \ldots, \text{NMG} \tag{1.27}$$

where (1.21) and (1.22) are the MGi's active and reactive power balance, respectively. Equations (1.23)–(1.25) are trading constraints. Since we use separate variables to denote the MG trading behavior, the variables in (1.23)–(1.25) are nonnegative. In the active power balance (1.21), buying variables are added and selling variables are subtracted. Constraint (1.26) is the limit on the MT generation capacity. Constraint (1.27) shows the trading balance among MGs. In other words, Eq. (1.27) represents that the total purchase by MGi is equal to its cumulative sale. The constraint (1.27) denotes the MG trading balance in the transactive energy market. Although MG can also buy from DSO using the location marginal price (LMP) in this section.

By relaxing (1.27), the transactive energy problem will be decomposed for each MG using the Lagrangian function,

$$L(\lambda) = \sum_{t=1}^{NT}\sum_{i=1}^{NMG}(C_i\left(P_{MT,i}^t\right) + \sum_{j=1}^{NMG}C_{ij}^{del} + \eta_{LMP}^t P_{pur,iD}^t$$
$$+\lambda_i^t\left(P_{pur,1i}^t + P_{pur,2i}^t + \cdots + P_{pur,NMG,i}^t - P_{sel,i}^t\right)) \tag{1.28}$$

The relaxed transactive energy problem can be decoupled into separate MG sub-problems. Accordingly, the objective function of the decoupled subproblem i is

$$\min \sum_{t=1}^{NT}\left(C_i\left(P_{MT,i}^t\right) + \sum_{j=1}^{NMG}C_{ij}^{del} + \eta_{LMP}^t P_{pur,iD}^t\right.$$
$$\left. + \lambda_1^t P_{pur,i1}^t + \lambda_2^t P_{pur,i2}^t + \cdots + \lambda_{NMG}^t P_{pur,i,NMG}^t - \lambda_i^t P_{sel,i}^t\right) \tag{1.29}$$
$$\text{s.t. } (1.21)-(1.25) \text{ for MG}i$$

Solve the subproblem (1.29) for each MG. The solutions of (1.29) may not satisfy the relaxed constraints (1.27). Therefore, we modify the multipliers in order to achieve the primal feasibility, i.e., satisfy the relaxed (1.27). The Lagrangian multipliers are updated by the subgradient method,

$$\lambda_i^t[k+1] = \lambda_i^t[k] + \left(\frac{1}{pk+q}\right)\left(\sum_{j=1}^{NMG}P_{pur,ij}^t[k] - P_{sel,i}^t[k]\right) \tag{1.30}$$

where k is the iterations.

Lagrangian multipliers represent the trading price in the proposed MG trading model. In (1.30), when the total power the MGi's sale exceeds its available supply, $\sum_{j=1}^{NMG}E_{pur,ji}^t > E_{sel,i}^t$, then $\lambda_i[k+1]$ will exceed $\lambda_i[k]$. Otherwise, $\lambda_i[k+1]$ will be smaller than $\lambda_i[k]$. The trend matches that of bulk market price which will be lower for buyers and higher for sellers. Hence, multipliers provide the transactive electricity price. Moreover, since λ_i is the sale price of MGi, the second line in (1.29) denotes the MGi's purchase cost from other MGs (purchase price multiplied by the quantity) minus the MGi revenue (sale price λ_i multiplied by the quantity $E_{sel,i}^t$). In this way, costs and revenues are considered in each MG's decision. Note that the trading price is decided by MG rather than DSO.

Once we find a feasible solution, the primal objective function value provides the upper bounds on the optimal value P^* of the problem (1.19). The optimal value of relaxed primal problem (1.28) yields lower bounds on P^*. The difference between the objective function value of the primal and dual problem is called the duality gap. Relative duality gap (RDG) is used to estimate how far the feasible solution is from the optimal solution. The RDG is calculated as follows:

$$\text{RDG}^{(k)} = \left| \frac{z^{(k)} - \phi^{(k)}}{\phi^{(k)}} \right| \leqslant \varepsilon \tag{1.31}$$

where $z^{(k)}$ and $\phi^{(k)}$ are the objective function values of problems (1.19) and (1.28) at iteration k, respectively. By setting prespecified tolerance ε of RDG, it is guaranteed that the final solution is close enough to the optimum. In this study $\varepsilon = 0.05$. If the RDG convergence criterion is not satisfied, the Lagrangian multipliers should further be adjusted to achieve a solution as close to the global optimality of the final solutions as possible.

Upper Level: Distribution System Reconfiguration

The optimization objective is

$$\min -\eta_{\text{LMP}}^t \cdot \sum_{i=1}^{N} PD_i \tag{1.32}$$
$$+ \eta_{\text{LMP}}^t \cdot \sum_{i=1}^{NS} P_{\text{inj},i} + \gamma \cdot \sum_{i=1}^{N} \left| \Delta E_i^t \right|$$

where PD_i is the constant load and equivalent load for non-MG nodes and MG nodes, respectively. P_{inj} is the power injection at the substation bus. $\mathcal{O}E_i^t$ is the adjustment requirement of trading amount for MGi. The absolute value function is needed since $\mathcal{O}E_i^t$ can be positive or negative. γ is a big constant to make sure that the trading adjustment is required only when the network constraints cannot be satisfied.

Constraints (1.33)–(1.36) guarantee that the distribution network has a tree structure. α_l is binary variable where $\alpha_l = 1$ if the line l is connected and 0 otherwise. β_{mn} is a binary variable where $\beta_{mn} = 1$ if bus n is the parent of bus m and 0 otherwise. The root node in each tree can be substation buses.

$$\alpha_l = \beta_{mn} + \beta_{nm}, \quad \forall l \in L \tag{1.33}$$

$$\sum_{n \in N(m)} \beta_{mn} \leqslant 1, \quad \forall m \in N \setminus N_s \tag{1.34}$$

$$\sum_{n \in N(m)} \beta_{mn} = 0, \quad \forall m \in N_s \tag{1.35}$$

$$\beta_{mn} \in \{0,1\}, \quad 0 \leqslant \alpha_l \leqslant 1, \quad \forall l \in L \tag{1.36}$$

Constraints (1.37) and (1.38) represent nodal real and reactive power balances. Constraints (1.39) and (1.40) denote real and reactive power flow from node m to node n, respectively. By defining auxiliary variable $u_m^l = V_i^2 / \sqrt{2}$, $r_i = V_i V_j \cos \theta_{ij}$,

$t_l = V_i V_j \sin \theta_{ij}$ where V_i is the voltage at node i, we calculate linearized power flow constraints (1.39) and (1.40). Constraint (1.41) represents the conic relaxation relationship of r_l, t_l, and u_m^l, u_n^l.

$$\Delta E_m - PD_m = \sum_{n \in N(m)} P_{mn}, \quad \forall m \in N \tag{1.37}$$

$$\varphi_m \cdot \Delta E_m - QD_m = \sum_{n \in N(m)} q_{mn}, \quad \forall m \in N \tag{1.38}$$

$$P_{mn} = \sqrt{2} g_l u_m^l - g_l r_l - b_l t_l, \quad \forall l \in L \tag{1.39}$$

$$q_{mn} = -\sqrt{2} \left(b + \frac{b_l^{sh}}{2} \right) u_m^l + b_l r_l - g_l r_l, \quad \forall l \in L \tag{1.40}$$

$$r_l^2 + t_l^2 \leq 2 u_m^l u_n^l, \quad \forall l \in L \tag{1.41}$$

Constraints (1.42) and (1.43) link the network configuration variable α_l to the auxiliary voltage variables u_m^l, u_m so that α_m^l can be set to 0 when α_l is 0 and u_m when α_l is 1. Equation (1.44) sets the current flow limit on each line. Equations (1.45)–(1.47) set upper and lower bounds for the auxiliary variables r_l, t_l and u_m.

$$0 \leqslant u_m^l \leq \frac{V_{m,max}^2}{\sqrt{2}} \alpha_l, \quad \forall m \in N \tag{1.42}$$

$$0 \leqslant u_m - u_m^l \leq \frac{V_{m,max}^2}{\sqrt{2}} (1 - \alpha_l), \quad \forall m \in N \tag{1.43}$$

$$\sqrt{2} A_l u_m^l - \sqrt{2} B_l u_n^l - 2 C_l r_l + 2 D_l t_l \leq I_{l,max}^2, \quad \forall l \in L \tag{1.44}$$

$$0 \leq r_l \leq V_{m,max} V_{n,max}, \quad \forall l \in L \tag{1.45}$$

$$-V_{m,max} V_{n,max} \leq t_l \leq V_{m,max} V_{n,max}, \quad \forall l \in L \tag{1.46}$$

$$\frac{V_{m,min}^2}{\sqrt{2}} \leq u_m \leq \frac{V_{m,max}^2}{\sqrt{2}}, \quad \forall m \in N \tag{1.47}$$

Constraints (1.48)–(1.52) ensure the number of switching status changes between every two adjacent hours is less than a certain threshold N_{switch} in order to reduce the switching cost and extend the life of switches. $\hat{\alpha}_l$ represents the last status of switch i. H_l is an auxiliary variable to apply the XOR operation of α_l and $\hat{\alpha}_l$.

$$\sum_l \left(\alpha_l + \hat{\alpha}_l - 2H_l \right) \leqslant N_{\text{switch}}, \quad \forall l \in L \tag{1.48}$$

$$H_l - \alpha_l \leqslant 0, \quad \forall l \in L \tag{1.49}$$

$$H_l - \hat{\alpha}_l \leqslant 0, \quad \forall l \in L \tag{1.50}$$

$$\alpha_l + \hat{\alpha}_l - H_l \leqslant 1, \quad \forall l \in L \tag{1.51}$$

$$H_l \geqslant 0, \quad \forall l \in L \tag{1.52}$$

By solving the problem stated in (1.32)–(1.52), the network topology as well as the trading quantity adjustment are obtained. If the trading adjustment amount ΔE_i is not 0, the following constraint is generated and added to MGs trading model.

$$\sum_{j \neq i, j \in \text{MG}} P_{\text{pur},ij}^t + P_{\text{pur},iD}^t - P_{\text{sel},i}^t + \Delta E_i \leqslant Pd_i - P_{\text{MT},i}^t \tag{1.53}$$

Flowchart for Multi-Microgrid Trading

The overall flowchart is shown in Fig. 1.7. At the multi-MG level, the process of the designed end-to-end trading process is shown in the lower dashed box of Fig. 1.7 which is described as follows:

(a) Each MG solves (1.29) for the given multipliers $\lambda_i^t, \forall i \in \text{MG}$. That is, each MGi calculates its buy $E_{\text{pur},ij}^t, \forall j$ and sell $E_{\text{sel},i}^t$ volumes according to other MG prices. Repeat Step (a) for all MGs.
(b) Submit trading results to transactive energy market. If the trading decisions for all MGs do not change much in two consecutive iterations, go to Step (c). Otherwise provide each MG's latest trading variables to other MGs and go to Step (d).
(c) The relative duality gap (RDG) ε is evaluated. In this study, $\varepsilon \leqslant 0.05$ is the convergence criterion. If the condition is satisfied, send the nodal equivalent load of MGs to DSO without publishing trading details. Otherwise go to Step (d).
(d) Each MG updates its selling price λ_i^t using the trading variables in (1.30). Repeat Step (d) for all MGs. Provide each MG's latest price $\lambda_j^t, \forall j$ to other MGs. Then go to Step (a).

For the distribution level, shown on the top section of Fig. 1.7, solve the reconfiguration problem (1.32)–(1.52) using the nodal equivalent load of MGs. Determine the latest network topology and evaluate the requirement of trading adjustment if the network security index is violated. Calculate the distribution path and the equivalent distance between MGs.

The basic processes in the proposed framework are: multi-microgrids make trading decisions and submit to DSO; then, DSO calculates the equivalent load of the

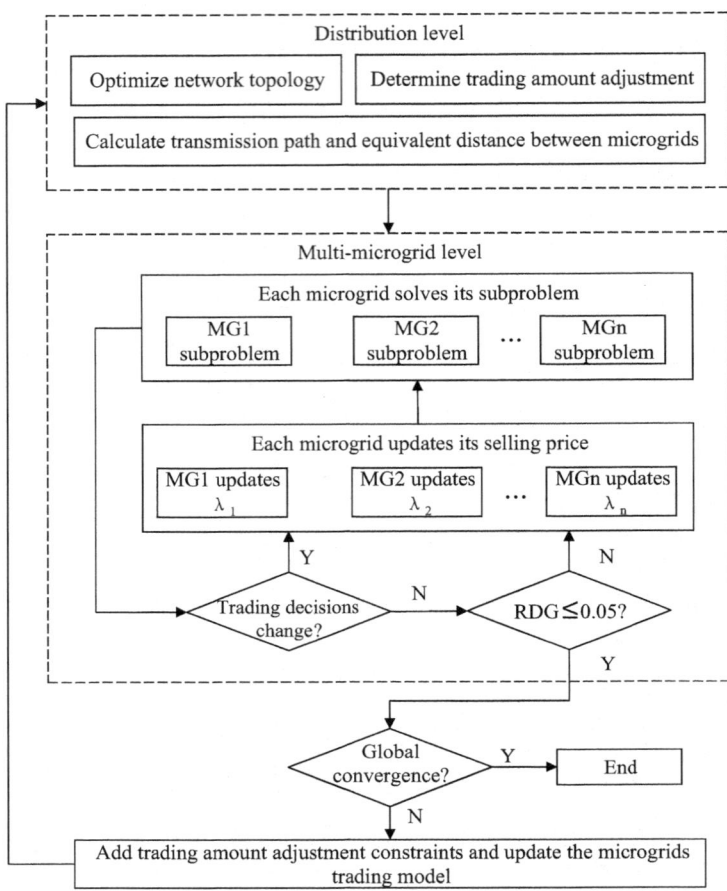

Fig. 1.7 Flowchart of proposed multi-microgrid trading

nodes where multi-microgrids locate; finally, DSO conducts the network reconfiguration to minimize the transmission loss of the distribution system. However, there are some existing problems: The network reconfiguration may change the network topology such that the equivalent distance between microgrids may change and affect the trading decisions of multi-microgrids. Therefore, two outcomes are possible after running the proposed bi-level optimization model: (1) the network reconfiguration does not change the network topology. It means the equivalent distance between microgrids does not change; (2) the network topology changes after network reconfiguration, but the equivalent distance between microgrids does not change. It means the transmission path between every two microgrids keeps the same; (3) the network topology changes after network reconfiguration and the equivalent distance between microgrids will also be changed.

For the first and the second situations, the trading decision of multi-microgrids will not be changed. The convergence criterion 1 is satisfied and the bi-level iteration process can be stopped.

For the third situation which is much more likely to occur, the trading decisions of multi-microgrids will be reconsidered due to the change of the equivalent distance between microgrids. The bi-level iteration process should be continued. Noted that for each line there are two states, i.e., on and off. It is known that the number of combinations of n lines is $2n$ which is very large but still limited. Especially for a distribution system with n lines, the number of combinations is less than $2n$. So we can make a conclusion that (1) in the proposed bi-level iteration process, the network reconfiguration result will be the same as the previous one after a certain number of iterations; (2) the convergence of the proposed bi-level iteration can always be achieved. It means the trading decisions of multi-microgrids will be repeated after a certain number of iterations. At this time, the convergence criterion 2 is satisfied and the bi-level iteration process can be stopped.

Note that the proposed bi-level programming problem can be transformed into an equivalent single-level mixed-integer SCOP problem by replacing the lower level optimization problem with its Karush–Kuhn–Tucker (KKT) optimality conditions. However, for the independency of each entity in the transactive energy market, multi-MG level, and distribution level are solved respectively in this situation.

Case Studies

To demonstrate the effectiveness of the proposed model and algorithm, the IEEE 33-bus [64] is used for simulation. The simulation is conducted using MATLAB R2014b with Gurobi solver. The distribution system topology is shown in Fig. 1.8. The three MG1, MG2, and MG3 are connected to nodes 9, 11, and 29, respectively. For simplicity, we assume each MG has one MT with the same quadratic generation cost function parameters $a = 0.0005$ $/(kWh)^2$, $b = 0.1809$ $/kWh, and $c = 1.223$$. In addition, $P_{\mathrm{MT},i}^{\min}$ is 0 kW for three MTs. The $P_{\mathrm{MT},i}^{\max}$ are 400, 500, and 500 kW, respectively for MG1, 2, and 3. The hourly LMPs at PJM are shown in Fig. 1.9.

Two cases are considered. Case 1 is to identify the impact of transactive energy trading among MGs without any network reconfiguration. Case 2 will add distribution reconfiguration to Case 1.

Case 1: MG Trading Without Network Reconfiguration

In Case 1, three scenarios are discussed as follows:

Scenario 1: MG participants use self-generation and do not trade any electricity, i.e., $E_{\mathrm{pur},ij}^{t} = 0$, $E_{\mathrm{sel},i}^{i} = 0$, and $E_{\mathrm{pur},iD}^{t} = 0$.

Scenario 2: MGs trade transactive energy but not trade with DSO, i.e., $E_{\mathrm{pur},ij}^{t} \geqslant 0$, $E_{\mathrm{sel},i}^{t} \geqslant 0$, and $E_{\mathrm{pur},iD}^{t} = 0$.

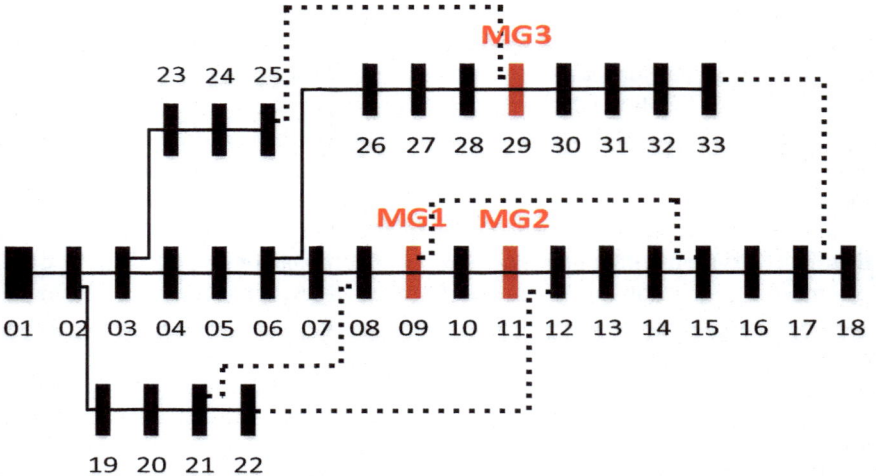

Fig. 1.8 Modified IEEE 33-bus distribution power system

Fig. 1.9 Hourly LMPs

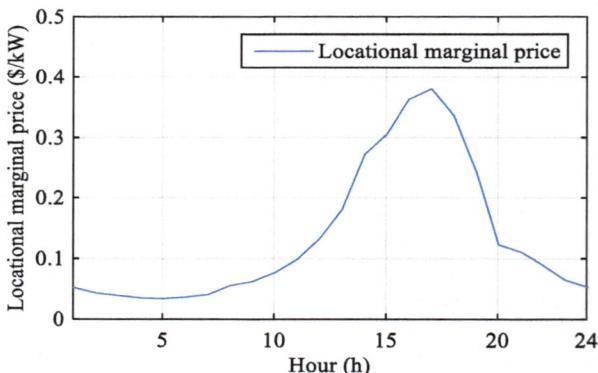

Scenario 3: MGs trade transactive energy with other MGs and DSO, i.e., $E_{\mathrm{pur},ij}^{t} \geqslant 0$, $E_{\mathrm{sel},i}^{t} \geqslant 0$, and $E_{\mathrm{pur},iD}^{t} \geqslant 0$.

In Case 1, Scenario 1 leads to the largest power consumption cost of \$1601.052 since no transactive energy between multi-MG is considered. Each MG supplies its load by own MT without any cooperation and coordination.

Scenario 2 leads to a lower consumption cost of \$1555.663 as transactive energy is considered among MGs. Take hour 19 for example. In the final transactive energy trading, Lagrangian multipliers are 0.2815, 0.2796, and 0.2872, representing sale prices for MG1, MG2, and MG3, respectively. To increase the payoff, MG3 buys 27.734 and 24.063 kW from MG1 and MG2, respectively. For MG1, it generates 100.579 kW, buys 11.405 kW from MG2, and sells 27.734 kW to MG3. That is,

11.405 kW of 27.734 kW sold by MG1 is generated by MG3. Accordingly, lowest generation and delivery cost can be obtained by dispatching the MG3 demand appropriately. If no delivery cost is considered, each MG will trade transactive energy based on its incremental generation cost and buys power from other producers. Otherwise, trading decisions will be based on generation cost and electrical distance among MGs.

In Scenario 3, shown in Fig. 1.10, MGs have more options to buy power as DSO supplies part of the demand. MGs trade with each other between hours 14:00 and 19:00 according to the hourly LMPs. Take hour 19:00 for example. MG3 with the heaviest load generates 91.900 kW and buys 11.549, 11.156, and 52.268 kW from MG1, MG2, and DSO, respectively. MG1 buys 9.800 kW from MG2 and sells 11.549 kW to MG3. MG2 generates 83.533 kW and sells 11.156 kW to MG3. The consumption cost is $895.909 in this case.

Case 2: MG Trading with Network Reconfiguration

DSO not only participates in transactive energy market but also reconfigures the distribution network as proposed in the bi-level coordination framework. Two scenarios are discussed.

Scenario 1: DSO reconfigures network and coordinates with three MGs located at bus 9, 11, and 29.

Scenario 2: MG3 is shifted from bus 29 to bus 33 to analyze the impact of MGs locations.

Figure 1.11 shows the hourly total cost of MGs and DSO with and without distribution reconfiguration which demonstrates the merits of reconfiguration. The cost of MGs is calculated by the primal objective function (1.19) and the cost of DSO is evaluated by the distribution network loss multiplied by LMPs. This observation indicates that the proposed framework can effectively reduce the social cost of distribution system with using MGs. It should be noted that network reconfiguration cannot always reduce MG trading costs which depend on the network topology and MG locations and operating conditions.

Figure 1.12 shows the MG total cost for different MG3 locations, which is impacted by network losses. An obvious reduction in the MG total cost is obtained when MG3 is relocated from bus 29 to 33. Although distribution losses are fairly close in most of the hours, Fig. 1.13 shows that the total power losses over 24 h are reduced after MG3 relocated. Hence, MG locations can have a major impact on trading and consumption costs. This conclusion also suggests that the MG operation and planning, market trading mechanism, and distribution system planning and operation, must all be coordinated in market operations.

Table 1.11 shows the interaction process between DSO and MGs at hour 14:00. At the initial iteration $k = 0$ transactive energy trading is set to 0 and the network is initiated. It takes three iterations to converge the bi-level solution. In each iteration, λ_i^t and trading behaviors are updated based on the last network topology. Accordingly, based on the transactive energy trading results, the network

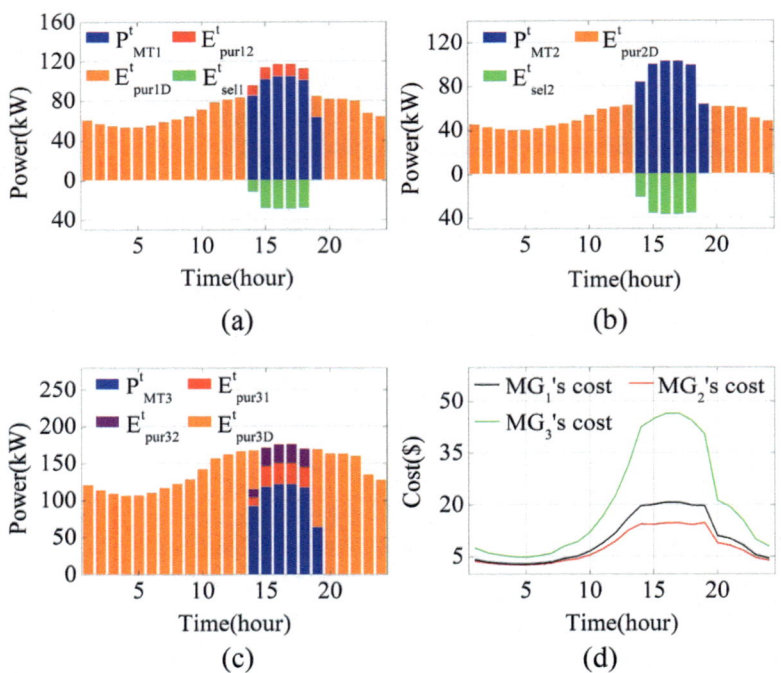

Fig. 1.10 Multi-MG trading behavior in Case 1 Scenario 3. (**a**) MG1 generation and trading. (**b**) MG2 generation and trading. (**c**) MG3 generation and trading. (**d**) MGs cost

Fig. 1.11 Total cost with and without distribution reconfiguration

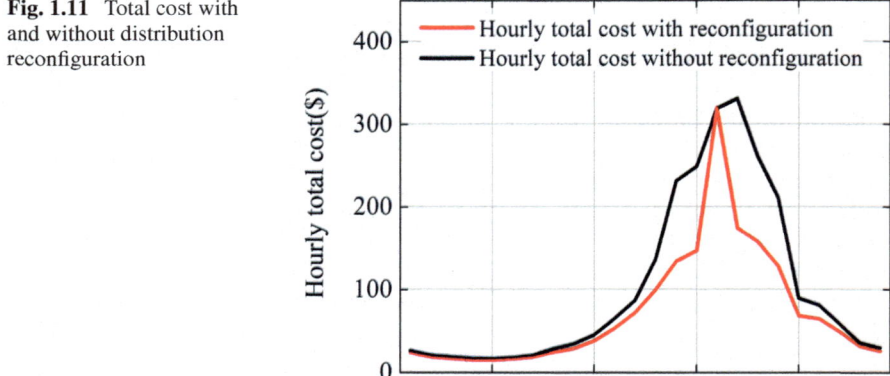

reconfiguration applies line switching and changes the electrical distance between MGs. The DSO cost is calculated by the first two terms in (1.32). In the column of MG trading behavior, variables that are zero are not shown for simplicity. As iterations continue, network losses will decrease further until DSO and MGs reach a stable point. At the last iteration $k = 3$, the electrical distance between MGs remains

Fig. 1.12 MGs total cost at different MG3 locations

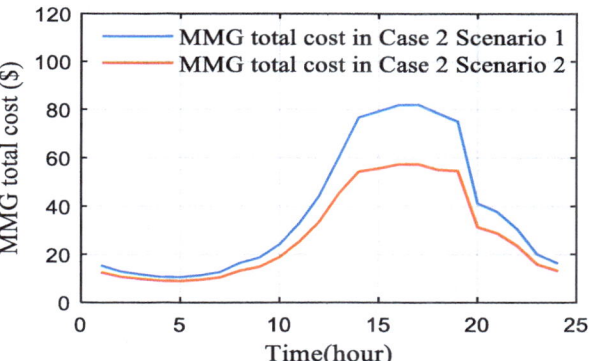

Fig. 1.13 Network loss at different MG3 locations

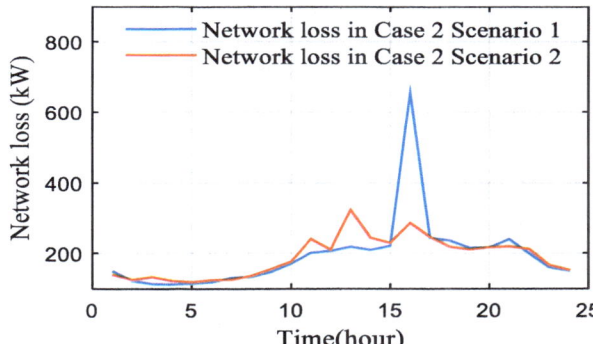

the same as that of $k = 2$ though the open switches are different. The convergence criteria are satisfied in the last transactive energy trading.

Table 1.12 shows the final results at hour 14:00 when MG3 is shifted from bus 29 to 33. Compared with Table 1.11, the MG relocation leads to different trading behavior and the reduction in the MG's total cost. The proposed model can help decision makers consider various scenarios for MG locations and the coordination and interaction among DSO and MGs.

Interpretation of Trading Behaviors

The MG's trading behavior is affected by marginal prices. The MGi's objective function (1.29) consists of four parts, i.e., MT generation cost, power purchase cost from DSO, power purchase from other MGs, and revenue for power sale to other MGs. For MGi, the derivation of each part is shown as (1.54)–(1.57):

$$M\left(P_{\mathrm{MT},i}^{t}\right) = C'\left(P_{\mathrm{MT},i}^{t}\right) = 2a_i P_{\mathrm{MT},i}^{t} + b_i P_{\mathrm{MT},i}^{t} \tag{1.54}$$

Table 1.11 Results at 14:00 in Case 2 Scenario 1

	MG trading behavior (kW)		$\begin{pmatrix} \lambda_1^t \\ \lambda_2^t \\ \lambda_3^t \end{pmatrix}$		
k	Buy	Sell		Open switches	DSO cost ($)
0	0	0	0	6, 8, 9, 17, 27	154.86
1	$E_{\text{pur}12} = 5.962$ $E_{\text{pur}31} = 8.342$ $E_{\text{pur}32} = 15.631$	$E_{\text{sel},1} = 8.342$ $E_{\text{sel},2} = 21.593$	0.267 0.265 0.267	6, 14, 17, 21, 29	64.81
2	$E_{\text{pur}12} = 10.057$ $E_{\text{pur}31} = 11.133$ $E_{\text{pur}32} = 11.010$	$E_{\text{sel},1} = 11.133$ $E_{\text{sel},2} = 21.066$	0.265 0.265 0.266	7, 10, 17, 29, 34	56.65
3	$E_{\text{pur}12} = 6.965$ $E_{\text{pur}31} = 9.850$ $E_{\text{pur}32} = 14.244$	$E_{\text{sel},1} = 9.850$ $E_{\text{sel},2} = 21.209$	0.267 0.265 0.267	6, 9, 12, 15, 37	63.37

Table 1.12 Case 2 Scenario 2 results at 14:00

	MG trading behavior (kW)		$\begin{pmatrix} \lambda_1^t \\ \lambda_2^t \\ \lambda_3^t \end{pmatrix}$		
k	Buying	Selling		Open switches	DSO cost ($)
3	$E_{\text{pur}12} = 7.443$ $E_{\text{pur}31} = 0.903$ $E_{\text{pur}32} = 4.977$	$E_{\text{sel},1} = 0.903$ $E_{\text{sel},2} = 12.420$	0.258 0.256 0.258	7, 8, 9, 16, 26	65.80

$$M\left(P_{\text{pur},ij}^t\right) = \sum_{j \in \text{MG}} \left(C_{ij}^{t\text{del}} + \lambda_j^t\right)$$
$$= \sum_{j \in \text{MG}} \left(\sigma_{ij}\left(\beta^2 + 1\right)d_{ij}P_{\text{pur},ij}^t + \lambda_j^t\right) \tag{1.55}$$

$$M\left(P_{\text{pur},iD}^t\right) = \eta_{\text{LMP}}^t \tag{1.56}$$

$$M\left(P_{\text{sel},i}^t\right) = \lambda_i^t \tag{1.57}$$

where (1.54)–(1.56) are the marginal price of microturbine generation, power purchase from other MGs, power purchase from DSO, respectively. Equation (1.57) is the marginal income for MGi sales. The above four parts will individually lead to different choices for MGi. Here, $M\left(P_{\text{MT},i}^t\right)$ and $M\left(P_{\text{pur},ij}^t\right)$ increase with $P_{\text{MT},i}^t$ and $P_{\text{pur},ij}^t$, respectively. Before $M\left(P_{\text{MT},i}^t\right)$ and $M\left(P_{\text{pur},ij}^t\right)$ grow to be equal to $M\left(P_{\text{pur},iD}^t\right)$, i.e., η_{LMP}^t, an MG will prefer to trade with other MGs. After that, the MG will buy from DSO. If $M\left(P_{\text{sel},i}^t\right)$ is higher than $M\left(P_{\text{MT},i}^t\right)$ and $M\left(P_{\text{pur},ij}^t\right)$, an MG will sell energy.

In Scenario 1 of Case 2, LMP is 0.273 $/kW at hour 14:00. The MG3 load is 166.873 kW which is supplied by purchasing 14.244, 9.850, and 50.880 kW, from

MG2, MG1, and DSO, respectively. We apply the final iteration results listed in Table 1.10 to (1.54)–(1.57) and depict the marginal cost of MG3 for suppling its load in Fig. 1.14. Here, the slope of marginal consumption cost has incorporated the impact of Lagrangian multipliers according to (1.54)–(1.57). When MT3 is dispatched at 83.800 kW, as shown by point 1 in Fig. 1.13, the incremental cost of MT3 is 0.265 $/kW which is the same as the price for buying power from MG2.

Next, MG3 will increase the dispatch of MT3 as the load increases and buys power from MG2 until MT3 reaches 86.300 kW and the purchase from MG2 reaches 2.500 kW, as depicted by point 2. Note that the incremental cost of MT3 and purchase price from MG2 have remained the same in order to minimize the cost as the load increases. At point 2, the incremental cost of MT3 and the purchase price from MG2 are 0.267 $/kW which is equal to that of purchasing from MG1.

The higher load will be supplied by MT3 dispatch and purchases from MG2 and MG1 simultaneously. When MT3 dispatch increases to 91.900 kW, purchases from

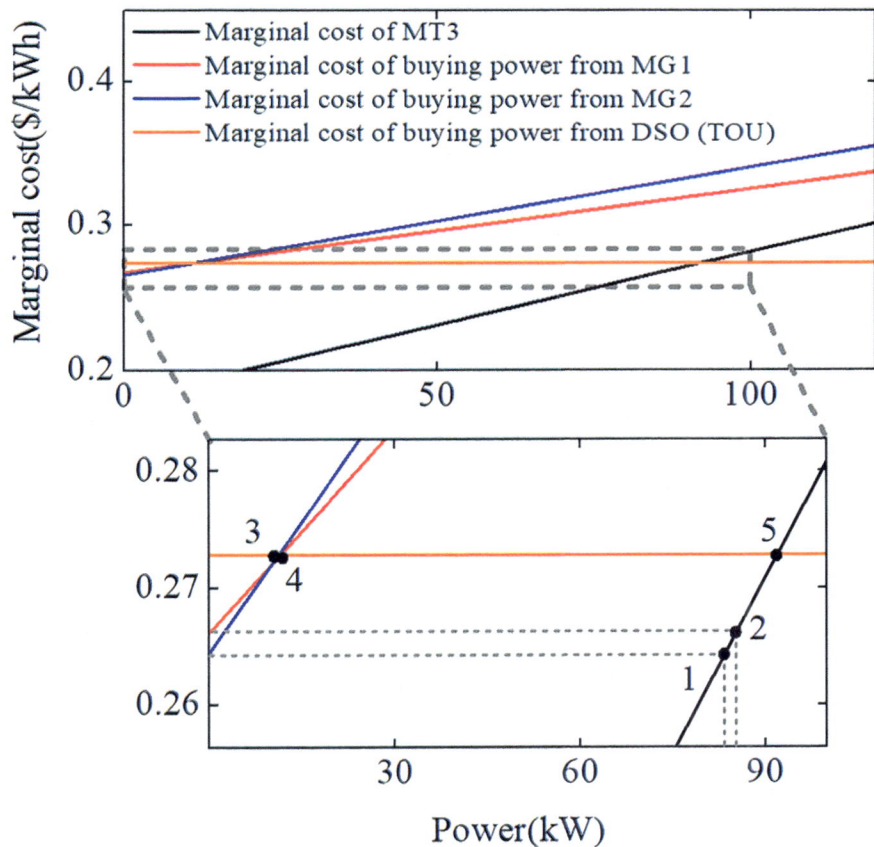

Fig. 1.14 Marginal consumption cost of MG3

MG1 and MG2 increase to 9.850 kW (point 3) and 14.244 kW (point 4). In this case, the incremental cost of MT3, as it purchases from MG2 and MG1, is equal to that of the power purchase from DSO, i.e., LMP at 0.273 $/kW. If the load continues to grow beyond 91.90 + 9.850 + 14.244 = 115.994 kW, MG3 will buy power from DSO to supply the load in excess of 115.994 kW.

Convergence Performance

Consider Case 2, Scenario 1, hour 14:00. Figure 1.15 shows the convergence trend and Fig. 1.16 shows the evolution of the Lagrangian multipliers in the last bi-level iteration. The proposed bi-level framework contains the distribution reconfiguration problem and multi-microgrid trading problem. The computer burden of each part and the whole framework is described as follows:

1. The reconfiguration problem is solved with a mixed integer second-order cone programming and takes about 2 s.
2. The Lagrangian relaxation iteration for MG trading problem converges in about 120 iterations and takes 100 s generally.
3. One bi-level iteration takes about 100 + 2 = 102 s. In general, it takes about three bi-level iterations to converge. Therefore, for each scheduling point, it takes about 102 × 3 = 306 s. And for a 24-h scheduling, it takes about 306 × 24 = 7344 s.

The main computation burden lies in the microgrids trading level. Note that the proposed end-to-end microgrids trading model has a distributed structure. If parallel computation is adopted to solve each Lagrangian subproblem, the total CPU time would be significantly reduced. Moreover, the proposed approach is implemented with a general solver in this test. The calculation efficiency can be further improved with algorithms particularly developed for this kind of problem. To sum up, with the development of fast computational and communication techniques, the computation burden would not be a challenge for the implementation of the proposed model.

Fig. 1.15 Objective function and relative duality gap

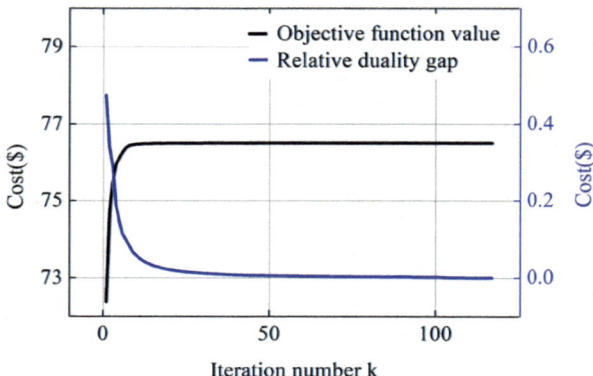

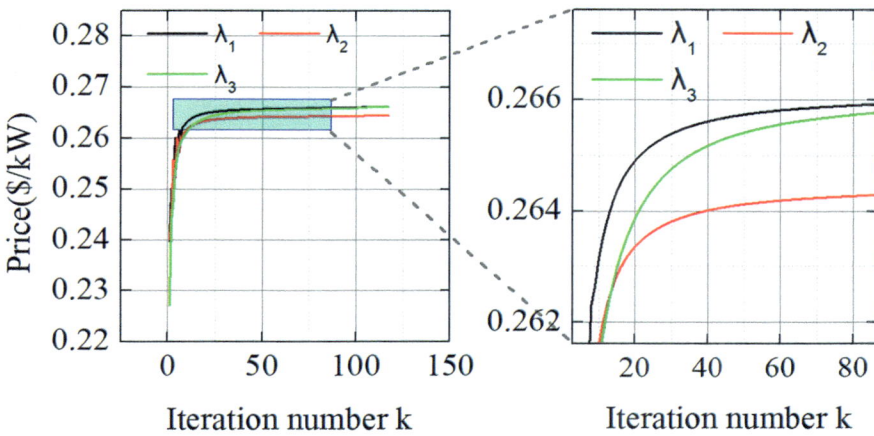

Fig. 1.16 Lagrangian multipliers (sale price) convergence

Conclusions

This section devises an end-to-end transactive energy market to highlight the cooperation among MGs and the coordination between the distribution system and MGs. The following conclusions can be drawn.

- Using the proposed bi-level transactive energy framework, MGs can trade energy to lower their costs and make more payoff by coordination their trades with DSO.
- Lagrangian relaxation and decomposition techniques provide an effective end-to-end transactive energy trading mechanism where Lagrangian multipliers provide price signals in transactive energy market.
- The MGs trading decisions are based on marginal consumption costs. The distribution path between MGs and Lagrangian multipliers determine the marginal costs and transactive energy trading behaviors.
- The locations of MGs in a distribution system pose a great impact on transaction costs and trading behaviors. The distribution reconfiguration should be considered effectively when multi-MGs are considered.
- The proposed bi-level framework can coordinate the DSO's operation with that of the multi-MG trading, reduce network losses, and enhance the MGs' payoffs effectively.

1.3.1.3   Resilience-Constrained Power Systems in Extreme Conditions

In this section, impact of cascading and common cause outages on resilience-constrained economic operation of power systems in extreme conditions will be explained.

Nomenclature

Variables and Functions	
$C_i(\cdot)$	Fuel consumption function of unit i
LC_j	Load curtailment at load j
i	Index for generator unit
j	Index for bus load
l, p, q	Index for transmission line
P_i	Generation of unit i
PL_l	Real power flow of line l
PL_l^{pc}	Line flow on l, due to an outage in line p
$\mathrm{PL}_l^{p-q,\mathrm{c}}$	Line flow on l due to common-cause outages in line p and q
p_k^{cf}	Weather-dependent outage probabilities of conductor k
p_k^{tf}	Weather-dependent outage probabilities of tower k
p_l^{wf}	Weather-dependent outage probabilities of transmission line l
$p_{p,q}^{\mathrm{cco}}$	Probability of common-cause outage of line p and q
p_l^{hf}	Hidden outage probability of line l
r_l	Absolute loading rate of line l
s_l, t_l	Auxiliary variables to linearize (1.73c)
u_l, v_l	Auxiliary variables to linearize (1.66a)
σ_l	Auxiliary binary variables to linearize constraint (1.66c)
$\varnothing\mathrm{PL}_l^{pc}$	Slack variable for line flow on l due to a single outage of p
$\Delta\mathrm{PL}_l^{p-q,\mathrm{c}}$	Slack variable for line flow on l due to common-cause outages of p and q
Constants and Sets	
α	Coefficient of generation cost
β	Coefficient of the first penalty term
γ	Coefficient of the second penalty term
PD_j	Real power demand of transmission load j
KD	Bus-load incidence matrix
KL	Bus-line incidence matrix
KP	Bus-generator incidence matrix
NL	Number of total transmission lines
NG	Number of generation units
ND	Number of transmission loads
NC_l	Number of conductors on line l
NT_l	Number of towers connected to line l
$P_{\min, i}$	Lower limit of real power generation of unit i

$P_{\text{max}, i}$	Upper limit of real power generation of unit i
$PL_{\text{max}, l}$	Capacity limit of line flow of line l
S_{AL}	Set of extreme weather affected lines

Introduction

An increasing number of cascading outages in extreme conditions indicate that power system vulnerabilities are continuously exposed to serious weather conditions which could culminate in extensive power blackouts. Resilience would evaluate the performance of an ecosystem affected by external changes and continually confronted by unexpected events. Similarly, power system resilience describes the capability of power systems to change itself to withstand major events with high impact and low probability [2, 5, 65].

Figure 1.17 shows a typical power system resilience curve which is divided into three development stages, i.e. adaptation, absorption, and restoration, with specific resilience indices [4, 66, 67]. For example, the BC slope denotes how fast the system deteriorates, CD segment denotes the system robustness, DF segment denotes how promptly the network recovers, and $BCDEF$ area denotes the system loss. According to the definition of power system resilience, the adaptation in Fig. 1.17 describes the power system capability to adapt to prevailing conditions in response to unexpected events. However, the adaptation stage lacks indices that describe the adaptive capacity of power systems in extreme events which may lead to cascading outages. In this study, a new resilience index is introduced to describe adaptation performance and establish a preventive resilience-constrained economic dispatch (RCED) strategy to improve power system adaptability.

Some previous studies proposed preventive strategies for enhancing the power system resilience. In [22], a sequential proactive operation strategy was proposed where the system state transition follows a Markov process. Reference [68] considered cascading outages and $N - k$ contingencies to establish a risk-based operation strategy. In [69], an $N - k$ contingency screening method for economic dispatch was proposed with multi-objective optimization where maximized system load shedding and minimized system load shedding are considered. In [70], the economic dispatch contingency set was constructed based on risk assessment under a bi-level framework. Reference [25] studied the proactive microgrid dispatch strategy to enhance resilience in which the islanded operation time is modeled as uncertainty set. To further enhance power system resilience, Reference [71] studied the power system economic dispatch integrated with microgrids in extreme conditions. Reference [72] proposed the optimal resilience operation in terms of line hardening. Different hardening methods were determined to reduce line outage probabilities and load shedding costs. Reference [73] studied the resilience enhancement strategy considering the line hardening and the formation of multiple islanded provisional microgrids. Reference [74] proposed an $N - k$ contingency screening method

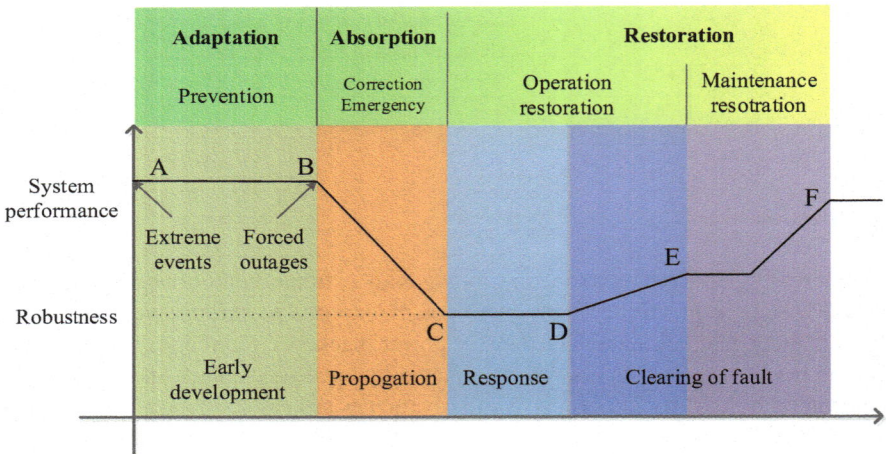

Fig. 1.17 Typical resilience curve

considering the hidden outages. Reference [75] proposed an approach to construct the constraints for contingency events using line outage distribution factors to reduce the computational burden. Reference [76] considered the extreme events and established a robust model to achieve optimal hardening strategy in integrated electricity and nature gas transportation systems.

Aforementioned references studied the power system resilience and security considering $N - k$ contingencies. Such resilience operation strategies could improve the system performance toward specific contingencies. These studies have had an implicit assumption that improving the power system reliability, such as implementing the $N - k$ contingencies, could lead to more resilient power system operations at certain circumstances. This view may be true when a power system is subject to typical outages. However, when a power system is subject to extreme conditions, such as severe weather with common-cause outages and cascading outages, the traditional $N - k$ reliability security strategies may not be effective.

In extreme situations, power system operation characteristics and forced outage modes could change [77]. Forced outages occur randomly with certain effects on reliability, but presumed outages could be much more profound if they demonstrate cascading effects and outage correlations in extreme circumstances. Moreover, contingencies representing typical power system outages would usually be more complicated in extreme events due to the extent of common-cause outages and cascading outages. Thus, if reliability-based operation strategies are instituted without considering the unique features of extreme events, the severe impacts of cascading outages, could more readily culminate the power system in blackouts (e.g., North America on August 14, 2003, Europe on November 12, 2006, Brazil on November 10, 2009, and India on July 30, 2012). To enhance the system resilience, the common-cause outages and cascading outages besides the typical outages should be considered in the operation strategy.

Common-cause outage refers to simultaneous outages of multiple components due to a common cause [78]. For example, the outage of two or more circuits on the same transmission tower can occur due to a single cure and the outage of multiple lines on the same substation due to lightning invasion wave overvoltage accident. Also, a major physical disturbance such as tornado can result in the outage of two or more transmission circuits on the same right-of-way. Such outages can be classified as common-cause outages since a single cause results in an outage of two or more elements [79, 80]. Cascading outages with inherently complex nature could have a compounded effect on power system operations. Some references linked power system operating conditions to cascading outages. Reference [27] demonstrated that self-organized criticality is an essential characteristic of large blackouts. Reference [28] illustrated that power system loading that is close to the system operating limits is the key contingency attribute that could lead to cascading outages. According to the system structure and operating states, Ref. [29] proposed an entropy-based metric to evaluate the power grid robustness with respect to cascading outages. Reference [32] showed the correlations between self-organized criticality and the heterogeneity of power flow distribution by introducing the power flow entropy index. Accordingly, the larger the power flow entropy, the more routinely a power system state can lead to self-organized criticality and eventually lead to cascading outages. Therefore, to improve the power system resilience, a comprehensive operation strategy is needed where the power flow distribution and customized contingencies under extreme events should be addressed simultaneously. This study fills the gap in which penalty terms and customized contingency constraints are established by considering extreme events to improve the uniformity of power flow distribution, reduce the impact of common-cause outages and cascading outages, and boost the system adaptability.

Reference [81] presented a resilience-constrained unit commitment model where the power flow entropy was considered to improve the power system resilience. The main difference between [81] and this study lies in the following points. Firstly, new penalty terms are proposed in this study. The required terms to be added are much less than that in [81], which leads to better computation performance. Secondly, the convexification method of penalty terms is different. An approximation method is established at the price of optimality loss in [81]. In this section, a transformed problem is established to solve original problem and prove that the two problems are equivalent when a necessary and sufficient condition is satisfied. Moreover, a new contingency set containing three types of contingency events is established. Furthermore, a new resilience evaluation index is proposed to reflect the adaptability of power systems under extreme conditions.

The main contributions of this work are summarized as follows:

- Considering common-cause outages and cascading outages, a resilience index is proposed to quantify the power system adaptability to extreme events. The evaluation index is utilized in the adaption stage.
- An RCED model for blackout prevention and resilience enhancement is presented in which the system security subjected to common-cause outages and

cascading outages is addressed simultaneously. Two penalty terms are introduced to improve the system resilience under hidden cascading outages. The common-cause outages and cascading outages types of contingencies are evaluated to improve the power system performance under reliability types of outages.
• A convexification method is proposed to linearize the RCED model without the loss of optimality. Although the linearized problem is not equivalent to the original one, a sufficient and necessary condition is introduced to ensure that the optimal value of linearized problem is the same as that of the original problem.

This session is organized as follows: Section "Resilience Evaluation for Cascading Outages" describes the proposed cascading-based resilience evaluation approach and index. Sections "The Proposed RCED Model" and "Solution Methodology of RCED Model" introduce the proposed resilience constrained economic dispatch and its convexification solution. Section "Case Studies" presents the case studies and the work is concluded in section "Conclusion".

Resilience Evaluation for Cascading Outages

A resilience index is proposed in this study for quantifying the adaptation performance, which is based on the probability distribution of blackout size. Blackout size in this study refers to the scale (severity) of blackout, which is defined as the percentage of load curtailment (LC), i.e., load curtailment/system load. A resilience evaluation approach considering the common-cause outages and cascading outages is established.

Random Outages in Extreme Conditions

The proposed process for considering random outages in power systems includes three stages. The first state considers extreme events, such as severe weather, to determine initial line outages. The second stage considers the common-cause outages of adjacent lines. The third stage considers cascading outages in which a simulation model is introduced to determine whether the remaining lines are subject to cascading outages. The details of each state are as follows:

Weather-Dependent Initial Line Outages

Without the loss of generality, we apply the generic wind-related fragility curves for transmission lines and towers [4]. For a real power system, the fragility function of each component in different weather conditions can be derived empirically from statistical analysis based on observed failures. The wind speed can get from meteorological monitoring system or derived from prevailing wind field model. Assume the failure probability of a single conductor and tower is p^{cf} and p^{tf}, respectively. Since individual failure of a conductor and transmission tower both will lead to outage of a transmission line, the outage probability of a transmission line structure is

$$p_l^{\text{wf}} = 1 - \prod_{k=1}^{Nc_l}\left(1 - p_k^{\text{cf}}\right) \cdot \prod_{k=1}^{Nt_l}\left(1 - p_k^{\text{tf}}\right) \tag{1.58}$$

Common-Cause Outages

In practice, common-cause outages occur when one event causes multiple outages which are not statistically independent. In this study, we only consider the common-cause outages of adjacent components in the weather-affected areas. The two lines could be either in a common right-of-way or connected with the same bus. The common-cause outages of transmission line p and line q can be derived from the following equation [82].

$$p_{p,q}^{\text{cco}} = \frac{\mu_p \mu_q \lambda_{p,q}^c}{\left(\mu_p + \lambda_p\right)\left(\mu_q + \lambda_q\right)\mu_{p,q}^c + \mu_p \mu_q \lambda_{p,q}^c} \tag{1.59}$$

where λ_p and λ_q are the failure rate (failures/year) of the transmission lines p and q, respectively, μ_p and μ_q are the repair rate (repairs/year) of the transmission lines p and q, respectively. $\lambda_{p,q}^c$ and $\alpha_{p,q}^c$ are the common-cause failure rate and repair of the transmission lines p and q, respectively.

Hidden Outages

A hidden outage remains undetected in normal operating conditions but exposed after the occurrence of a power system disturbance, which may cause relays to trip erroneously [83]. According to [84, 85], the components connected to tripped lines would also be exposed to incorrect tripping. The hidden outage probability, which has an approximate linear relationship with the line loading when the corresponding flow exceeds its limit [83], is stated as

$$p_l^{\text{hf}} = \begin{cases} p_0, 0 \le r_l \le 1 \\ k_0 \cdot r_l + b, 1 \le r_l \le r_l^t \\ 1, r_l \ge r_l^t \end{cases} \tag{1.60}$$

where p_0 is the initial hidden outage probability which depends on line parameters, k_0 and b are the coefficients of the linear function, and r_l^t is the thermal limit of line l.

Proposed Resilience Evaluation Indices

Figure 1.19 shows two different complementary cumulative distribution function (CCDF) of blackout size distribution of a system in different operating strategies, where x could be any blackout measure, such as load curtailment percentage, tripped

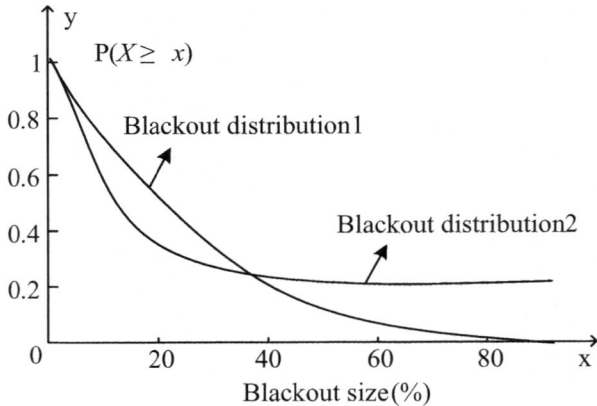

Fig. 1.18 Two blackout distributions

lines percentage, y is the probability of $X \geq x$. Initially, curve 1 drops more sharply as blackout gets larger, which indicates that curve 1 has higher proportion of small accidents but fewer major accidents. On the contrast, curve 2 has a relatively flat tail which means that curve 2 has a higher proportion of large accidents but fewer small accidents. Therefore, from the perspective of preventing large blackout in extreme conditions, the first operational strategy is more resilient.

However, the reliability indices of the two strategies, such as the expected load curtailment, may be similar. Therefore, traditional reliability evaluation indices underestimate the risk of large blackout and are not suitable to evaluate system resilience. Resilience considers the performance when power system suffers from extreme events, conventional reliability index cannot reflect the resilience characteristics effectively. To fill the gap, we propose a resilience index

$$\text{RI} = \sum_{k=1}^{\text{NX}} x_k \cdot P\left(X \geq x_k\right) \tag{1.61}$$

where NX is the number of points to evaluate (1.61). In this study, NX = 100. RI is evaluated by changing x_k from 0% to 100%, which ensures that large blackouts, which pose a greater influence on resilience, are represented by larger RI (though the two expected load curtailments are similar). The incremental resilience index considering two different strategies is stated as

$$\Delta RI = \text{RI}_2 - \text{RI}_1 = \left(\sum_{k=1}^{\text{NI}} x_k \cdot \left(P_2\left(X \geq x_k\right) - P_1\left(X \geq x_k\right)\right) \right) \tag{1.62}$$

The blackout size x in (1.61) and (1.62) is defined as load curtailment/system load which is dimensionless. Thus, RI is dimensionless. To calculate the RI, the input information includes system parameters, forecasted weather-affected areas, dispatch condition, and failure characteristics. The output information includes sampled scenarios, probability distribution of load curtailment, and resilience index.

To calculate x, we apply the Monte Carlo technique to the proposed resilience evaluation method. The overall flowchart of the simulation process is shown in Fig. 1.20. For each Monte Carlo simulation:

1. Sample weather-related initial line outages according to (1.58). The initial tripped transmission lines are simulated by comparing p_l^{wf} with a uniformly distributed random number $\rho_1 \sim U(0, 1)$. Trip line l if $p_l^{wf} > \rho_1$.
2. Sample common-cause outages adjacent to outage lines according to (1.59). Compare $p_{p,q}^{cco}$ with $\rho_2(\rho_2 \sim U(0, 1))$, trip line i and line j if $p_{p,q}^{cco} > \rho_2$.
3. Check network connectivity, calculate the blackout size in each island if the network is partitioned, and end the process. Otherwise, go to Step 4.
4. Calculate power flow and check thermal limit violations. Trip the lines with load rates exceeding r_l^t.
5. Sample hidden outages. Identify lines connected to tripped lines and calculate hidden outage probability according to (1.60). Trip individual lines when $p_l^{hf} > \rho_3 \left(\rho_3 \sim U(0,1) \right)$. Go to Step 3.

The Proposed RCED Model

Penalty Terms Based on Power Flow Entropy

The power flow entropy provides a measure of power flow distribution uniformity. Reference [32] showed that the power flow entropy has a close relation with the blackout size in cascading outages. When the entropy is high, transmission lines which carry heavy loads can fail and trigger cascading outages more easily. However, it is difficult to optimize the power flow entropy directly in a mathematical programming model. Therefore, to reduce the power flow entropy and homogenize the power flow distribution, we consider two penalty terms as

$$
pn_1 = \sum_{l \in S_{AL}} \frac{PL_l}{PL_l^{max}} \tag{1.63}
$$

$$
pn_2 = \sum_{l=1}^{NL} \left\| \frac{PL_l}{PL_l^{max}} \right| - \frac{\sum_{l=1}^{NL} \left| PL_l / PL_l^{max} \right|}{NL} \right| \tag{1.64}
$$

The first penalty term pn_1 denotes the absolute loading rate of weather affected lines. This term is introduced to adjust the power flow and avoid the weather affected lines undertaking heavy loads. The second penalty term pn_2 denotes the mean absolute deviation (MAD) of lines load ratio which is a measure of statistical dispersion defined as $\frac{1}{n} \sum_{k=1}^{n} |y_k - \bar{y}|$ where $\bar{y}$ is the mean of $\{y_1, y_2, ..., y_n\}$. pn_2 is established to

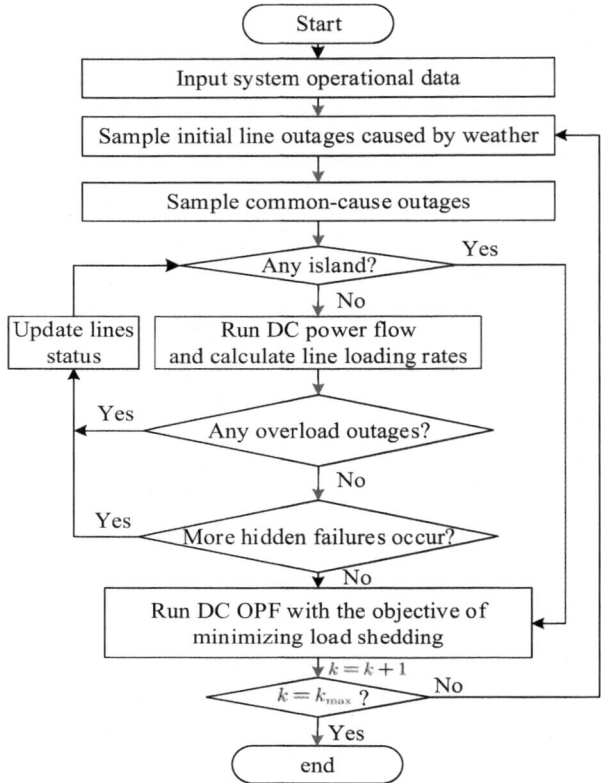

Fig. 1.19 Overall flowchart of the proposed resilience evaluation method

reduce the power flow heterogeneity and the cascading risk when lines are subject to extreme events. By including the above two penalty terms, the transmission network is operating at a more resilience loading level. That is, each component has a minimum impact upon failure and larger operating margin. Note that the term is nonconvex where its convexification and linearization methodology are proposed in the next subsection.

Reference [81] presented a resilience-constrained unit commitment model where the power flow entropy was considered to improve the power system resilience. However, the two penalty terms proposed in [81] are approximated to make them solvable at the price of optimality loss. In this work, new penalty terms and solution methodology are proposed without loss of optimality. Moreover, the penalty terms have much less terms than before which will lead to faster computation performance.

Contingency Set for RCED

In here, common-cause outages of two adjacent lines and cascading outages are considered to construct contingency set. For common-cause outages contingency, assuming that bus i has k connected lines and C_k^2 contingencies are added to this bus. Traverse all buses in the weather-affected areas and establish the contingencies accordingly. Practically, there are few lines connected to one bus and the proposed common-cause outage contingencies will not lead to the curse of dimensionality. For cascading outages, the heavy loading lines upon the weather-induced initial outage are identified and added into contingency set in case of hidden outages.

Constraints corresponding to each contingency are constructed using line outage distribution factor [75]. The corresponding contingency constraints for initial outages are shown in (1.65) and (1.66). Given line p is failure, the lines adjacent to line p and the lines undertaking heavy load upon p failure are denoted as q. The constraints for common-cause contingency and cascading contingency are shown in (1.67) and (1.68).

$$\text{PL}_l^{pc} = \text{PL}_l + \text{LODF}_{l,p}^{pc} \cdot \text{PL}_p \quad \forall l, p \tag{1.65}$$

$$\left| \text{PL}_l^{pc} \right| \leq \text{PL}_{\max,l} + \Delta\text{PL}_l^{pc} \tag{1.66}$$

$$\text{PL}_l^{p-q,c} = \text{PL}_l + \text{LODF}_{l,p}^{p-q,c} \cdot \text{PL}_p + \text{LODF}_{l,q}^{p-q,c} \cdot \text{PL}_q \quad \forall l, p, q \tag{1.67}$$

$$\left| \text{PL}_l^{p-q,c} \right| \leq \text{PL}_{\max,l} + \Delta\text{PL}_l^{p-q,c} \tag{1.68}$$

where PL_l^{pc} is line flow on l, due to an outage in line p. PL_l, PL_p, and PL_q are line flow on l, p, and q in steady state, respectively. $\text{PL}_l^{p-q,c}$ is the line flow on l due to common outages in lines p and q. $\varnothing\text{PL}_l^{pc}$ and $\Delta\text{PL}_l^{p-q,c}$ are slack variables for line flow on l due to the outage of p and outages of p and q, respectively. $\text{LODF}_{l,p}^{pc}$ is line outage distribution factor between the flow of line l and flow of line p when line p is on outage. $\text{LODF}_{l,p}^{p-q,c}$ and $\text{LODF}_{l,q}^{p-q,c}$ are line outage distribution factor between the flows of line l and line p, line l and line q, respectively.

The line outage distribution factor of the single outage and multiple outages can be calculated by the following equation referred to Reference [86].

$$\text{LODF}_{M,O} = \text{PTDF}_{M,O}^0 \left(\mathbf{E} - \text{PTDF}_{O,O}^0 \right)^{-1} \tag{1.69}$$

$$\text{PTDF}_{M,O}^0 = \mathbf{X}_M^{-1} \Phi^{\mathsf{T}} \left[\mathbf{B} \right]^{0-1} \Psi \tag{1.70}$$

$$\text{PTDF}_{O,O}^0 = \mathbf{X}_O^{-1} \Psi^{\mathsf{T}} \left[\mathbf{B} \right]^{0-1} \Psi \tag{1.71}$$

where $\mathbf{E}$ is an identity matrix of $\nu \times \nu$, ν is the number of outage lines. **PTDF** is power transfer distribution factor which determines a change in the power flow at

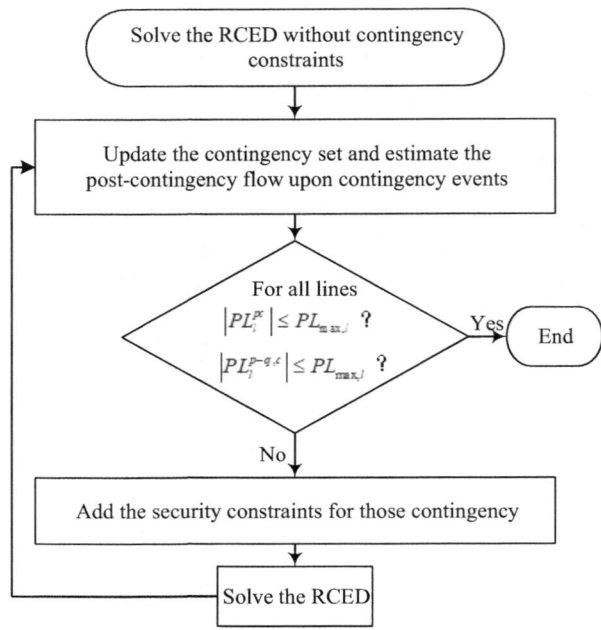

Fig. 1.20 Flowchart of the LODF post-contingency filter for the proposed RCED problem

each branch when one unit of power is transferred from on bus of the network to another. $\mathbf{X_M}$ and $\mathbf{X_O}$ are diagonal matrices with elements representing the reactance of lines that are monitored and those on the outage, respectively. $\boldsymbol{\Phi}$ is a bus to monitored line incidence matrix and $\boldsymbol{\Psi}$ is bus-to-outage line incidence matrix.

The contingency screening and constraints construction process are shown in Fig. 1.20 and described as follows. The method starts by calculating the proposed RCED without contingency constraints and obtaining the pre-contingency line flows. Then the contingency filter uses (1.65) and (1.67) based on LODF to calculate all the post-contingency line l power flow PL_l^{pc} and $\mathrm{PL}_l^{p-q,c}$ when the line p is on outage or line p and q are failure simultaneously. Then, if all post-contingency power flow is under the maximum value and it is no need to include more pre- and post-contingency combinations and the final solution is obtained. If the above condition is not fully filled, then every combination between the failed line p and overloaded line l are stored and afterward added to the RCED model using the corresponding constraints (1.65) and (1.66) or (1.67) and (1.68). After solving the RCED with the binding combination of (1.65) and (1.66) or (1.67) and (1.68), the new pre-contingency flows are analyzed again to verify if additional combinations should be added. This iterative approach checks all the contingency of lines in extreme events area and common-cause contingency of any adjacent line which are in a common right-of-way or connected the same substation or tower located in extreme area at each iteration. Then if the estimated power flow on line l due to an outage in line p or line simultaneous outage of p and q is higher than its maximum capacity, those contingencies are added to the RCED in the next iteration. In a real power system, it is expected, from the experience of the ISO, that only a short list of active line outages is required to establish a secure operation.

The Proposed RCED Model

The proposed resilience-constrained economic dispatch model is formulated as follow with established penalty terms and contingency constraints.

$$
\text{min.} \quad \alpha \cdot \sum_{i=1}^{\text{NG}} C_i \left(P_i \right) + \beta \cdot \sum_{l=S_{\text{AL}}} \left| \frac{\text{PL}_l}{\text{PL}_l^{\max}} \right|
$$

$$
+ \gamma \cdot \sum_{l=1}^{\text{NL}} \left| \left| \frac{\text{PL}_l}{\text{PL}_l^{\max}} \right| - \frac{\sum_{l=1}^{\text{NL}} \left| \text{PL}_l / \text{PL}_l^{\max} \right|}{\text{NL}} \right| + C_{\text{DR}} \cdot \sum_j^{\text{ND}} \text{LC}_j \tag{1.72a}
$$

$$
\text{s.t.} \quad C_i \left(\cdot \right) = a_i + b_i P_i + c_i P_i^2, \quad \forall i \tag{1.72b}
$$

$$
\sum_i^{\text{NG}} P_i = \sum_j^{\text{ND}} \left(\text{PD}_j - \text{LC}_j \right) \tag{1.72c}
$$

$$
P_{\min,i} \le P_i \le P_{\max,i}, \quad \forall i \tag{1.72d}
$$

$$
0 \le \text{LC}_j \le \text{PD}_j, \quad \forall j \tag{1.72e}
$$

$$
\sum_l^{\text{NL}} \text{KL}_{b,l} \cdot \text{PL}_l = \sum_i^{\text{NG}} \text{KP}_{b,i} \cdot P_i - \sum_j^{\text{ND}} \text{KD}_{b,j} \cdot \left(\text{PD}_j - \text{LC}_j \right), \quad \forall b \tag{1.72f}
$$

$$
\text{PL}_l = \frac{\theta_{fl} - \theta_{tl}}{x_l}, \quad \forall l \tag{1.72g}
$$

$$
-\text{PL}_{\max,l} \le \text{PL}_l \le \text{PL}_{\max,l}, \quad \forall l \tag{1.72h}
$$

$$
\text{PL}_l^{pc} = \text{PL}_l + \text{LODF}_{l,p}^{pc} \cdot \text{PL}_p, \quad \forall l, p \tag{1.72i}
$$

$$
\left| \text{PL}_l^{pc} \right| \le \text{PL}_{\max,l} + \Delta \text{PL}_l^{pc}, \quad \forall l \tag{1.72j}
$$

$$
\text{PL}_l^{p-q,c} = \text{PL}_l + \text{LODF}_{l,p}^{p-q,c} \cdot \text{PL}_p + \text{LODF}_{l,q}^{p-q,c} \cdot \text{PL}_q, \quad \forall l, p, q \tag{1.72k}
$$

$$
\left| \text{PL}_l^{p-q,c} \right| \le \text{PL}_{\max,l} + \Delta \text{PL}_l^{p-q,c}, \quad \forall l \tag{1.72l}
$$

The model constraints are shown in (1.72b)–(1.72l) which the common-cause outages and cascading outages contingency constraints are included. Constraint (1.72c) ensures the power balance. Constraint (1.72d) limits the upper/lower bounds of generation output. Constraint (1.72e) limits the upper/lower bounds of load curtailment. Constraints (1.72f)–(1.72h) represent line power flows and capacity limits. Constraints (1.72i) and (1.72j) are constructed for $N-1$ contingency in the weather-affected areas. Upon the weather-induced initial outages, constraints (1.72k)–(1.72l) are constructed for the common-cause outages and cascading outages.

Solution Methodology of RCED Model

Note that the non-convexity of objective function (1.72a) is due to the nested absolute function of the third term. We substitute $r_l = \left| \dfrac{\mathrm{PL}_l}{\mathrm{PL}_{\mathrm{max},l}} \right|$ into (1.72a), the problem (1.72a)–(1.72l) yields to

$$\min. \quad \alpha \cdot \sum_{i=1}^{\mathrm{NG}} C_i\left(P_i\right) + \beta \cdot \sum_{l \in S_{\mathrm{AL}}} r_l + \gamma \cdot \sum_{l=1}^{\mathrm{NL}} \left| r_l - \frac{\sum_{l=1}^{\mathrm{NL}} r_l}{\mathrm{NL}} \right| \tag{1.73a}$$

$$\text{s.t.} \quad (1.72\mathrm{b})-(1.72\mathrm{l}) \tag{1.73b}$$

$$r_l = \left| \frac{\mathrm{PL}_l}{\mathrm{PL}_{\mathrm{max},l}} \right|, \quad \forall l \tag{1.73c}$$

A general optimization method is proposed in [81] to linearize the objective function with absolute value functions. Accordingly (1.73a)–(1.73c) yields to (1.74a)–(1.74e) while the optimal solution remains the same.

$$\min. \quad \alpha \cdot \sum_{i=1}^{\mathrm{NG}} C_i\left(P_i\right) + \beta \cdot \sum_{l \in S_{\mathrm{AL}}} r_l + \gamma \cdot \sum_{l=1}^{\mathrm{NL}} \left(u_l + v_l\right) \tag{1.74a}$$

$$\text{s.t.} \quad (1.72\mathrm{b})-(1.72\mathrm{l}) \tag{1.74b}$$

$$r_l - \frac{\sum_{l=1}^{\mathrm{NL}} r_l}{\mathrm{NL}} + u_l - v_l = 0, \quad \forall l \tag{1.74c}$$

$$u_l \geq 0, \quad v_l \geq 0, \quad \forall l \tag{1.74d}$$

$$r_l = \left| \frac{PL_l}{PL_{\max,l}} \right|, \quad \forall l \tag{1.74e}$$

We further linearize constraints (1.74e) which include absolute functions without the loss of optimality.

Consider a more general form of the proposed mathematical programming model as shown in (1.75a)–(1.75c).

$$\min. \quad F(x) \tag{1.75a}$$

$$\text{s.t.} \quad g(x) + \sum_{k=1}^{K} \left| \sum_{i=1}^{n} d_{k,i} x_i \right| = b \tag{1.75b}$$

$$\text{and other model constraints} \tag{1.75c}$$

Define auxiliary variables s_k, t_k, $k = 1, 2, \ldots, K$, to construct the model in (1.76a)–(1.76e).

$$\min. \quad F(x) \tag{1.76a}$$

$$g(x) + \sum_{k=1}^{K} (s_k + t_k) = b \tag{1.76b}$$

$$\sum_{i=1}^{n} d_{k,i} x_i = s_k - t_k, \quad k = 1, 2, \ldots, K \tag{1.76c}$$

$$s_k \geqslant 0, \quad t_k \geqslant 0, \quad k = 1, 2, \ldots, K \tag{1.76d}$$

$$\text{and other model constraints} \tag{1.76e}$$

Theorem 1 If (x, s, t) is the feasible solution of the model in (1.76a)–(1.76e), the sufficient and necessary condition for x as the feasible solution of the model in (1.75a)–(1.75c) is $s_k t_k = 0$, $\forall k$.

Proof of Necessity If (x, s, t) is the feasible solution of the model in (1.76a)–(1.76e) and x is the feasible solution of (1.75a)–(1.75c), then

$$\sum_{k=1}^{K} (s_k + t_k) = b - g(x) = \sum_{k=1}^{K} \left| \sum_{i=1}^{n} d_{k,i} x_i \right| = \sum_{k=1}^{K} (s_k - t_k) \tag{1.77}$$

The three equal signs in (1.77) are due to (1.76b), (1.75b), and (1.76c), respectively. Since $s_k \geqslant 0, t_k \geqslant 0$, we have

$$\left| s_k - t_k \right| \leqslant \left| s_k \right| + \left| t_k \right| = s_k + t_k \tag{1.78}$$

According to (1.77) and (1.78), we have $|s_k - t_k| = s_k + t_k$, $\forall~k$. Thus, at least one term in a pair of s_k, t_k, is 0, i.e., $s_k t_k = 0$.

Proof of Sufficiency If (x, s, t) is the feasible solution of (1.76a)–(1.76e) and $s_k t_k = 0$, $\forall~k$, then

$$s_k + t_k = \left| s_k - t_k \right| = \left| \sum_{i=1}^{n} d_{k,i} x_i \right| \tag{1.79}$$

The second equal sign is true because of (1.76c). Substitute (1.79) into (1.76b), then (1.76b) yields to (1.75b). Therefore, x is the feasible solution of (1.75a)–(1.75c) when $s_k t_k = 0$ and (x, s, t) are the feasible solution of (1.76a)–(1.76e). ∎
Add the sufficient and necessary condition to (1.76a)–(1.76e) to get (1.80a)–(1.80e). The big-M method is used to relax $s_k t_k = 0$, $\forall~k$.

$$\text{min.} \quad F(x) \tag{1.80a}$$

$$\text{s.t.} \quad (1.76a) - (1.76e) \tag{1.80b}$$

$$0 \leqslant s_k \leqslant M \times \delta \tag{1.80c}$$

$$0 \leqslant t_k \leqslant M \times (1 - \delta) \tag{1.80d}$$

$$\delta \in \{0,1\} \tag{1.80e}$$

Theorem 2 The optimal objective function value of (1.80a)–(1.80e) is equal to that of (1.75a)–(1.75c).

Proof If x is the feasible solution of (1.75a)–(1.75c), it is also feasible in (1.76a)–(1.76e) since we can always assign s_k, t_k to make (1.76b) yield to (1.75b). Recalling Theorem 1, we conclude that the feasible region of x in (1.80a)–(1.80e) is the same as that of (1.75a)–(1.75c). So, the optimal objective function value of (1.80a)–(1.80e) is exactly equal to that of (1.75a)–(1.75c), although the two models are not essentially equivalent. ∎

According to Theorem 2, problem (1.81a)–(1.81e) can be constructed with the same optimal solution as that of (1.74a)–(1.74e) which is equivalent to (1.73a)–(1.73c) and (1.72a)–(1.72l).

$$\text{min.} \quad \alpha \cdot \sum_{i=1}^{NG} C_i(P_i) + \beta \cdot \sum_{l \in S_{AL}} r_l + \gamma \cdot \sum_{l=1}^{NL} (u_l + v_l) \tag{1.81a}$$

$$\text{s.t.}\quad \text{Original constraints}: (1.72\text{b})-(1.72\text{l})\tag{1.81b}$$

$$\text{Linearize}(1.73\text{a}):\quad (1.74\text{c})-(1.74\text{d})\tag{1.81c}$$

$$\text{Linearize }(1.73\text{c}):\quad \begin{cases} r_l = s_l + t_l \\ \dfrac{PL_l}{PL_{\max,l}} = s_l - t_l \end{cases},\quad \forall l\tag{1.81d}$$

$$\text{Ensure optimality}:\quad \begin{cases} 0 \le s_l \le M\cdot\sigma_l \\ 0 \le t_l \le M\cdot(1-\sigma_l),\quad \forall l \\ \sigma_l \in \{0,1\} \end{cases}\tag{1.81e}$$

Case Studies

Case Studies on IEEE-30 Bus Test System

To verify the effectiveness of proposed model, the modified IEEE 30-bus system is introduced and tested in MATLAB 2016a using the Gurobi solver on a personal computer with a 3.20 GHz i5 processor and 8 GB RAM. The IEEE 30-bus system is composed of 6 generators, 21 loads, and 41 transmission lines. Without the loss of generality, all transmission lines are assumed to be exposed to the same weather conditions. We generate 1000 scenarios in order to calculate blackout performance distributions.

The following four cases are discussed.

Case 1: Solve the model in (1.69) without contingencies. This case is to study the regulating effects of penalty terms on the uniformity of power flow distribution.

Case 2: Based on the dispatch solution in Case 1, perform a resilience evaluation to calculate the proposed resilience index. This case is to verify the effectiveness of penalty terms and the rationality of proposed resilience index.

Case 3: The introduced contingencies constraints are added to Case 2. The system resilience is further studied under the proposed RCED strategy.

Case 4: Evaluate the impact of weather severities on the system resilience.

For the sake of brevity, in the following discussion, NCED denotes the networked-constrained economic dispatch without penalty terms and contingencies constraints. SCED denotes the NCED with $N-1$ contingency constraints. RCED denotes the dispatch model with penalty terms. C-RCED denotes the dispatch model with both penalty terms and the contingencies constraints. We vary the coefficients in the objective function, to demonstrate two RCED models: RCED I with $\alpha = 1$, $\beta = 100$, $\gamma = 1000$; RCED II with $\alpha = 1$, $\beta = 1000$, $\gamma = 10,000$. The results for four cases are presented as follows:

Case 1

Case 1 is performed to verify the feasibility of linearization method and the effectiveness of the proposed penalty terms embedded in the RCED model. The optimization preference between generation cost and power flow distribution can be adjusted by varying β, γ. The two penalty terms are studied separately and the results are shown in Table 1.13. Table 1.14 shows the results with both penalty terms in place. Since the penalty terms are introduced to adjust the power flow, four indices are calculated including the average line loading rate (Ave. r_l), maximum line loading rate (Max. r_l), number of heavy loaded lines (Num. $r_l > 0.7$ for $r_l > 0.7$), and mean absolute deviation of line loading rate (MAD r_l).

In Table 1.13, four power flow distribution indices are improved when β is larger than 100. This is because when $\beta = 1$, pn_1 is 16.669 and the corresponding generation cost is \$8495.18. As β increases, pn_1 becomes larger and makes up a higher proportion of the objective function. Thus, larger β which increases the generation cost, results in a more effective penalty term and more homogeneous power flow. When β is 10,000, the average loading rates are lowered to 0.304. However, the adjustment in power flow distribution is limited by generation and line capacities, network topology, system loading, etc. The results corresponding to $\beta = 100, 000$ are the same as those of $\beta = 10, 000$. In addition, MAD of r_l which represents the homogeneity of power flow distribution decreases first as β increases and then increases when β becomes very large. This outcome indicates that although the first penalty term can reduce the average loading rate, its impact on power flow distribution is not uniform.

The results with varying γ are also shown in Table 1.13. Both MAD of r_l decreases as γ increases. Compared with the results with β varies, the smallest MAD is reduced by about 46%. Both the maximum loading rate and number of heavy loaded lines

Table 1.13 Results when β, γ vary

	β varies, $\alpha = 1$, $\gamma = 0$				γ varies, $a = 1$, $\beta = 0$			
	Ave. r_l	MAD r_l	Max. r_l	Num. of $r_l > 0.7$	Ave. r_l	MAD r_l	Max. r_l	Num. of $r_l > 0.7$
1	0.407	0.220	1.00	5	0.406	0.218	1.00	5
100	0.388	0.190	1.00	4	0.407	0.137	1.00	1
1000	0.315	0.173	0.87	3	0.367	0.097	0.75	1
10,000	0.304	0.179	0.87	3	0.357	0.096	0.64	0
100,000	0.304	0.179	0.87	3	0.357	0.096	0.64	0

Table 1.14 Results when both β and γ vary

Model	NCED	RCED I	RCED II
Gen. cost (\$)	8495.17	8982.11 (+5.73%)	9219.44 (+8.52%)
Ave. r_l	0.407	0.366 (−10.07%)	0.356 (−12.50%)
Max. r_l	1.00	0.74	0.64
Num. of $r_l > 0.7$	5	1	0

are also lower than that with β varies. When $\gamma = 10,000$, line loading rates are all below 0.7 with a maximum loading rate of 0.64. This is because heavily loaded lines will have a higher priority to be optimized in our case which would lower pn_2. However, the average loading rate is deteriorated which indicates that the two penalty terms fulfill different tasks for adjusting power flows.

Table 1.14 shows the comparison of conventional NCED and the proposed RCED in which the proposed RCED model performs better than the traditional NCED. In Table 1.14, there are five lines with a loading rate that exceeds 0.7, while there are only one in RCED I and none in RCED II. Besides, the maximum r_l of NCED is 1, which indicates that certain lines are operated at their capacity, which is more prone to hidden outages when power flows fluctuate, especially when line outages occur in extreme weather conditions. Even at higher dispatch costs, the operation security remains to be the primary consideration in extreme conditions.

Case 2

In this case, the three models, NCED, RCED I, and RCED II, are tested without contingencies. Accordingly, the reliability and resilience performances are compared and discussed. The weather condition is assumed to be a major storm at an average speed of 35 m/s. The wind-dependent line and tower outage probabilities are calculated according to fragility curves. The hidden outage probability is $p_0 = 0.02$ when $r_l \leq 1$ which increases to 1 when $r_l = 1.4$. The generation ramping limit is set at 10% in each island. The number of cascading simulation scenarios is 1000, i.e., $k_{max} = 1000$ in the resilience evaluation process. The blackout size is denoted by load curtailment amount/system load × 100%.

Figure 1.21(a) shows the probability distribution of blackout size. Both RCED I and RCED II perform better than the traditional NCED. The maximum load curtailment percentage of NCED is 60%, while it is only 40% in RCED I and 30% in RCED II. Moreover, the probability distribution of NCED in Fig. 1.21(a) shows a relatively flat tail when load curtailment percentage is over 30. To further investigate this situation, the log–log plot of blackout size is shown in Fig. 1.21(b). The NCED curve in Fig. 1.21(b) shows the characteristics of power law distribution with the power tails. A long flat tail generally implies a higher risk of large blackouts [83]. The two RCED curves drop exponentially with the blackout size in Fig. 1.21(b). Therefore, the RCED model reduces the risk of large blackouts effectively by improving the power flow distribution.

Table 1.15 shows the outage results for NCED and RCED. The weather-induced outage scenarios are similar since the weather conditions are assumed identical for triggering initial outages in the three models. However, different dispatch strategies could lead to various hidden outage scenarios and blackout sizes. There are 186 hidden outage scenarios in NCED, while there are only 78 and 74 in RCED I and RCED II. Furthermore, the average numbers of lines with hidden outages are 3.02, 1.15, and 0.90, respectively. This indicates that both numbers of overload and hidden outage scenarios and lines are improved in the RCED models, even though the initial triggers are the same. Moreover, the proposed resilience index distinguishes

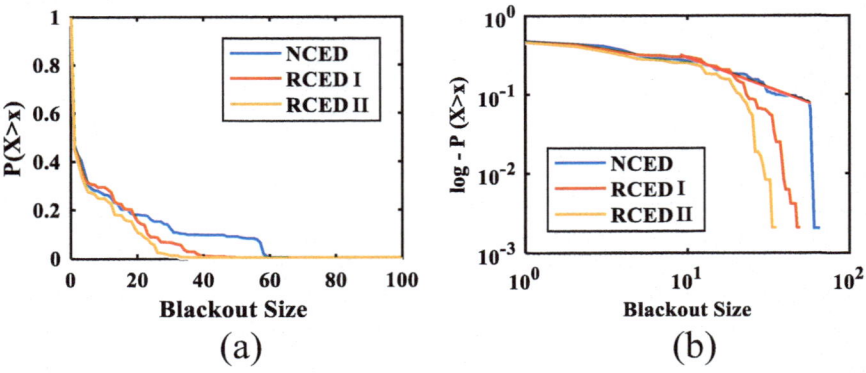

Fig. 1.21 Distribution of blackout size in NCED and RCED. (**a**) In linear plot. (**b**) In log–log plot

Table 1.15 Results in different dispatch models

Model	NCED	RCED I	RCED II
Generation cost ($)	8495.17	8982.11	9219.44
Weather-induced outage scenarios N_0	554	582	572
Hidden outage scenarios, N_h	186	78	74
Average hidden outage lines	3.02	1.15	0.90
Expected load curtailment (%)	9.65	6.91	5.35
Resilience index RI (%)	194.19	81.96	50.29

the three models more effectively than expected load curtailment does. The expected load curtailment of RCED I and II is similar. However, their resilience indices are very different.

The traditional NCED aims to determine the least production operation cost of power systems but pays little attention to the power flow distribution. In this way, there could be a few dangerous states in which some of the transmission lines are heavy loaded in which any minor power flow fluctuations or transfers would lead to cascading outages. That situation will be magnified further under extreme events. The proposed RCED model can be adopted as a more resilient operation strategy under normal and extreme conditions for blackout prevention.

Case 3

In Case 3, the $N - 1$, common-cause contingency and cascading outage contingency security constraints are further added to illustrate its effect on blackout prevention. Based on the generation dispatch plan in Case 2, we follow the contingency check with an optimal load shedding model. It is not surprising that neither RCED I nor RCED II model used in Case 2 satisfies the contingency constraints. The load curtailed in the two models is 24.9 and 20.97 MW, respectively, which indicate that an improved uniformity in power flow distribution cannot always ensure a higher reliability in response to typical outages. Hence, contingency for typical outages and penalty terms for power flow adjustment should be considered simultaneously in power dispatch strategies.

Figure 1.22 shows the blackout size distribution in different models. Comparing NCED (in blue) and SCED with $N - 1$ contingency (in red), the power tail of blue curve gets improved in the red case. However, the resilience in both NCED and SCED cases is still worse than that in RCED models due to higher proportion of large blackouts. Comparing RCED I (in yellow) with C-RCED I (in purple), we can see that the purple curve is lower than yellow, which means the CCO contingency constraints lead to less blackout and could improve the system resilience. On the contrary, the resilience of C-RCED II (in black) is worse than that of RCED II (in green). In this case, contingency constraints pose negative effect on resilience. This observation indicates that the impact of contingency constraints on resilience is uncertain. The results show that the addition of contingency constraints might not always improve resilience and sometimes it could even make worse under certain circumstances.

This interesting phenomenon can be explained in terms of the power flow distribution as shown in Table 1.16. Comparing the line flow distribution of different models, we encounter that the power flow distribution of RCED I is improved when contingency constraints are added. However, the distribution indices of RCED II are deteriorated when contingency constraints are added. Accordingly, contingency constraints can improve the power system resilience if they improve the power flow

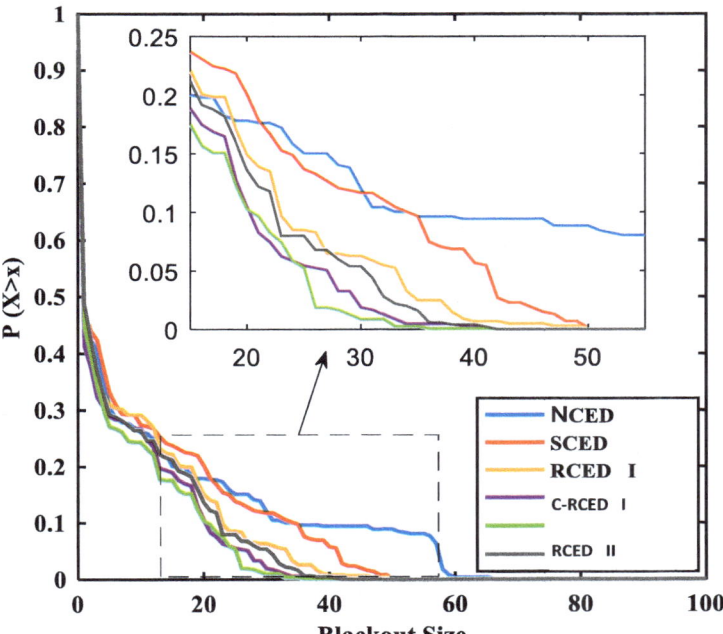

Fig. 1.22 Blackout size distribution in different models

Table 1.16 Power flow distribution and RI in different models

Model	Ave. r_l	MAD r_l	Num. of $r_l > 0.7$	LC (MW)	RI (%)
RCED I	0.366	0.097	1	19.604	81.96
C-RCED I	0.348	0.098	0	15.938	56.96
RCED II	0.356	0.096	0	15.168	50.29
C-RCED II	0.350	0.098	1	18.331	70.79

uniformity. This observation further demonstrates that the power system resilience has a close relationship with power flow distribution and the traditional SCED cannot prevent blackouts effectively with only the $N - k$ reliability strategy is pursued.

Case 4

Case 4 studies the influence of wind speed on resilience. This case helps find the effectiveness of dispatch strategies when weather condition varies. Figure 1.23 shows the expected LC percentage over all scenarios for different wind speeds in which the system is resilient (i.e., LC percentage is under 10%) when wind speed is below 30 m/s. The LC percentage has a sharp increase as wind speed increases. For wind speed below 40 m/s, the LC percentage in the proposed C-RCED is obviously below that of NCED and SCED. As wind speed increases, the gap between C-RCED and SCED becomes smaller which indicates that the operational strategy would have a weaker influence on the system resilience. When the wind speed is over 60 m/s, there is no difference among dispatch strategies. That is, the resilience cannot be improved by enhancing operational strategies. However, the infrastructural improvements, such as hardening of lines and towers will have a more profound impact on resilience when extreme events become destructive.

Case Studies on Large Test Systems

The added penalty terms and contingency constraints are suitable for large power systems. To show the computational efficiency and validity, we apply the model to more complicated test systems, including RTS-96, IEEE 118-bus, and Polish 2383wp test case. The parameters of IEEE 118-bus test system and Polish 2383wp test system are from Matpower 5.0.

Table 1.17 demonstrates the computation time of the proposed method in different test systems and the simulation time for the resilience index evaluation.

As seen from Table 1.17, the proposed penalty terms and common-cause contingency constraints will result in a longer computing time, but even for the Polish 2383wp test system, the time extension is still within the acceptable range of the scheduling department. However, the improvement of system resilience of the proposed method is obvious. Tables 1.18 and 1.19 demonstrate the load flow distribution and resilience index of the proposed strategy and conventional NCED in RTS-96 and Polish 2383wp test system, respectively. Figures 1.24(a), (b) show the blackout size distribution of different strategies in RTS-96 and Polish 2383wp test

Fig. 1.23 Influence of wind speed on blackout size

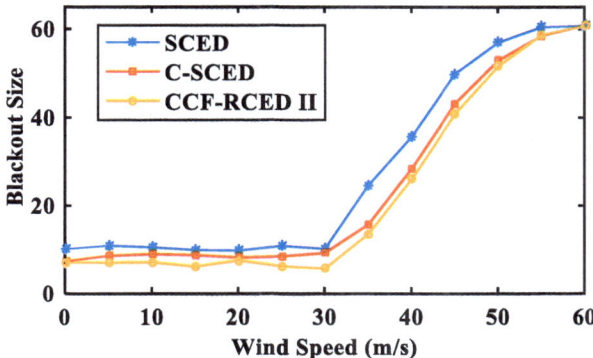

Table 1.17 Computational time of different dispatch strategy and resilience evaluation

Time(s)	RTS-96	IEEE 118	Polish 2383wp
NCED	0.25	0.39	4.06
RCED	100.34	150.16	322.07
C-RCED	496.45	500.93	856.35
Resilience index calculation	134.17	153.99	14,469.02

system, respectively. From Tables 1.18 and 1.19, we can see that the load flow derived from the proposed strategy RCED and C-RCED is better than that of NCED. The maximum loading rate in NCED is up to 1 in both RTS-96 test system and Polish 2383wp test system while it is only 0.675 and 0.741 in RCED, respectively. The heavy loaded lines of RCED and C-RCED are both less than that of NCED. It can be seen from Fig. 1.24, the blackout size distribution of the proposed strategies is better than that of NCED. The C-RCED is better than RCED, with the consideration of preventive contingency constraints. The resilience index of NCED, RCED, and C-RCED in Polish 2383wp test system is 42.42, 17.26, and 9.85, respectively.

Conclusion

This study proposes a resilience-constrained economic dispatch and corresponding set of resilience indices for blackout prevention. The following conclusions can be drawn.

- The proposed resilience indices demonstrate the power system adaptability to extreme events and distinguish different dispatch strategies even their LC are similar. The proposed indices can serve as adaptation indices in resilience evaluation effectively.
- The $N - k$ reliability strategy cannot always obtain a better resilience, especially when a power system is subject to extreme events. On the contrary, the $N - k$

Table 1.18 Load flow distribution and resilience index of different strategies in RTS-96 test system

Model	SCED	RCED	C-RCED
Gen. cost ($)	10,258	16,311	13,021
Ave. r_l	0.4956	0.2998	0.3233
Max. r_l	1	0.6754	0.741
Num. of $r_l > 0.7$	51	5	6
Max. LC (%)	61.64	37.08	30.39
Resilience index	317.04	102.84	45.16

Table 1.19 Load flow distribution and resilience index of different strategies in Polish 2383wp test system

Model	SCED	RCED	C-RCED
Gen. cost ($)	1,799,400	1,885,100	1,883,980
Ave. r_l	0.3524	0.3382	0.321
Max. r_l	1	1	1
Num. of $r_l > 0.7$	584	523	528
Max. LC (%)	22.94	17.68	13.87
Resilience index	42.42	17.26	9.85

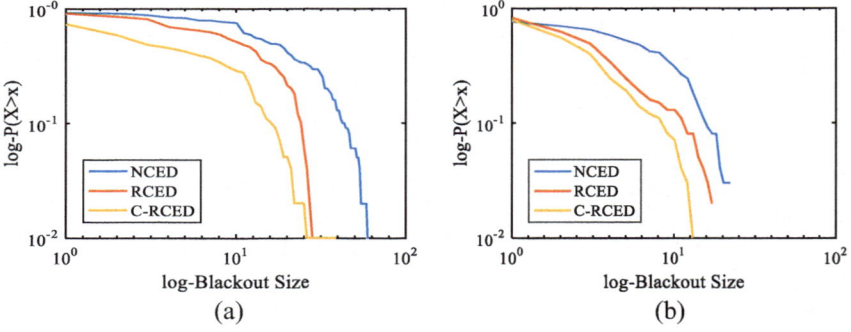

Fig. 1.24 Blackout size distribution of different strategies in (**a**) RTS-96 and (**b**) Polish 2383wp test system

reliability strategy could make system resilience performance worse in certain cases. The impact of $N - k$ strategy on resilience may also depend on power flow distribution.

- The proposed RCED simultaneously considers set and power flow distribution uniformity. The synergy of penalty terms and contingency constraints can achieve higher system resilience in extreme conditions.
- When extreme weather conditions land, the effectiveness of operation strategies becomes less critical than those of infrastructural for resilience.

1.3.1.4   Reliability Evaluation of Communication-Constrained Protection Systems

In this section, reliability evaluation of communication-constrained protection systems using stochastic-flow network models will be given.

Introduction

The relay protection systems in service generally use object-oriented layout to trip associated circuit breakers in response to faults. These relays utilize local measured signals provided by the cable and add hardware redundancy to ensure reliability. Long-term practice shows that the failure rate of protective relays caused by hardware and software is very low and such failure events usually can be detected by the built-in monitoring and self-checking facilities. Considerable works have been done to examine different reliability aspects of conventional protection systems, such as the routine test, self-checking intervals, and the redundant configuration, or even hidden failures [87–89].

Conventional protection devices adopt closed and independent arrangement and their working state can be described as being in operation or failure. In the smart grid, protection and control systems will be widely constructed upon wide area or local area networks. These communication-based protection layouts are decentralized, while devices are linked via the information flow. Under the new circumstance, the communication network architecture and the available information resources are changing heavily. The information flow becomes the important part in maintaining the high level of system reliability [90–92].

Modern protection schemes in a typical IEC61850-based substation have more components than the conventional one that is mainly composed of several intelligent electronic devices (IEDs) such as merging units, Ethernet switches, intelligent terminals of breakers, digital protective relays, and Ethernet communication media [93, 94]. Some progress have been made recently in the reliability evaluation of IEC61850-based substation protection systems. Reference [95] introduces the Markov model to calculate the reliability indices of all-digital protection systems including the impact of repair. Lei et al. present cyber-physical interface matrix to implement the protection system reliability analysis [96]. These reliability studies are meaningful; however, the reliability models are built under Markov state space theory or reliability block diagram in which each unit has only two states, that is, the flawless state or the completely unavailable state. This general binary-state description cannot reflect the performance degradation of the information flow that may cause protective function failure and cannot completely express the complexity of protection reliability based on communication networks.

In an IEC61850-based substation relay protection system relies on the Ethernet network, whose performance is more subject to the information flow fluctuation. The influence of various information disturbances can be simply classified as the decline in the level of transmission capacity. The capacity level that can be run up to each component in the network may be changeable. If the maximum capacity can be reached, it indicates the component is in an intact state that can operate properly. And if the operating capacity can only be zero, it means that the component is in a complete failure state. If the maximum capacity cannot be reached due to performance degradation such as transmission delay increasing and network throughput decreasing and so on, it can be defined as some middle states in which the component is not disable and still in service. Hence it is necessary to consider the multistate of components along the information flow path while analyzing the protection reliability in intelligent substations.

In fact, almost all networks are stochastic in nature and they can be modeled as stochastic-flow networks in which the performance of network devices varies from working properly to complete failure. A lot of algorithms based on minimal cut (MC) or minimal path (MP) have been developed to evaluate the reliability of a stochastic-flow network with multivalued random capacities [97–103]. An MP/MC is a path/cut set such that if any edge is removed from this path/cut set, then the remaining set is no longer a path/cut set. The reliability of stochastic-flow networks can be computed in terms of level d where a lower boundary point for d means the maximum flow passing through the network is not less than d units. And the reliability for level d is the probability that d units of flow can be transmitted from the source to the sink.

Stochastic-flow network model exploits multistate classification for the system components to describe the communication network under the degradation and even congestion circumstances. Hence it is suitable for reliability analysis of communication-based protection systems. The reliability of communication-constrained protection systems based on stochastic-flow network models can be defined as the probability satisfied the demand flow by the sink while ensuring network connectivity between the source node and the sink node.

A new and practicable reliability analysis method for future protection layout is presented in this section that can preferably take into account the influence of information flow. Firstly, the multiple states for components are put forward using continuous-time discrete-state Markov Chain. Secondly, the stochastic-flow network model of protection system is established and the system availability and reliability are defined. Thirdly, the improved depth-first searching method is proposed to optimize the search process. And all lower boundary points for d are judged by the maximum flow calculation and all valid system states for required demand can be determined. Finally, the reliability analysis approach upon SFN models is built and the specific computing procedure of reliability indices of protective systems is discussed.

The remainder of this section is organized as follows. Section "Stochastic-Flow Network Modeling" introduces the stochastic-flow network model with limited demand and variable edge capacities into reliability analysis for communication-based protection systems and takes multistate dynamic division of components via Markov process. Section "Reliability Calculation of Stochastic-Flow Network Model" presents the computational procedures of the proposed method, also gives the availability and reliability definition and calculation, and the specifics of improved depth-first state-tree searching. Section "Reliability Analysis of Protective System in Intelligent Substation" discusses the reliability index calculation of the typical protective structures in intelligent substations using the proposed algorithm. And it also gives a comparison of the performance of different structures. Section "Conclusions" gives the conclusion.

Stochastic-Flow Network Modeling

Mathematical Statement

According to Graph Theory, relevant definitions of stochastic-flow network models are stated as follows:

Definition 1 Let $G = (V,E,C,P)$ be a stochastic-flow network, where $V = \{v_i | 1 \le i \le n\}$ is a n-node-set with s and t defined as the source node and the sink node respectively, $E = \{e_i | 1 \le i \le a\}$ is an a-edge set, all nodes and edges are the components of G, $C = \{c_i^j | 1 \le i \le n + a |, 0 \le j \le h_i\}$ is a set of multistate limited-capacity with h_i being the largest state of component i, and $P = \{p_i^j(t) | 1 \le i \le n + a |, 0 \le j \le h_i\}$ is the set of the probability function of each component in different states.

Note that the capacities of different components are statistically independent. And there are $h_i + 1$ independent states for component i including 0-state, and the maximum capacity of components is an integer-valued random variable that takes values $0 < 1 < 2 < \cdots < h_i$ according to a given distribution. In particular, it is assumed that the set of stationary distribution is $P_\pi = \{\pi_i^j | 1 \le i \le n + a |, 0 \le j \le h_i\}$.

Definition 2 Let $Y = (y_1, y_2, \ldots, y_{a+n})$ be system state vector, $X = (x_1, x_2, \ldots, x_{a+n})$ be the capacity vector under current system state Y, then the universal set of system state vector constitutes the system state space $\Omega = \{Y_1, Y_2, \ldots, Y_M\}$ of a stochastic-flow network system where M is the total number of system states, $y_i \in \{0, 1, \ldots, h_i\}$ and $x_i \in \{c_i^0, c_i^1, \ldots, c_i^{h_i}\}$ are the current state and the corresponding capacity of component i, respectively.

For any given Y, a reliable stochastic-flow network can be generally denoted by

$$\left[f := \varphi(Y)\right] \ge d \tag{1.82}$$

where f is the maximum flow transmitted successfully from the source node to the sink node under current state Y, φ is an operator to solve the maximum flow which can satisfy the flow-conservation law, and d is the demand level required at the sink node.

Multistate Dynamic Division of Components

In the former related literatures, multistate division of components adopts static partitioning method, which can be assumed that reasonable ranges are evenly divided according to the historical data, and then the number of data samples falling in certain intervals is counted, so the probability under each state of components can be obtained and $h_i + 1$ states are allocated eventually.

This classification method uses samples to establish the discrete distribution of components, only reflects the static probability distribution, and only the static probabilistic reliability index such as availability can be derived. The time-varying reliability index, such as reliability degree and mean time to first failure (MTTFF) are unable to calculate with this model. For comprehensive assessing the reliability of protection systems, Markov process is firstly introduced to form the components' multistate model of stochastic-flow network, which is more reasonable than the conventional static partitioning.

State-space models are always used to represent time-varying systems, and the multi-state of the random flow passing through the components can also be described by Markov chain because the future state is independent of the past, but only given by the current state. It is assumed that the state process of components is Markovian. Supposed for component i, including a sequence of random state vectors $\left\{ c_i^j \right\}_{j=1:h_i}$, there is a related observation flow sequence F_i. Based upon the sequence of known observations, the sequence of unknown latent states can be inferred.

Let Δt be sampling interval, T_m be observation duration, $f_{\max}$ and $f_{\min}$ be the observed maximum and minimum flow respectively. To be discretized into h_i flow states, the total flow interval $[f_{\min}, f_{\max}]$ is divided into h_i nonoverlapping partitions. Assume the capacity of component i in state j be the mean flow over jth interval $\left(f_i^{j-1}, f_i^j \right]$ as follows:

$$c_i^j = \frac{f_i^j + f_i^{j-1}}{2\Delta t} \tag{1.83}$$

where $j = 1, 2, \ldots, h_i$. When considering the physical failure of component i, the zero state c_i^0 can be included and it means that the capacity is limited to zero and could not pass flow through the component. So the state space of component i can be written as $C_i = \left\{ c_i^0, c_i^1, c_i^2, \ldots, c_i^{h_i} \right\}$.

Let T_g^k be the kth sojourn time of target component in g-state which obeys the exponential distribution, K_g be the number of occurrences in g-state, and K_{gu} be number of state transitions from g-state to u-state. Then, the transition rate matrix A can be obtained, and among them, the unbiased estimation of state transition rates is calculated by

$$\hat{a}_{gu} = \frac{K_{gu}}{\sum_{k=1}^{K_g} T_g^k}, \quad g \neq u; \quad \hat{a}_{gg} = -\sum_{u=1}^{h_u} a_{gu}, \quad u \neq g \tag{1.84}$$

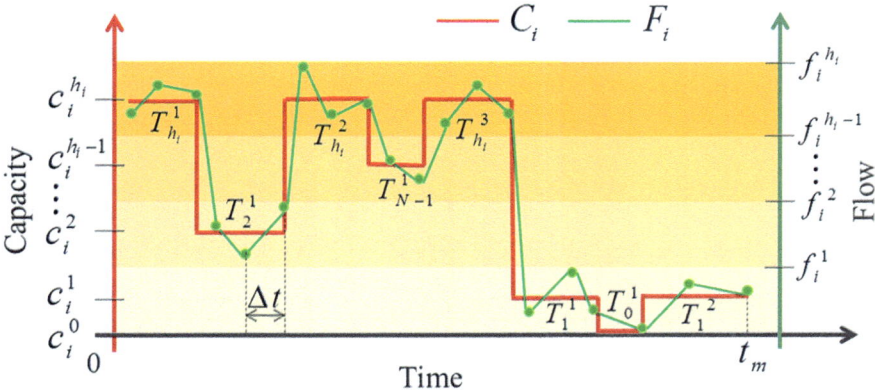

Fig. 1.25 Schematic diagram of states classification

A general process of states dividing is shown in Fig. 1.25. The Markov state transition diagram can be formed by counting the number of state transitions and the staying time of the target component in each state, as shown in Fig. 1.26.

The probability of component i being in state j has already defined as $p_i^j(t)$ and the probability vector as $P_i = \left[p_i^1(t), p_i^2(t), \ldots, p_i^{h_i}(t) \right]^T$. So according to the Kolmogorov backward equations, the linear differential equations $\dot{P}_i = AP_i$ can be solved by setting the initial value and $\left\{ p_i^j(t) \mid 1 \le i \le n + al, 0 \le j \le h_i \right\}$ can be obtained. Then, the steady-state probability is as follows:

$$\pi_i^j = \lim_{t \to \infty} p_i^j(t); \quad j = 0, 1, \ldots, h_i \tag{1.85}$$

and there is $\sum_{i=1}^{h_i} \pi_i^j = 1$.

Table 1.20 provides the allocation of component i in each state.

Topology Construction of a Stochastic-Flow Network

When the practical system architecture is transformed into an SFN model, it is necessary to consider the actual devices as the nodes and edges in the network. Different topological models can be constructed by the corresponding equivalent ways, but the results are consistent for reliability analysis [97]. In this study, the stochastic-flow network models are built, in which each edge has several capacities and may fail, but nodes are considered permanently reliable and are used only for forwarding, whose delivered flow is equal to the sent flow.

In order to highlight the impact of network performance degradation on protection reliability, the devices in communication-based protection system that may be invalid or degraded are equivalent to the edge components, while the connection of devices without flow-constraint are considered as nodes. Therefore only the

Fig. 1.26 A state-transition diagram of component i

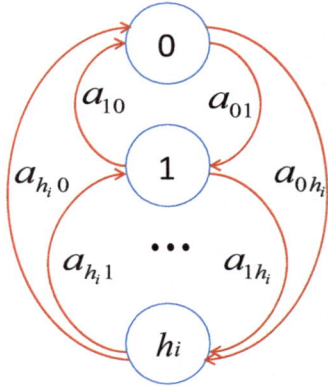

Table 1.20 Allocation of component i in each state

State no.	Capacity	Probability function	Steady-state probability
0	c_i^0	$p_i^0(t)$	$\neq_i^0$
1	c_i^1	$p_i^1(t)$	$\neq_i^1$
...	...	...	...
h_i	$c_i^{h_i}$	$p_i^{h_i}(t)$	$\neq_i^{h_i}$

multistates of edges need to be discussed, and the dimension of system state vector can be reduced from $n + a$ to a consequently. Consider the merging units as edges that connected with the virtual source node directly, protection and intelligent terminal IED also act as edges and linked with the virtual sink node on the other side. Thus, the stochastic-flow network model is constructed for the studied protection structure. The process translating a real system to an SFN model with reliable nodes and unreliable edges is shown in Fig. 1.27.

Reliability Calculation of Stochastic-Flow Network Model

Protection Reliability Index

From the definition of a reliable stochastic-flow network as given in Eq. (1.82), relay protection system reliability based on information-flow can be defined as the probability of protection and terminal IEDs receiving data no less than d units within the prescribed time while ensuring the system topology connectivity. In other word, every protection IED must receive a certain amount of data to issue a tripping instruction in time, and terminal IED to act effectively.

Definition 3 If $\varphi(Y) \geq d$, then Y is a valid system state, and if $\varphi(Y) < d$, then Y is an invalid system state. Let Y_X be all valid states of the system. If there are two system state vectors $Y_i, Y_j \in Y_X$ satisfying $\varphi(Y_i) \leq \varphi(Y_j)$, then the relation between these

Fig. 1.27 Translate a real
system to an SFN model

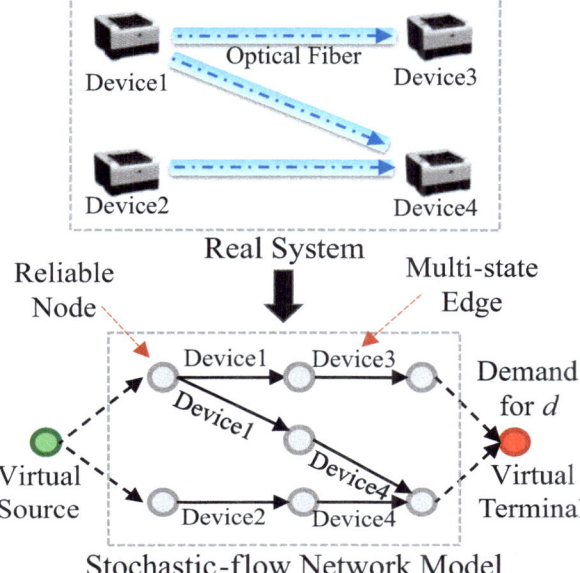

two state vectors is simply denoted as $Y_i \le Y_j$. In particular, if there is a system state $Y \in Y_X$ satisfying $Y \le Y_X$, then this state is a critical state, denoted as Y_d. Let Y_d be a lower boundary point for d.

Obviously, if the current system state $Y \in \Omega$ satisfies $Y \ge Y_d$, the system is reliable. Therefore once all the lower boundary points for d are located, which means that all the valid states of the system can be learned accordingly. The reliability of the whole system can be evaluated by computing the probability of the valid states. Assume that there are l lower boundary points for d on the system state space Ω, the set of all lower boundary points is shown as $L = \left\{ Y_d^1, Y_d^2, \dots, Y_d^l \right\}$. Then the availability and reliability of the protection system under stochastic-flow network model can be defined with (1.86) and (1.87) respectively as below:

$$A_d = \lim_{t \to \infty} \Pr\left(\Upsilon\left(t \right) = 1 \right) \tag{1.86}$$

and

$$R_d\left(t \right) = \Pr\left(T_{\text{sys}} > t | \Upsilon\left(t \right) \equiv 1 \right) \tag{1.87}$$

where $\Pr(\cdot)$ is the probability operator, T_{sys} is a random variable that represents the continuous running time of the system, $\Upsilon(t)$ is a two valued function, if the system satisfies $\varphi(\Omega) > d$ at the t moment, its value is equal to 1, otherwise its value is equal to 0. The system availability is the probability that the system can work at time t, which covers the process of failure and repair. When t tends to infinity, the availability only reflects the probability of the system being in steady state and it can be solved with the stationary distribution. However, the reliability of the system is the probability that the system can work normally in the specified time period $[0, t]$, which is a measure of the system being in good condition continuously while the system changes with time. And there is no need to take into account the repair period, instead, the probability function of components should be incorporated.

Reliability Computation Method

In order to compute system availability and reliability under the SFN model, the set of all lower boundary points for d should be found out. There are two things that are needed to do, the first is to determine whether a system state is valid or invalid by identifying a d-lower boundary point. The second is to find all d-lower bound points in the state space Ω that ensures the completeness of the state set.

Maximum Flow Algorithm

The maximum flow $\varphi(Y)$ through a stochastic-flow network needs to be computed, in order to detect whether such a transmitted flow from the source node to the sink node can satisfy the minimum demand d at the sink or not, and judge whether the current system state Y is valid or not. Ford–Fulkerson algorithm, based on Max-flow Min-cut Theorem, is used to solve the maximum flow of network [104]. The algorithm can be generally described as two parts, namely, the first part is locating, finding out paths that are suitable to augment; the second part is augmenting, increasing network flow–volume along the paths located in the first part.

Improved Depth-First State-Tree Searching

The following describes how to search the state space Ω to have all the d-lower boundary points. The obvious difficulty in searching the target space is its direct dependence on the size of the network, and needless to say the complexities are enormous while dealing with multistates, that easily get into the "Curse of Dimensionality." Therefore, the heuristic algorithms are generally adopted to traverse the state-tree based on depth-first search (DFS) or breadth-first search (BFS) that both can work effectively.

For a given state tree, suppose that depth-first searching starts from zero-state node $Y = (0, 0, \ldots, 0)$, it searches from left to right on the same state layer for valid state identification and moves downward to the deeper layer after the test of this layer is completed. Once reach to the deepest one, then trace back to its parent node and start searching the next branches on the right. And finally going through the whole state tree without skipping any node and obtain all the valid states. However, if too many branches exist, there will be a heavy burden during the searching which leads to low efficiency.

The improved depth-first state-tree search is presented in this section to reduce the workload during traversing. Because the crucial kernel of depth-first search is backtracking, two theorems are given below in order to set up reasonable backtracking conditions. While combining the searching order with the characteristics of stochastic-flow network, it can perform a trace-back operation at more upper level, which can improve the efficiency of the algorithm.

Theorem 1 Let MP be the set of minimal paths between the source node and the sink node. If there is a state $Y = (y_1, y_2, \ldots, y_k, 0, \ldots, 0), y_k \neq 0$, existing in which k is the maximum index of nonzero state element in Y. Mark those elements with zero-state on the left of y_k as set B including failure elements. And if $MP \cap B = \emptyset$, there is no need to go through the sub-tree generated by Y.

Theorem 1 states that if there is no minimal path between the source and the sink, it shows that the topology of the network is without connectivity under Y and no need to calculate the maximum flow to testify Y and its generated sub-tree are invalid states or not, so just stopping going deeper and implementing the backtracking operation directly. For example, in Fig. 1.32 the states with green strikethrough and their generated sub-tree are invalid as checked by Theorem 1.

Theorem 2 During the depth-first searching, if the current state vector Y first satisfies $\varphi(Y) \geq d$, then this state is denoted as a generalized d-lower boundary point. Let $\Gamma(Y)$ be the state set extended from current Y, and the states in $\Gamma(Y)$ must be valid, and there is no need to continue searching $\Gamma(Y)$.

Since the states number of components is allocated according to the ascending order, therefore, the state nodes in the state-tree on deeper layer have more capacities than those on upper layer. Once the first appearance of $\varphi(Y) \geq d$, the current state Y must be a lower boundary point and the capacity of the state extended by Y will be greater than that of under Y, so there is no need to search down, then performing backtrack.

A further supplement is a generalized d-lower boundary point from Theorem 2 is not precisely a d-lower boundary point given by Definition 3, but the two are consistent in nature that they and their subsets are valid states. From the whole state space, the generalized lower boundary points cannot guarantee accurate because the maximum flow under these states is not exactly equal to d, which is greater than or equal to d, so they must be valid. For example, in Fig. 1.32, the states with red horizontal line below the numbers represent the generalized lower boundary points and two horizontal lines of red and blue for exact lower boundary points. By Theorem 2, these d-lower boundary points and their subsets are valid states.

By using Theorems 1 and 2, searching can be performed more effectively. Main steps of improved state-tree searching are as below:

Step 1. Initialize the state-tree and regard $Y = (0, 0, \ldots, 0)$ as the top node that is denoted as the starting point and the first parent node. Define $Y(r) \in \{0, 1, \ldots, h_r\}$ as element y_r in the state vector Y representing rth edge of the model and for $1 \leq r \leq a$. Initialize the state variable and let $r = 1$. Set up STACK to record the order of depth extending in which top data is noted as TOP.

Step 2. Perform the depth extension. If $Y(r) < h_r$, then $Y(r) = Y(r) + 1$; otherwise, let $r = r + 1$, and then $Y(r) = Y(r) + 1$. Push the current value of r into STACK and determine whether the current state meets MP $\cap$ B $= \varnothing$ or not. If satisfied, then backtracking; otherwise, go to Step 3.

Step 3. Calculate the maximum flow of current state. If $\varphi(Y) \geq d$, that means the current state Y is a d-lower boundary point Y_d, so mark this state into the state set L, then backtracking; otherwise, go to Step 4.

Step 4. If STACK is empty and $r = a$, the searching algorithm is ended. Meanwhile, the set L stores all of the d-lower bound points of the stochastic-flow network. Otherwise, go to Step 2.

The specific backtracking operation is as follows:

- If $r < a$, then pop TOP out stack, let $Y(\text{TOP}) = Y(\text{TOP}) - 1$, trace back to the parent node on upper layer, and let $r = \text{TOP} + 1$.
- If $r = a$, then pop TOP out stack, let $Y(\text{TOP}) = Y(\text{TOP}) - 1$, and let $r = \text{TOP} + 1$. And If $r > a$, then repeat the popping operation till $r < a$.
- Finally, recall back to Step 4.

System Availability Calculation

The availability index represents the probability for the system being in steady-state taking into account the devices' maintenance. So the availability of the system defined in (1.86) can be calculated based on all the available states accessed by the searching algorithm. If Y_d is used as a parent node, by Theorem 2, the states in its downward-extended sub-state set $\Gamma(Y_d)$ are also effective. Then the collection of the state set generated by all the d-lower boundary points in set L contains all the valid states in the whole state space, marked as $W = \left\{ \Gamma\left(Y_d^1\right), \Gamma\left(Y_d^2\right), \ldots, \Gamma\left(Y_d^l\right) \right\}$. Thus,

---

Algorithm 1.1 Improved Depth-First State-Tree Searching
Input: h_r, a, d
Output: *L*
Main
1: $STACK \leftarrow$ null; $Y \leftarrow zeros(1, a)$; $r \leftarrow 1$; $i \leftarrow 1$;
2: DO{
3:   if $Y(r)<$ hr do
4:     $Y(r) \leftarrow Y(r) + 1$
5:   else
6:     $r \leftarrow r + 1$; $Y(r) \leftarrow Y(r) + 1$
7:   end if
8: $STACK \leftarrow r$ [push]
9: if *Theorem_1*(Y) is null do
10:    Backtrack(Y)

11: else if $\varphi(Y) \geq d$
12:    $\{L(i) \leftarrow Y; i \leftarrow i + 1; \text{Backtrack}(Y)\}$
13:    else
14:       goto 3
15:    end if}
16:    WHILE STACK is null and $r = a$
17:       return L
18: END Main

Backtrack
1: if $r < a$ do
2:    $\{TOP \leftarrow STACK[\text{pop}]; Y(TOP) \leftarrow Y(TOP) - 1; r \leftarrow TOP + 1\}$
3: else
4:    $\{TOP \leftarrow STACK[\text{pop}]; Y(TOP) \leftarrow Y(TOP) - 1$
5:    if $TOP + 1 > a$
6:       goto 4
7:    end if
8:    $r \leftarrow TOP + 1\}$
9: end if
10: END Backtrack

the calculation formula of the protection system's availability under the stochastic-flow network model can be refined into

$$A_d = \Pr(W) = \Pr\left(\bigcup_{i=1}^{l} \Gamma(Y_d^i)\right) \tag{1.88}$$

Because the state tree is arranged in order, the states in the tree are mutually disjoint. And the states generated by a d-lower boundary point also have this property and there is no need to apply inclusion-exclusion rule to delete the duplicated states. So, Eq. (1.88) can be further expressed as

$$A_d = \sum_{i=1}^{l} \Pr\left(\Gamma(Y_d^i)\right) \tag{1.89}$$

where

$$\Pr\left(\Gamma(Y_d^i)\right) = \prod_{r=1}^{k-1} \pi_r^g \prod_{r=k}^{a} \sum_{j=g}^{h_r} \pi_r^j \tag{1.90}$$

in which, let $g = Y_d^i(r)$ be the serial number for current state of component r, k be the maximum serial number corresponding to nonzero element in Y_d^i, $\neq_r^g$, and $\neq_r^j$ be the steady-state probability in state g and j of component r.

System Reliability Calculation

If considering the performance of the system continuous operation without failure, the reliability index should be adopted for the system evaluation. In order to get the continuous probability distribution, the zero state of components is set to the absorbing state firstly, means that the state transition rate of the certain line where the zero state is located in the transition rate matrix A is set to 0, that is

$$\hat{a}_{01} = \hat{a}_{02} = \cdots = \hat{a}_{0h_i} = 0 \tag{1.91}$$

In solving the Kolmogorov backward equation $\dot{P}_i = AP_i$, by setting the initial value of component i being in the best condition that there is no congestion or degradation for passing flow, the obtained probability function of component i in each state can be expressed as

$$P_i = \left[p_i^1(t), p_i^2(t), \ldots, p_i^{h_i}(t) \right]^T \tag{1.92}$$

Because of the topological complexity caused by the large states number of components in the system, it is difficult to derive the analytic expression of $R_d(t)$ directly. In this study, an indirect method is used to get the reliability curve by observing the reliability at discrete time points. Let T be the total observation period, and Δt be the observation interval, at t_k (for $t_k \in [0, T]$), the reliability can be calculated by

$$R_d(t_k) = \sum_{i=1}^{l} \Pr\left(\Gamma\left(Y_d^i \right) \right) \tag{1.93}$$

where

$$\Pr\left(\Gamma\left(Y_d^i \right) \right) = \prod_{r=1}^{k-1} p_r^g(t_k) \prod_{r=k}^{a} \sum_{j=g}^{h_r} p_r^j(t_k) \tag{1.94}$$

Then, the unreliability of system is presented by

$$Q_d(t_k) = 1 - R_d(t_k) \tag{1.95}$$

Based on all observed values of R_d over the period T, a fitting curve denoted as $\hat{R}_d(t)$ can be formed, and the mean time to first failure is

$$\text{MTTFF} = \int_0^\infty \hat{R}_d(t)\,dt \tag{1.96}$$

Component Importance

In the reliability analysis of binary-state system, the Birnbaum importance is generally used to identify the vulnerable points of the system and determine the maintenance priority of the components. For multistate system, the Birnbaum importance with extended definition is adopted [105]. And for component i, the Birnbaum importance is calculated as

$$I_i^B = \frac{\sum_{j=0}^{h_r} \left| \Pr(\varphi(\Omega) \geq d | y_i = j) - \Pr(\varphi(\Omega) \geq d) \right|}{h_r - 1} \tag{1.97}$$

Reliability Calculation Procedure

After translating the physical structure of protection system into a stochastic-flow network model with reliable nodes and unreliable edges, using adjacent matrix to indicate the connected relation among components, setting the demand level at the sink node, the reliability evaluation of the target system can be started. The procedure for system assessment is shown in Fig. 1.28.

Reliability Analysis of Protective System in Intelligent Substation

Typical Protection Structure and Its SFN Model

The relay protection structures based on Ethernet in intelligent substations are used as the study cases. The process bus in the intelligent substation undertakes the time-critical messages transmission, such as raw data messages and trip messages, which directly determines the reliability of protection systems. According to pilot projects of smart substations in recent years, six typical protective structures are developed [106–108], as shown in Fig. 1.29, including merging units (MUs), Ethernet switches (SWs), intelligent terminal of circuit breakers (ITs), and protective relays (PRs), in particular, SVA/B and GSA/B refer to switches for SV (sampled value) and GOOSE (generic object-oriented substation event) packets transmission respectively, and SV communication is shown with solid lines whereas dashed lines show GOOSE path.

Structure 1 adopts the architecture with double-star topology, networked sampling, and networked tripping, SV and GOOSE decoupled transmission; Structure 2 with double-star topology, networked sampling, and networked tripping, SV and GOOSE coupled transmission; Structure 3 with star topology, networked sampling, and networked tripping, SV and GOOSE coupled transmission; Structure 4 with point-to-point sampling and networked tripping; Structure 5 with networked sampling and point-to-point tripping; and Structure 6 with point-to-point sampling and point-to-point tripping.

Fig. 1.28 Reliability
calculation procedure

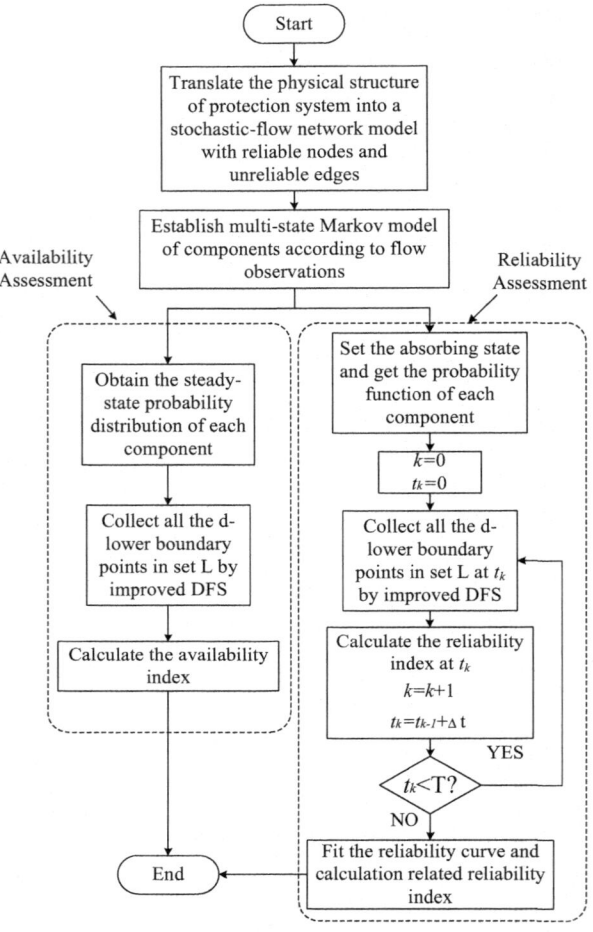

In order to highlight the influence of the sampling and tripping data flow on the reliability of protective scheme, the GOOSE upgoing flow from the intelligent terminal of circuit breakers and synchronizing signal flow over process bus are ignored. There is no consideration for the potential failure of the instrument transformer and circuit breaker operating mechanism. The storage and forwarding process of the switch are also without concern. Packets are transmitted in time since Ethernet medium with enough capacity can always work properly, for it is more reliable than the other components. So these medium lines are regarded as nodes in the stochastic-flow network models. Accordingly, the other devices are as edges. Therefore stochastic-flow network models corresponding to the six structures are shown in Fig. 1.30, in which dashed lines represent the virtual links with infinite capacity so as to distinguish redundant devices in adjacency matrix.

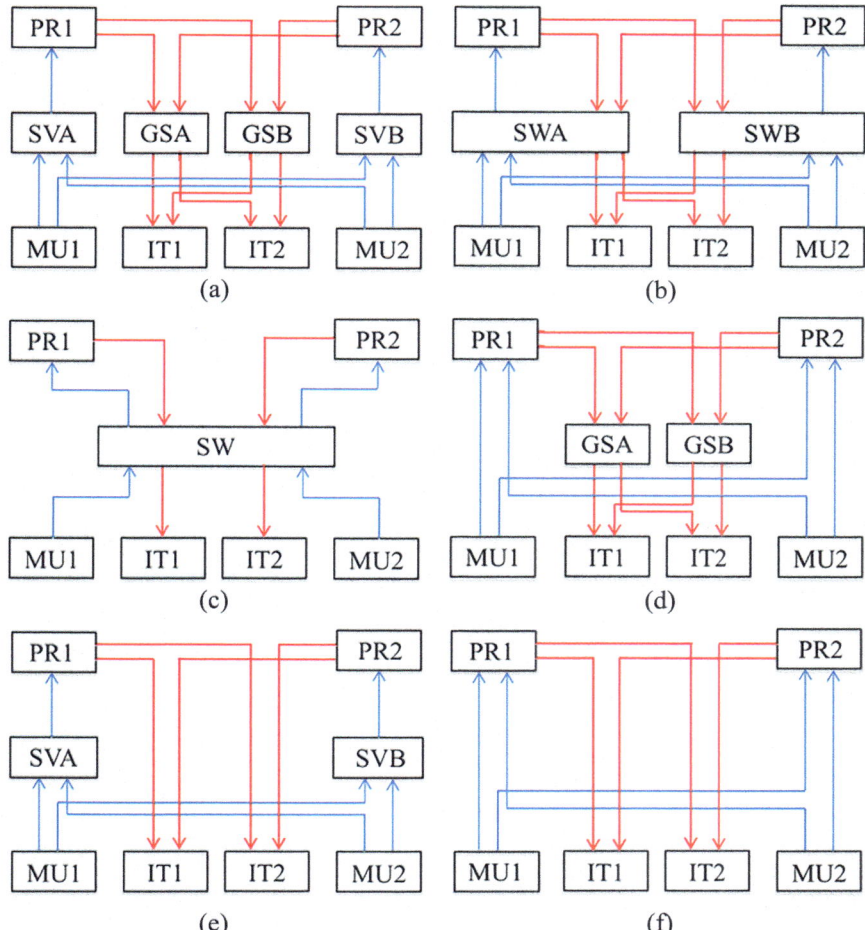

Fig. 1.29 Typical structures of protection system in intelligent substations: (**a**) Structure 1, (**b**) Structure 2, (**c**) Structure 3, (**d**) Structure 4, (**e**) Structure 5, and (**f**) Structure 6

Apparently there are two different types of information-flow in stochastic-flow network model recorded as SV flow and GOOSE flow. According to the developed custom switching technology in smart substation sharing the same Ethernet network to realize decoupling transmission based on MPLS (multiple protocol label switching) labels [109], those stochastic-flow network models can be separated into SV flow and GOOSE flow subnet models, as shown in Fig. 1.29. By calculating the reliability of the two subnets, the reliability index of the whole network can be obtained.

Let A_{SV} be the availability of SV flow subnet, A_{GS} be the availability of GOOSE flow subnet, then the reliability of the whole network can be worked out by

Fig. 1.30 Stochastic-flow network models (with reliable nodes): (**a**) Structure 1, (**b**) Structure 2, (**c**) Structure 3, (**d**) Structure 4, (**e**) Structure 5, and (**f**) Structure 6

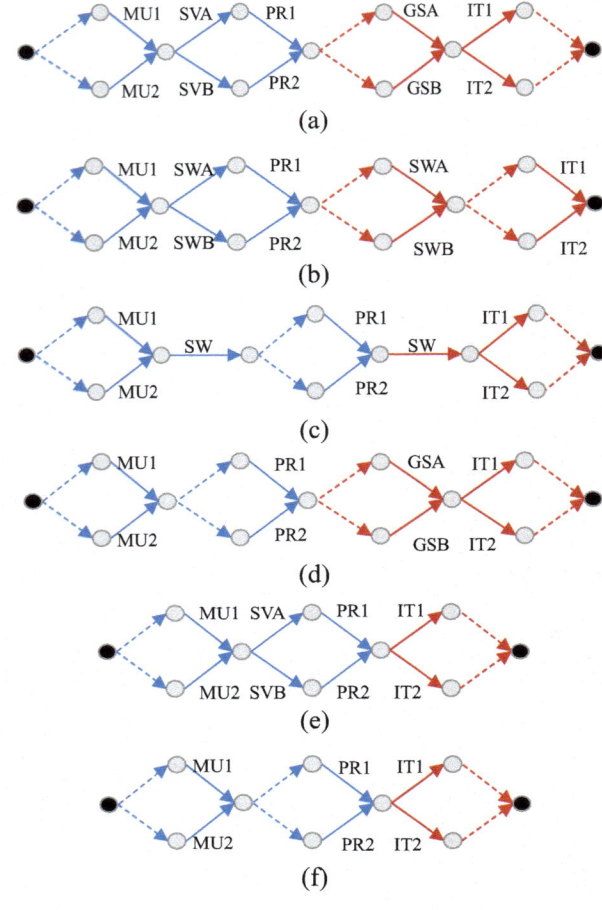

$$A_{\text{SYS}} = A_{\text{SV}} \times A_{\text{GS}} \tag{1.98}$$

Similarly, the reliability of the system is replaced by

$$R_{\text{SYS}}(t) = R_{\text{SV}}(t) \times R_{\text{GS}}(t) \tag{1.99}$$

And the unreliability is

$$Q_{\text{SYS}}(t) = 1 - R_{\text{SYS}}(t) \tag{1.100}$$

Fig. 1.31 Decomposition of the stochastic-flow network model

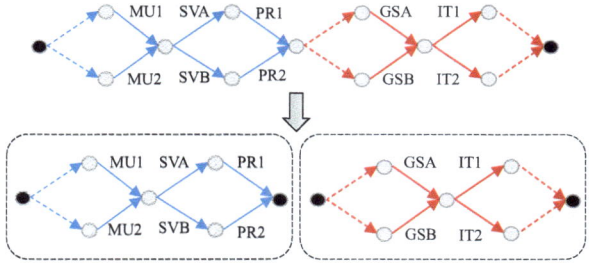

States Division of Components

A lot of aspects can cause the switch performance degradation, such as throughput capacity, data saturation, broadcast strategy, and collision conflict, etc. In this study, all these aspects affecting the work performance of the switch are taken into consideration with various capacity constraints. Usually, it is a random variable whose distribution can be determined through continuous observation and forecasting. For component i, if the historical data F_i are known, the transition rate matrix A can be obtained using the multistate dynamic division stated earlier, and Kolmogorov backward equations $\dot{P}_i = AP_i$ can be solved by Runge–Kutta iterative method. Then, the steady-state probability can be obtained by (1.85).

In fact, due to the smart substation is currently in the pilot stage, it is hard to get the traffic flow statistics. According to the network structure and operating characteristics, multiple levels along with steady-state probability values for switches are assumed as shown in Table 1.21. The level defined as 100% capacity means packets sent into switches can be processed and forwarded at the prescribed time. And 90% means the switch is overloading and can only deal with the 90% data in time while the remaining 10% of the data needs additional delay to complete the transmission. So it can be seen that the representation of multistate constrained capacity can reflect not only the traffic throughput of switches directly but also indicate the response time and network latency indirectly.

The steady-state probability values are used as the original input data. Then the Markov Chain, whose stationary distribution is consistent with the known assumed distribution, can be constructed based on the idea of Markov chain Monte Carlo [110]. And the transition rate matrix A can be obtained, as shown in Table 1.22. Furthermore, the reliability can be calculated based on A.

This study focuses on the switches with a variety of constrained state and the rest components in models other than switches just have two capacity levels with the failure state 0 and the operating state 1. The annual failure rates of MU, PR, and IT are set to 0.0067, 0.0067, and 0.01 respectively [96] and repair time is 24 h.

Table 1.21 The levels and parameters of switches

Components	State no.	Capacity (%)	Steady-state probability (%)
SVA/B	0	0	0.003
	1	60	5.095
	2	90	9.402
	3	100	85.50
GSA/B	0	0	0.003
	1	80	9.047
	2	100	90.95
SW	0	0	0.003
	1	60	7.147
	2	90	17.10
	3	100	75.75

Table 1.22 Transition rates of switches (hour-1)

Components	State no.	0	1	2	3
SVA/B	0	−0.0402	0.0141	0.0129	0.0132
	1	1.1435×10^{-6}	−0.9490	0.0943	0.8547
	2	1.0760×10^{-6}	0.0506	−0.9060	0.8533
	3	1.2710×10^{-6}	0.0509	0.0940	−0.1450
GSA/B	0	−0.0343	0.0187	0.0156	−
	1	1.1432×10^{-6}	−0.9095	0.9095	−
	2	1.1760×10^{-6}	0.0905	−0.0905	−
SW	0	−0.0381	0.0131	0.0108	0.0142
	1	1.0415×10^{-6}	−0.9285	0.1712	0.7573
	2	1.1680×10^{-6}	0.0711	−0.8290	0.7578
	3	1.1520×10^{-6}	0.0716	0.1709	−0.2425

State-tree Searching

After constructing the stochastic-flow network model, d-lower boundary points can be gained by searching the state tree in order to study its reliability analysis. Although protection structures have some degrees of redundancy, lack of information may also be caused by equipment failure or switch performance degradation. The protective function can be realized only receiving enough data, embodied in the algorithm is the demand d transmitted. If the minimum demand flow d is set to be 60 units of data, then the corresponding steps of the state-tree generation and verification process can be expressed partly as shown in Fig. 1.32, which takes the SV flow subnet of Structure 3 shown in Fig. 1.30 as an example. The figure only shows the front steps of the search process.

The sequence with five numbers represents the current system state, in which each number indicates the state of MU1, MU2, SW, PR1, and PR2. The number in brackets means the maximum flow sending under the current state. And the number on the arrows shows clearly the search order. The state with a strikethrough

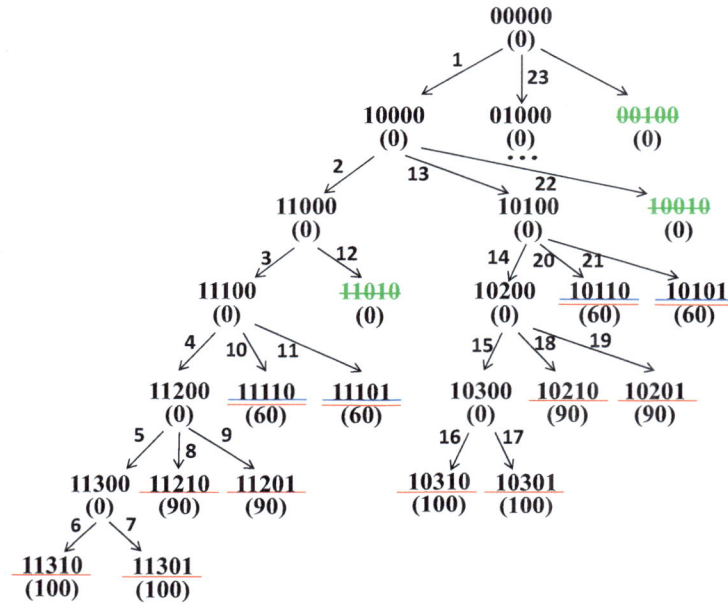

Fig. 1.32 The state-tree of SV subnet of the third structure under $d = 60\%$

indicates all child nodes generated from it and brother nodes at the right side are invalid. Finally, all lower boundary points can be found out. The 18 lower boundary points for 60 units of flow are obtained: 11310, 11301, 11210, 11201, 11110, 11101, 10310, 10301, 10210, 10201, 10110, 10101, 01310, 01301, 01210, 01201, 01110, and 01101. The last six states are generated by 01000 and cannot be fully displayed in Fig. 1.32.

Results Analysis

Availability Analysis

The availability index of each structure with the different flow requirements at the sink is shown in Fig. 1.33, in which Fig. 1.33(b) is the partial enlarged drawing of Fig. 1.33(a). It can be seen that six structures' availability are relatively perfect without considering the influence of information flow. When considering the influence of the capacity constraints of switches, the availability of the protective architecture on communication network is decreased in varying degrees. Therefore, it is necessary to take into account the information flow performance of the network when analyzing the reliability of the protection system in intelligent substations. Otherwise, the reliability evaluation will tend to be optimistic.

Specifically, when the demand is 60%, the availability of each structure is almost consistent with that of the traditional binary-state one which does not consider the

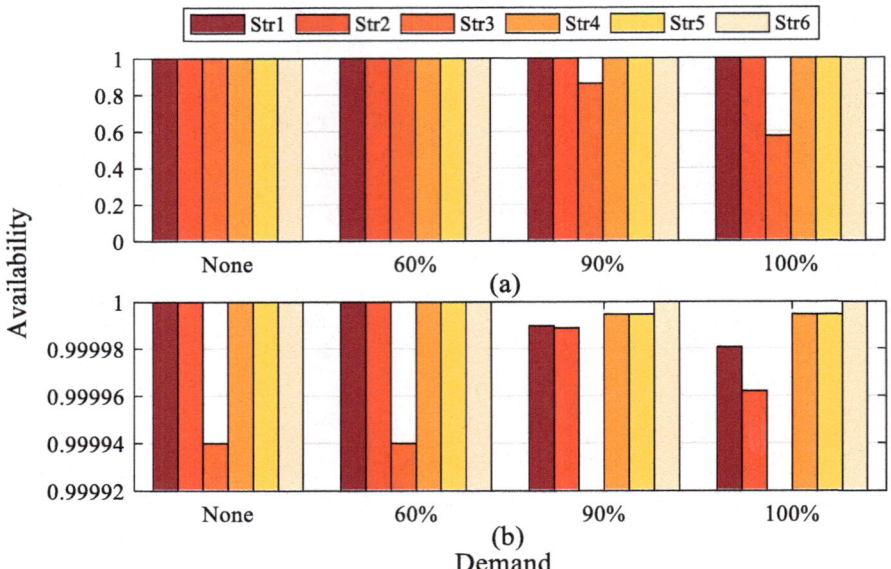

Fig. 1.33 Availability degree of six protective structures

flow-constraint. However, with the increase of the demand, the index of all but Structure 6 is a certain degree of decline, but the availability of Structure 3 has been reduced rapidly, which is mainly due to its SV and GOOSE coupled transmission using one common network while lacks of network redundancy.

In practice, the performance requirements for SV and GOOSE flow dominated by IEC61850 standards are quite strict. If no packet loss is allowed, that is, the sink point such as the protection IED will be able to gain timely access to all the data sent by the source device. In other words, the system still has a high availability when the d is set to 100%. In Structure 3, the availability is only about 50% under $d = 100\%$, obviously this cannot meet the requirements, whereas the other structures are basically available. Structure 6 adopts the point-to-point connection for sampling and tripping. It does not rely on network transmission and therefore it is not subject to the variations of d.

Structures 1 and 2 adopt the architecture with SV and GOOSE decoupled and coupled transmission respectively, when d is lower, the height of the two structures is very close, and when d is increased from 60% to 100%, the height of Structure 2 reduces obviously. It shows that if coupled transmission for SV and GOOSE flow is used, for some design decisions, such as enhancement of equipment redundancy or capacity, it might reasonably be updated in response to network congestion, even network storm and other adverse events. Generally speaking, suppose the network transmission should be implemented, the performance of SV and GOOSE decoupled transmission would be better.

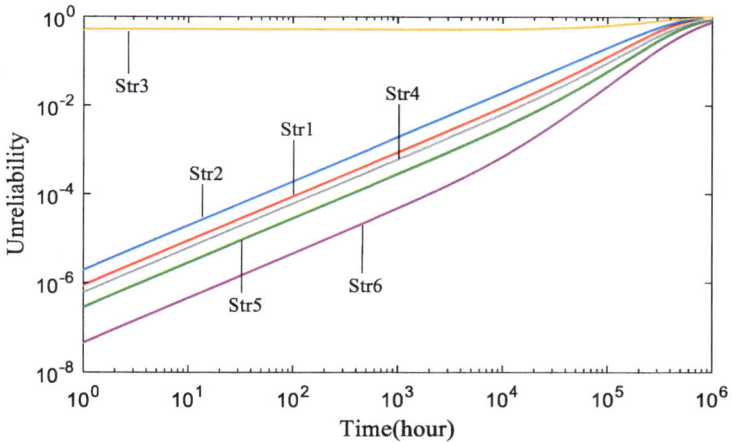

Fig. 1.34 Unreliability over time of six structures under $d = 100\%$

Table 1.23 MTTFF of six structures under $d = 100\%$

Structures	1	2	3	4	5	6
MTTFF (year)	44.39	37.53	17.53	59.90	52.12	73.82
Qsys (1000 h)	8.42×10^{-4}	1.88×10^{-3}	5.01×10^{-1}	6.09×10^{-4}	2.79×10^{-4}	4.62×10^{-5}

Reliability Analysis

The time-varying curves of unreliability of six protective structures for $d = 100\%$ are shown in Fig. 1.34. Correspondingly, the mean time to first failure (MTTFF) of these structures is shown in Table 1.23.

Therefore the results show that, under the same demanded flow, all structures except Structure 3 are comparatively highly reliable with descending order listed as Structure 6, Structure 5, Structure 4, Structure 1, Structure 2, and Structure 3. Among them, Structure 3 with single-star topology is the least reliable because it lacks network redundancy and deals with large amount of information in the common network that easily leads to traffic overloading. Structures 1 and 2 with double-star decoupled and coupled transmission respectively are much more reliable than Structure 3 because of the redundancy configuration in the communication network and they are not too inferior if compared with Structures 4–6 which partly or entirely depend on point-to-point communication. However, the arrangement with point-to-point sampling and/or point-to-point tripping follows the traditional protection design ideas. As compared with the double-star structures of networked sampling and networked tripping, there is no significant improvement in reliability, but they have lost a lot of opportunities to realize more intelligent power grid advancements.

Table 1.24 Birnbaum importance of Structures 1, 2, and 6

Structure	Rank	Component	Birnbaum importance
1	1	PR	1.450×10^{-1}
	2	SV	4.482×10^{-2}
	3	GS	4.453×10^{-2}
	4	IT	2.740×10^{-5}
	5	MU	1.826×10^{-5}
2	1	PR	2.425×10^{-1}
	2	SW	1.421×10^{-1}
	3	IT	2.740×10^{-5}
	4	MU	1.826×10^{-5}
6	1	IT	2.739×10^{-5}
	2	PR	1.826×10^{-5}
	2	MU	1.826×10^{-5}

Importance Analysis

Taking Structures 1, 2, and 6 as examples, the results of component importance are shown in Table 1.24. It can be seen that PR is the most important component for network transmission, that is, whether with SV and GOOSE decoupled or coupled transmission, the change of the working state of PR has the greatest impact on the reliability of the whole protection system. This is mainly because PR is the sink of SV flow, and can only work in the operation or failure state, so once PR fails, the information cannot be used, and protection system cannot work properly. In addition, for Structures 1 and 2, the next important components ranking are switches, because the performance of the switches has a direct impact on whether or not the sink can receive the demand traffic flow in a given time.

Structure 6 uses point-to-point connect without depending on network transmission and its importance order is basically proportional to the failure rate of components. Generally, there are obvious differences in the ranking of components' degree of importance between the communication-based and the traditional protection schemes. When the maintenance strategy for the communication-based protection systems is formulated, it is recommended to give priority to maintain the protection IED and switches.

Discussion

As for the low reliability problem of Structure 3, if broaden the switch capacity and replaced by the Gigabit switch, and demand level d is still set to 100%, the unreliability curve of Structure 3 is shown in Fig. 1.35. With respect to 100M-Ethernet networks, the data-exchange capability of Gigabit switch can be improved significantly with few performance deficiencies, the simulation results can be equivalent to that with measurements using a binary-state model. And the reliability for Structure 3 based on Gigabit switch is almost equal to that of Structure 2 with double-star arrangement and 100M-Ethernet-coupled transmission.

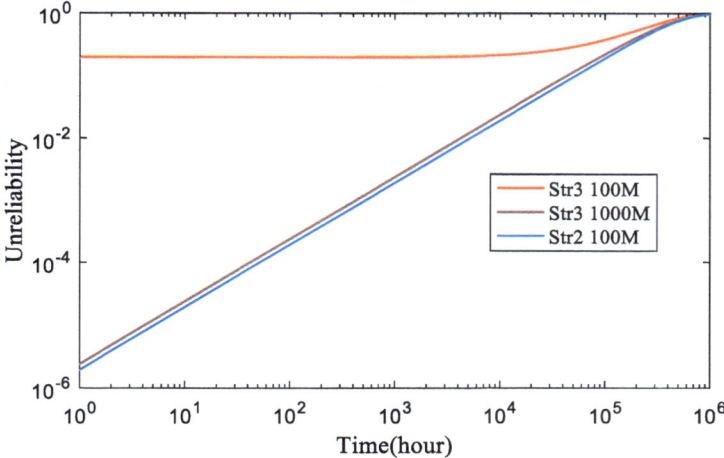

Fig. 1.35 Unreliability of the different switch capacity

Above all, from the reliability level of relay protection and fully sharing of information to achieve interoperability, it is recommended to use Structures 1 and 2 in intelligent substations. For current substation layout, the double-star architectures of process bus constructed on 100 M-Ethernet network with networked sampling and networked tripping, can be separated or use a common network transmission for SV and GOOSE flow. If only Structure 3 can be adopted, then at least Gigabit Ethernet networking should be adopted.

Conclusions

In the future, the application of relay protection in power grid is more concerned with the functional constraints under whatever circumstances due to network performance degradation. Effective analytical model is a key design under such unexpected circumstance and the properties of stochastic-flow networks are available to the activities executing in decision-making by communication-based protection system.

   The presented model mainly considers the influence of the information flow into the constrained capacity, by improved measures such as multistate division, statetree searching, index description, and calculation. Further developing the reliability computation method for next-generation protection and control technologies based on information transmission mechanism has been conducted. As compared with other ways with only considering topological connectivity, the proposed approach can comprehensively include the network adequacy, flow characteristics, and the protective business requirements.

1.3.1.5   Power Market Load Forecasting

In this section, power market load forecasting based on neural network with beneficial correlated regularization will be presented.

Nomenclature

$\odot$	Matrix doc product operator
$\times$	Matrix cross product
$Fj()$	Activation function in hidden layer neurons of NN
$Fp()$	Activation function in output layer neurons of NN

Introduction

The Background

Deregulation of power market is a slow but solid trend in power system development. In power market, pricing is agreed between energy buyers and sellers instead of official formulation. Cost variation of power generation and transmission is able to be transferred to consumer side. Power system deregulation is also recognized as a platform for competition enhancement so that asset utilization in power system is optimized and efficiency is improved [111].

In deregulation environment, electricity markets are intricate systems that consist of multiple market modes. There are three typical dominant conceptual modes, which are pool model (day-ahead and intraday markets), bilateral transactions (forwards, futures, options, and contracts for difference) and mix model [112, 113]. Nord pool spot market in Nordic Europe [114], market in Austria, early British power market [115], PJM, CAISON, and ERCOT in North America electricity markets adopt the pool-based auction mechanisms [114]. Bilateral market is utilized in most of the Nordic electricity markets and California [116]. The Nordpool in Scandinavia, the MIBEL in Portugal and Spain are examples of mixed trading system [117]. Other than modes mentioned above, China has constructed a special deregulated mechanism. Since 2015, China has initiated its second power market revolution. In the document of this reform which is published by Chinese government in Several Opinions on Further Deeping Electricity System Reform, China put forwards price differences as products for power trading [118–120].

The competition in the electricity market is mostly realized by concentrated spot trading on electric power. This spot trading system consists of the day-ahead market and real-time market. In the day-ahead biding transaction, the generator and the purchaser quote the respective prices through the dispatch of electrical network and then the day-ahead market form. The quotation is based on the forecast about uncertain electric power demand, so there may be some deviations [121, 122]. The calculation of day-ahead prices and real-time prices is based on locational marginal pricing (LMP), which contains system energy, congestion, and loss cost [122].

Electrical power load forecasting is vital to the operation and planning of a utility company influencing decisions such as purchasing and generating electric power in power market, load switching, and infrastructural development. Classified in time span, load forecasts can be divided into short-term forecast, medium-term forecast, and long-term forecast [123]. Nearly all load forecasting techniques are trying to improve the prediction accuracy by attempting on new feature space, new models, or new application domain. Accuracy seems to be the only target from implementation for decision maker to verify performance of different models, such as artificial neural network (ANN) [124–126], fuzzy logic [127–129], support vector machine (SVM) [130–132], and other time series forecasting models [133–135], etc. For example, authors of [136] deploy an ANN-based model to improve load forecasting accuracy in PJM market and ISO New England market. An accuracy-based prediction model using hybrid adaptive fuzzy neural system has been proposed in [137]. Reference [138] presents a hybrid method using period refinement scheme and adaptive strategy, which is also focusing on accuracy. Reference [139] has introduced another accuracy-based model on daily schedule behavior pattern analysis and context information for load forecasting, which is different from traditional model based on pure historical load data and weather data.

In historical load forecasting methods, accuracy nearly becomes the only pursued target for load forecasting models. Though in general, a better accuracy will achieve a better decision making. But in research on Load Serving Entities' (LSEs) behavior in day-ahead power market, we find that the accuracy of load forecast in day-ahead power market schedule submission is not always synchronized with the LSEs benefit, though the accuracy and benefit are both computed by output of forecasting model. The optimal point of accuracy and that of LSEs' power purchasing cost do not coincide. This phenomenon is named inconformity between load accuracy and LSEs' benefit. Thus, a more accurate performance based on pure accuracy-based model may not lead to optimal benefit of decision maker. Output of pure accuracy-based load forecasting techniques may decrease benefit of decision maker. In this case, a new load forecasting technique is required for LSEs' benefit improvement other than pure accuracy-based methods.

Original Contribution

In fact, accuracy and benefit nonlinearly relate to each other. They are both computed from output of load forecasting model. But the optimal point of accuracy and LSEs' benefit do not coincide together. This phenomenon is named inconformity in this study. So when traditional accuracy-based model gets results closer to the optimal accuracy in training, pure accuracy-based model would provide a solution with the best accuracy instead of a solution with the best benefit.

Facing this inconformity of accuracy and benefit, this section puts forward a regularization model named beneficial correlated regularization (BCR) for feed-forward neural network's (FFNN) prediction to improve LSEs' benefit with accuracy insurance. In this model, a market-based regularization term with virtual neuron is created for objective function optimization. Also, this section initiates a modified Levenberg–Marquardt algorithm for non-quadratic optimization in back-propagation network training of FFNN with BCR. This model provides a load fore-cast solution which ensures the required accuracy and improves the power purchasing benefit. Numerical study is carried out with PJM historical data from 2014 to 2016. FFNN with BCR provides up to 2–3% cost reduction.

The content of this section is summarized below. The overview of problem is given in section "Problem Definition: Inconformity Between Accuracy and Benefit on Load Forecast in Day-Ahead Power Market"; detailed introduction of FFNN with BCR is presented in section "Feed Forward Neural Network with Beneficial Correlated Regularization"; section "Modified Levenberg-Marquardt Training" introduces details of modified Levenberg–Marquardt (MLM) algorithm; and sec-tion "Numerical Study" presents a numerical study to support the performance of FFNN with BCR.

Problem Definition: Inconformity Between Accuracy and Benefit on Load Forecast in Day-Ahead Power Market

Power Market Introduction

Two settlement market structure is one of the most accepted market pattern world-wide, consisting of day-ahead market and real-time balancing market. It encourages market participants to preschedule their multiple-day future operation in day-ahead market. This market matches the trading by considering optimal power system oper-ation (unit commitment and dispatch) with network stability constraints. Uncertainty and fluctuation between one-day future operation and real-time operation are settled in real-time market, whose information is released much closer to the actual opera-tion. Locational marginal pricing (LMP) is the pricing model for pricing processes in both markets [140].

The day-ahead market (DAM) is a forward market in which hourly clearing prices are calculated for each hour of a future operating day. Before daily account-ing deadline, each LSE should submit their hourly consumption schedule on the target day to market operator. Before submission, LSEs use load forecast techniques to predict their hourly load in the target day. After receiving all load schedules and power generation bids, market operator follows procedure of LMP to find out the selected power generation with bid-winning capability and the price of each net-work node for LSEs in DAM [140]. Taking PJM as an example, Table 1.25 intro-duces the daily accounting deadline of PJM market for LSEs in the United States [141].

Real-time market (RTM), named balancing market, calculates the clearing prices at minute's level based on the actual system operations. Real-time LMPs are calcu-lated based on actual system operating conditions as described by the market

Table 1.25 Daily accounting deadline of PJM market for LSEs

General	4:00 PM for schedule changes from two business days prior.
Load schedule	Monday–Thursday Operating Days due two business days later, by 4:00 PM.
	Friday–Sunday Operating Days due on the following Tuesday, by 4:00 PM.

operator. LSEs will pay real-time LMPs for any demand that exceeds their day-ahead scheduled quantities and will receive revenue for demand deviations below their scheduled quantities. All spot purchases and sales in the balancing market are settled at the real-time LMPs [141].

Inconformity Between Accuracy and Benefit

In hourly consumption schedule submission in DAM, LSEs need to utilize load forecast techniques to predict the future load. The aim of LSEs is to purchase sufficient power for their downstream consumers with least cost. This cost consists of two components. The first component is the fee paid in DAM and the other is the fee paid/received in RTM. In fact, a more accurate schedule submission may not lead to a lower cost level. The following case study is performed to verify this phenomenon.

The load of a large consumer in Pittsburgh USA, DAM price and RTM price in the same area are selected for analysis. Data from 2014 to 2015 are adopted for the study. A traditional FFNN is used to predict the load two days ahead. Within the training process, sensor of cost is inserted to each iteration for observing the variation of cost while the accuracy is improving. Figure 1.36 shows the training process of the load forecast model.

From Fig. 1.36, the cost of load output from FFNN does not always reduce while the error of load is decreasing. After the 13th iteration, the variation directions of accuracy and benefit appear to be completely different. This phenomenon can also be revealed by Eq. (1.101) and Table 1.26. In Eq. (1.101), mse_i and cst_i represent the mean square error (MSE) and the cost at the ith iteration. Thus, variables mse_var_i and cst_var_i describe the accuracy changing direction and cost changing direction of FFNN output. If an FFNN training step improves both accuracy and benefit together, changing direction of MSE and cost will be the same and thus correlation between mse_var_i and cst_var_i is strictly positive 1. Oppositely, correlation becomes negative if changing direction of MSE is different from that of cost. Table 1.26 shows that the correlation between iterations 1 and 13 is positive but not strictly 1. It represents that a small section of sample pairs still appears to have a diverse changing direction, such as the third iteration and the seventh iteration. Between iterations 14–29, correlation becomes strong negative correlation. It means that accuracy and cost change oppositely when FFNN is close to its accuracy optimal. Thus, traditional accuracy-based load forecast model does not guarantee the optimal of power cost. This phenomenon reflects the inconformity between accuracy and benefit.

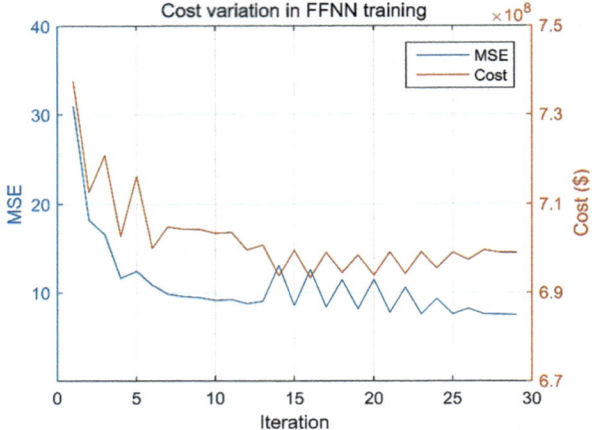

Fig. 1.36 Cost variation while accuracy improvement in FFNN training

Table 1.26 Correlation between mse_var*i* and cst_var*i* on traditional FFNN

	Iteration 1–13	Iteration 14–29
Pearson correlation	0.7748	−0.9871
Kendall correlation	0.5758	−0.9000
Spearman correlation	0.6014	−0.9794

$$\begin{cases} \mathrm{mse\_var}\,i = \{\mathrm{mv}_i\}, \quad \mathrm{mv}_i = \dfrac{\mathrm{mse}_{(i+1)} - \mathrm{mse}_i}{\left| \mathrm{mse}_{(i+1)} - \mathrm{mse}_i \right|} \\[4mm] \mathrm{cst\_var}\,i = \{\mathrm{cv}_i\}, \quad \mathrm{cv}_i = \dfrac{\mathrm{cst}_{(i+1)} - \mathrm{cst}_i}{\left| \mathrm{cst}_{(i+1)} - \mathrm{cst}_i \right|} \end{cases} \qquad (1.101)$$

The aim of LSEs is to achieve a lower cost of power purchase with acceptable accuracy. But accuracy pursuing prediction model may miss the cost reduction requirement. Thus, a model that considers both power cost and accuracy satisfaction is much more preferred for LSEs load schedule submission.

Feed-Forward Neural Network with Beneficial Correlated Regularization

The Aim of LSE's Load Schedule Submission

For the aim of LSEs, cost of energy purchase and the accuracy of load forecast should both be considered. The accuracy and cost should both be included in training process, either in objective function or constraints. Equation (1.102) introduces the aim-satisfied objective function for model training.

$$\text{min}: \text{obj} = \text{Err} + \mu \cdot \text{PC} \tag{1.102}$$

where Err is a term to reflect the differences between actual load and predicted load. PC is the power cost of the predicted load, which is the term of BCR. μ(MIU) is the penalty coefficient, which controls the weight of cost. In Eq. (1.102), "Err" is not removed, leaving "Err" together with "PC" and "MIU" in objective function to adjust the importance between accuracy and benefit. Naturally, market operator prefers better accuracy of submitted load schedule. An extreme low accuracy of submitted load will deeply interrupt market operation and will be punished by market operator. Different markets will have different capability of error acceptance and have different minimum accuracy requirements. For this reason, "Err," "PC," and "MIU" are combined into the same objective function to provide a function to adjust training importance between accuracy and benefit for minimum accuracy requirements. When facing a market with strict accuracy requirement, decision maker may decrease the "MIU" so that a more accurate solution can be obtained. Oppositely, decision maker may increase the "MIU" to obtain a more economical solution. If "Err" is removed, this adjusting function is lost. On the other hand, effect of "Err" removal can be obtained by selecting a very large "MIU."

Term PC is computed by predicted load, price of DAM, and price of RTM. In fact, when submitted load schedule is changed largely, the price of both markets will not stand still. Also, price in DAM will possibly be larger than price in RTM. For example, price of DAM is larger than price of RTM at 64.37% time between 2010 and 2016. Only 36.6% time appears to be a reversing case. This phenomenon reflects that generation resources are comparatively abundant in RTM. Thus, price variation of buying in RTM is smaller than selling under a similar trading quantity unit. So this model assumes the price will not change with two insuring schemes:

Scheme 1 is that the final accuracy of load prediction by Eq. (1.102) should not be
   far from the traditional model. Because price changes will be sufficient small
   with sufficient prediction accuracy.
Scheme 2 is that LSEs will not get paid if its submitted load schedule exceeds its
   actual load. This scheme not only contains more price inertia, but also ensures
   the amount of LSEs benefit. Any result under this scheme will be less beneficial.
   If this scheme can still reduce cost, then the actual cost will be even smaller.

FFNN Training with BCR

Back-propagation training is selected as training process for FFNN with BCR. In forward calculation in each training iteration, FFNN with BCR performs the same as general FFNN. In back-propagation, the objective function is shown in Eq. (1.103) below.

$$\min: \begin{cases} \text{obj} = \text{Err} + \mu \cdot \text{PC} \\ = \text{mean}\left(\text{Acc}^2\right) + \text{sum}\left(\text{DAMC} + \text{RTMC}\right) \\ \text{DAMC} = \text{PDA} \odot Y \\ \text{RTMC} = \text{Acc} \odot \text{PRT} \odot \varepsilon\left(\text{Acc}\right) \\ \text{Acc} = T - Y \end{cases} \qquad (1.103)$$

In NN with BCR, assume dimension of feature space is I, dimension of output space is P, number of neurons in hidden layer is J, number of data samples in training data is N.

In Eq. (1.103), T is a P by N matrix representing the training target, which is the actual load. Y is a P by N matrix representing the output of neural network, which is the predicted load. PDA is a P by N matrix representing price of DAM. PRT is a P by N matrix representing price of RTM. DAMC and RTMC are the cost in DAM and RTM, which are new terms other than those in traditional FFNN. Function sum() receives a matrix or vector input and compute the summary of all input elements. This summery is output of this function.

Following the second scheme mentioned above, $\varepsilon(\text{Acc})$ is a step function as shown by the blue curve in Fig. 1.37, which ensures that no payment is received when schedule load is larger than the actual load.

Back-propagation is a derivation-based optimization method. But $\varepsilon(x)$ is not derivable, therefore a substitute function in Eq. (1.104) can be used instead of $\varepsilon(\text{Acc})$. Figure 1.37 shows the function output with coe = 0.5, 0.8, and 1.3. The substitute function can obtain a better approximation with larger coe.

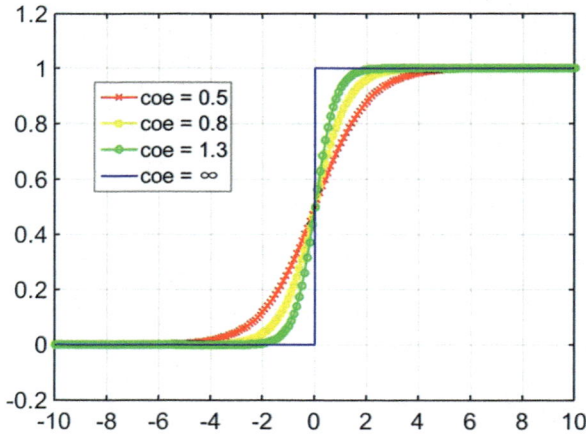

Fig. 1.37 Step function and its approximation function

$$\begin{cases} \varepsilon\left(\text{Acc}\right) \approx \dfrac{\left(\tanh\left(\text{coe}\cdot\text{Acc}\right)+1\right)}{2}, \quad \text{coe} > 0 \\[2ex] \tanh\left(\text{Acc}\right) = \dfrac{e^{\text{Acc}} - e^{-\text{Acc}}}{e^{\text{Acc}} + e^{-\text{Acc}}} \end{cases} \tag{1.104}$$

The substitute function is similar to the activation function for neurons in FFNN. But this function does not take effect in forward calculation. It only appears in back-propagation procedure. Thus, it is called virtual neuron in network training, which is specified for power cost computation.

Considering Eqs. (1.103) and (1.104), the structure of FFNN with BCR is given in Fig. 1.38. For practical requirements, network output should be positive as load is always a positive number. Scheme 1 should be considered too. Thus solution selection range should satisfy requirements in Eq. (1.104).

$$\text{selection range}: \begin{cases} Y > 0 \\[1ex] \text{mean}\left(\dfrac{|\text{Acc}|}{T}\right) \leqslant K \end{cases} \tag{1.105}$$

where K is the limit for prediction error set by network user. Function mean() receives a matrix or vector input and compute the mean value of all input elements. This mean value is output of this function.

Modified Levenberg–Marquardt Training

Levenberg–Marquardt (LM) algorithm is used to solve nonlinear least squares problems, which is widely used in FFNN training. It uses Jacobian matrix to construct the approximation of Hessian matrix in Newton method. LM algorithm is widely used in FFNN training. In FFNN with BCR, the objective function in Eq. (1.103) is not pure least square problems because of the cost term. Thus, this section introduced a modified Levenberg–Marquardt (MLM) algorithm for network training. Figure 1.39 provides the flow chart of MLM. In Fig. 1.39, procedure of back-propagation is constructed by two components. The first component is the accuracy contributed adjustment (ACA) of weight and bias. The second component is the cost contributed adjustment (CCA) of weight and bias, which is different from traditional FFNN.

Forward Calculation in MLM

Assume that dimension of FFNN's input space is I and that of output space is P. Number of neurons in hidden layer is J. There are N data samples in training data set. Thus, the output of the whole network in forward calculation is the same as traditional LM algorithm. Details can be found in [144].

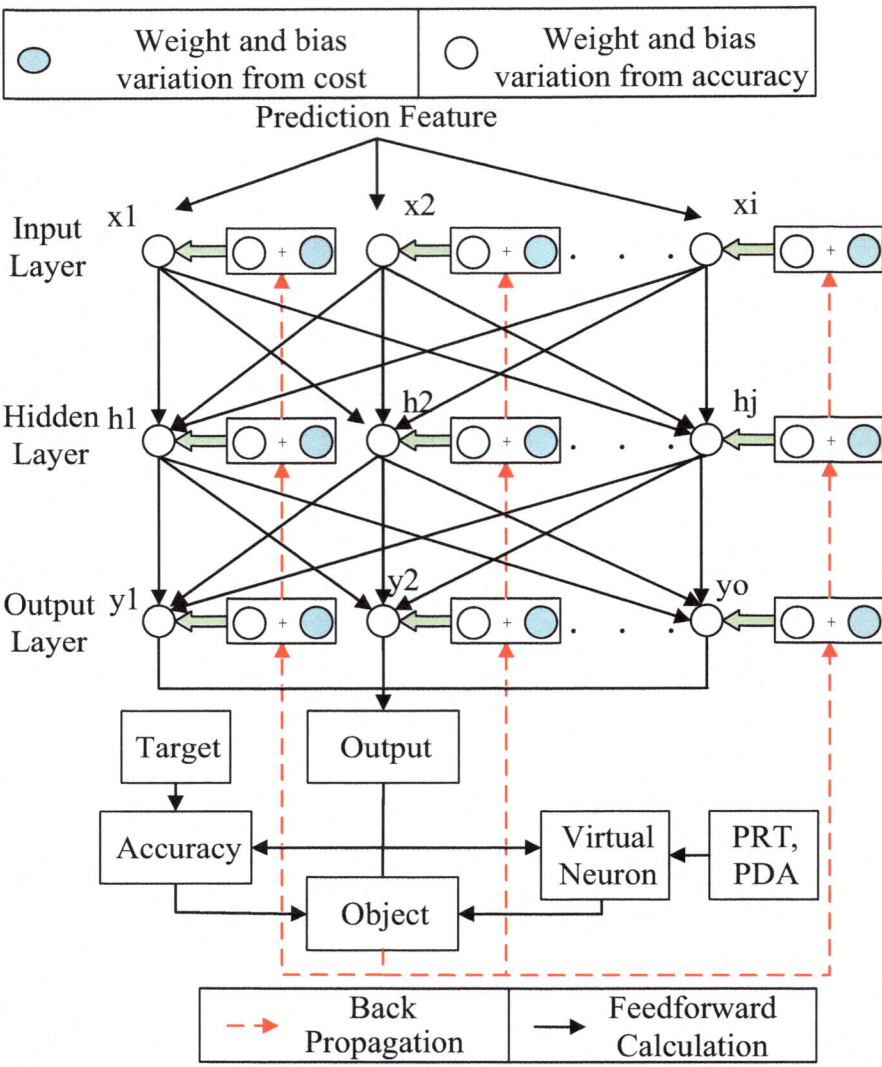

Fig. 1.38 Structure diagram of FFNN with BCR

Using the same forward calculation with traditional LM algorithm represents that network operation after training does not require participation in market price. This feature means that implementation of FFNN with BCR does not require to obtain price information ahead.

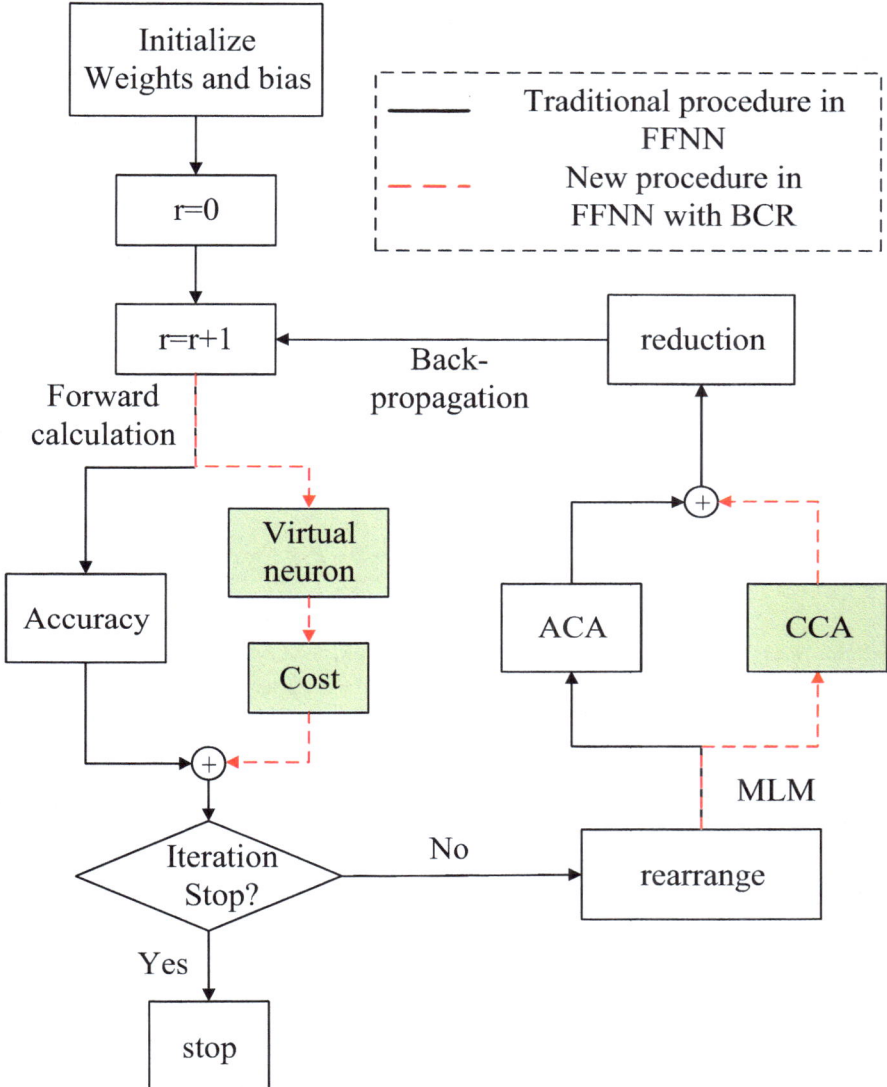

Fig. 1.39 MLM flow chart for FFNN with BCR

Variable Rearrangement in MLM

This step is specified for back-propagation preparation. It computes the derivation and transfer variables into suitable format. In FFNN with BCR, training process is to adjust the weights and bias to achieve the optimization. Weights and bias include W_{jp}, W_{ij}, B_j, B_p.

Rearrangement step 1: Derivation of W_{jp}

As Jacobian matrix is constructed by derivation for error only, deviation of each
argument should be computed first. The equation can be found in [144].

Rearrangement step 2: Flattening the variables

For computation with Jacobian matrix in MLM, arguments and relevant variables
are preferred to be of vector structure. So matrices model is needed to be
transferred into vector format. Equation (1.111) shows the vector format
transformation.

$$
\begin{cases}
X = \left[L\left(W_{ij}\right)^T, L\left(W_{jp}\right)^T, b_j^{\,T}, b_p^{\,T} \right]^T \\
Yf = L(Y) \\
Tf = L(T) \\
E = Tf - Yf = L(\text{Acc})
\end{cases}
\tag{1.106}
$$

$$
\begin{cases}
H = I \times J + J \times P + J + P \\
Q = P \times N
\end{cases}
\tag{1.107}
$$

where X is the vector of optimization arguments whose elements number is
H. Equation (1.107) gives the computation of H. Yf is the vector of output in
training data set whose elements number is Q. Tf is the vector of target in training
data set. Vector E represents the error. Vectors E and Tf are Q by 1.

In Eq. (1.106), function $L()$ is called flattening function. This function receives a
matrix input and sequentially moves each column of input matrix to the end of the
first column. Finally, all columns in the input matrix are put together into the first
column, establishing a column vector that contains all elements in input matrix.
This column vector is the function output. If dimension of input matrix is i by j, then
dimension of output vector is $i * j$ by 1.

Back-Propagation in MLM

MLM algorithm adopts Newton method to obtain the variation of arguments. Back-
propagation in MLM uses four steps to complete the arguments variation.

Step 1. Objective Function Transformation

After flattening the network output, target and the optimization arguments, the
objective function can be transformed into Eq. (1.108).

$$\begin{cases} \min : \mathrm{obj} = \mathrm{Err} + \mu \cdot \mathrm{PC} \\ \quad = \dfrac{\sum_{q=1}^{Q}\left(e_q^{\,2}\right)}{Q} + \sum_{q=1}^{Q}\left[\mathrm{pdaf}_q \cdot \mathit{yf}_q + e_q \cdot \mathrm{prtf}_q \cdot \varepsilon\left(e_q\right)\right] \end{cases} \tag{1.108}$$

where pdaf_q is the qth element in vector $L(\mathrm{PDA})$. prtf_q is the qth element in vector $L(\mathrm{PRT})$. yf_q is the qth element in vector Yf. Equation (1.108) shows that objective function of MLM is not least square problem, which is different from traditional LM algorithm.

Step 2. Gradient Computation

Deriving from Eq. (1.108), the gradient of objective function on argument can be shown in Eq. (1.109).

$$\begin{cases} \nabla \mathrm{obj} = \dfrac{2}{Q} \cdot J^T \times E + \mu \cdot J^T \times \Theta_1 \\ \Theta_1 = L(\mathrm{PRT}) \odot \left[\varepsilon(E) + E \odot \varepsilon'(E) - L(\mathrm{PDA})\right] \end{cases} \tag{1.109}$$

Different from traditional LM algorithm, a new term constructed with Θ_1 is introduced for cost representation.

Step 3. Hessian Computation

Derived from Eq. (1.109), Hessian matrix is given in Eq. (1.110) with order of argument greater than or equal to 2.

$$\begin{cases} \mathrm{Hess} = J^T \times \mathrm{COPY}\left(\Theta_2\right) \times J \\ \Theta_2 = \dfrac{2}{Q} + L(\mathrm{PRT}) \odot \left[2 \cdot \varepsilon'(E) + E \odot \varepsilon''(E)\right] \end{cases} \tag{1.110}$$

where function COPY() receives a vector with any dimension and outputs a matrix in which the dimension of rows and column are both equal to the dimension of input vector. Also, each column of output matrix is the same as that in input vector.

Comparing to use $J^T \times J$ in traditional LM algorithm, Hessian matrix approximation in MLM selects Eq. (1.110) instead. In Eq. (1.110), a new term COPY(Θ) is introduced, which is constructed by virtual neurons in Eq. (1.105) and the price data of DAM and RTM.

Step 4. Argument Updated

With gradient in Eq. (1.109) and Hessian in Eqs. (1.110) and (1.111), this produces the expression of argument updates.

$$X^{(r+1)} = X^{(r)} - \left[\mathrm{Hess}^{(r)} + \lambda I\right]^{-1} \times \nabla \mathrm{obj}^{(r)} \tag{1.111}$$

where λI is a term ensuring that Hess$^{(r)} + \lambda I$ is positive. After computation of $X(r+1)$, the update of W_{jp}, W_{ij}, B_j, B_p can be obtained from the corresponding elements from $X(r+1)$.

Numerical Study

Background

Load prediction for weekdays on Pittsburgh, USA is selected for numerical study to verify the capability of FFNN with BCR and MLM training algorithm. Load data of network training and testing set are selected from PJM's "Metered Load Data" [142]. DAM and RTM price data are selected from PJM's "Hourly Real-time & Day-ahead LMP" [143]. Training set of FFNN with BCR is constructed by data from 2014 to 2015 and testing set is constructed by data from 2016.

Feature space of network input is established by historical load section and target day weather section. In historical load section, the latest weekday's 24 hourly load data within data acquisition capability is selected as network's input. In target day weather section, hourly temperature and humidity are selected as network's input, which assumes the weather prediction accuracy is sufficiently high. Data of weather are selected from "Local Climatological Data" in National Oceanic and Atmospheric Administration (NOAA) [144].

Due to the rules of PJM schedule submission of DAM in Table 1.25, the data acquisition capability for each weekday is different. Weekday load forecast in this case study is a mix step ahead load forecast problem. The historical load input for each weekday is shown in Table 1.27.

FFNN with BCR is a non-convex model, therefore, the multiple extremums will cause training instability. To improve the training stability, a framework containing multiple networks is selected for each training set. This framework selects C different networks for each training set. The predicted load is the average value of output from each network. In this study, number of hidden layer neurons is 10. Matlab is used for model programming and simulation. Table 1.28 provides initial condition for network training.

Table 1.27 Historical load input of FFNN with BCR for each weekday

Weekday	Historical load input	Day step ahead (days)
Mon	4:00 PM in last Wed to 3:00 PM in last Thur.	4
Tue	4:00 PM in last Thur to 3:00 PM in last Fri.	4
Wed	4:00 PM in Sun to 3:00 PM in Mon.	2
Thur	4:00 PM in Mon to 3:00 PM in Tue.	2
Fri	4:00 PM in Mon to 3:00 PM in Tue.	3

Table 1.28 Parameters for network training

Stop condition	Epochs	100
	MSE goal	10^{-4}
	Cross-validation continuously miss time	3
	Minimum updates	10^{-10}
λ		10^{-2}
Number of hidden layer neurons		10
Activation function for hidden layer neurons		Log-Sigmoid
Activation function for output layer neurons		Linear
Arguments initialization method [145]		Random with normal distribution

Load Forecast on FFNN with BCR

With training data and initial condition given in Table 1.28, the training process is shown in Fig. 1.40.

In Fig. 1.40, FFNN's converging level on accuracy is slightly better than FFNN with BCR. But FFNN with BCR performs better in benefit converging. It shows that BCR term in Eq. (1.101) is effective. Also, adding cost in objective function will reduce the focus of accuracy. This effect is verified by the accuracy reduction in Fig. 1.40. This phenomenon can also be revealed by the benchmark between FFNN with BCR and other traditional load forecast techniques in Table 1.29. FFNN with BCR can achieve lower cost with small accuracy reduction.

In Table 1.29, the same training and testing data set are used for different load forecasting models, including radial basis function network (RBF), feedforward neural network (FFNN), binary decision tree for regression (BDTR), and linear regression (LR). The negative value in "cost relative diff" means that benefit of FFNN with BCR is better than any other prediction model with "a not bad" accuracy level. When MIU = $10^{-4.4}$ and coe = $10^{-2.2}$, FFNN with BCR can receive a maximum 0.8% accuracy reduction with up to 2.8% benefit improvements. The reason for best cost reduction in new model is that the cost of schedule load is added into the objective function as BCR term in model training. The BCR terms add consideration of cost into model learning so that cost of predicted load will be considered in FFNN's prediction. Decision maker of LSEs should consider if the reduced accuracy is acceptable. For further analysis, Fig. 1.41 shows the forecasting load in a typical day.

In Fig. 1.41, predicted load of FFNN with BCR is smaller than target load when price of DAM is higher than price of RTP. The largest load bias between target load and FFNN with BCR in Fig. 1.41 is up to 70 MW, which is about 4.52%. Figure 1.41 reveals that the new training objective function is trying to reduce load at period when price of DAM is higher than RTM with accuracy consideration. In this situation, LSEs may purchase more power at a lower price.

Fig. 1.40 Training process of FFNN and FFNN with BCR

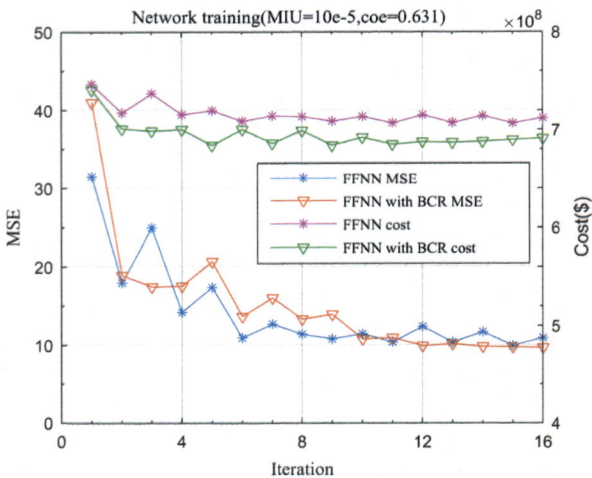

Table 1.29 Benchmarking between FFNN with BCR and other methods ($\mu = 10^{-4.4}$, coe $= 10^{-0.2}$)

Network	Training set accuracy (%)	Test set accuracy (%)	Test set cost relative diff (%)	Training set annual cost ($)	Test set annual cost ($)
FFNN with BCR	92.07	90.11	0	5.2992×10^8	4.0409×10^8
BDTR	97.05	90.12	−2.38	5.4850×10^8	4.1646×10^8
FFNN	95.83	90.92	−2.84	5.5406×10^8	4.1589×10^8
LR	93.89	89.33	−2.34	5.5551×10^8	4.1628×10^8
RBF	95.92	88.68	−2.50	5.5121×10^8	4.1446×10^8

Fig. 1.41 Load forecast comparison in a typical day

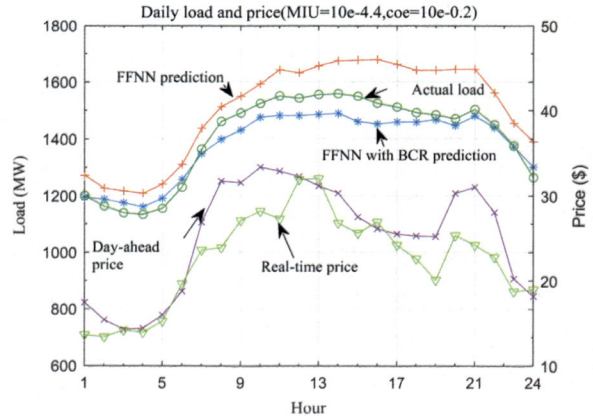

MIU Analysis of Cost Sections

MIU is a weight coefficient controlling the relative importance of power cost in Eq. (1.101). The variation of MIU deeply influences the training effect. Theoretically, the cost of predicted load will decrease further with larger MIU as PC term in Eq. (1.101) is paid larger attention in model training. But the accuracy will be worse. Table 1.30 shows the effect of new network training under different MIU. As the result of highly nonlinearity of training effect under different value, the value MIU is set shown with a log expression. Figure 1.42 shows the variation of accuracy.

From Fig. 1.42, the error of FFNN with BCR increases when MIU increases. Because MIU has increased the importance of cost and decreased the impact from accuracy in optimization, the accuracy of FFNN with BCR decreases. Decision maker can select any value whose error is below the error accepting line.

Reversely, Fig. 1.43 shows the cost reduction under different MIU value. From Fig. 1.43, cost reduction tends to decrease when MIU increases. This is the opposite effect of accuracy decreasing. Because the increasing proportion of cost in objective function will directly lead to higher cost reduction level. Considering the error accepting line, decision maker may select an MIU value with maximum cost reduction. For example, if decision maker's error accepting line is 85% as shown in Fig. 1.42, the maximum test set cost reduction can be found as 4.0145×10^8 in Fig. 1.43.

In implementation of FFNN with BCR, MIU is an important variable that deeply influences the model performance. So a procedure of MIU selection is necessary for decision maker. The spirit of MIU selection is trying to obtain lower cost within accuracy constraints. But the relationship between MIU and accuracy constraint deeply depends on constraint level selection and data set selection. Figure 1.44 introduces the MIU selection procedure.

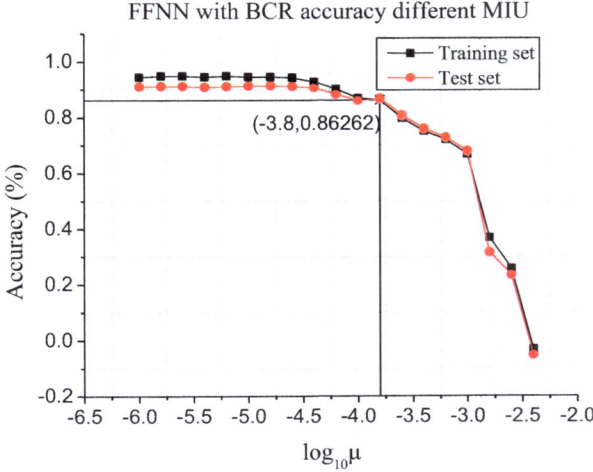

Fig. 1.42 Accuracy variation of FFNN with BCR under different MIU

Table 1.30 Comparsion of FFNN and FFNN with BCR at difference MIU (coe = 1, FFNN test set annual cost is 4.1503 × 10⁸, training set annual is 5.5365 × 10⁸)

$\log 10(\mu)$	FFNN training set mape (%)	FFNN test set mape (%)	FFNN with BCR training set mape (%)	FFNN with BCR test set mape (%)	Training set benefit improvement (%)	Test set benefit improvement (%)	Training set annual cost ($)	Test set annual cost ($)
-6	94.53	90.60	94.47	90.72	0.02	-0.05	5.5355×10^8	4.1525×10^8
-5.8			94.89	90.90	0.34	0.29	5.5175×10^8	4.1382×10^8
-5.6			94.90	90.84	0.50	0.33	5.5090×10^8	4.1364×10^8
-5.4			94.59	90.59	0.59	0.18	5.5040×10^8	4.1427×10^8
-5.2			94.87	90.81	0.93	0.31	5.4850×10^8	4.1373×10^8
-5.0			94.58	91.02	1.21	0.66	5.4695×10^8	4.1228×10^8
-4.8			94.59	91.10	1.87	1.01	5.4330×10^8	4.1085×10^8
-4.6			94.31	90.87	3.30	1.78	5.3540×10^8	4.0766×10^8
-4.4			92.75	90.45	4.05	2.04	5.3125×10^8	4.0655×10^8
-4.2			90.18	88.19	5.63	2.91	5.2250×10^8	4.0296×10^8
-4.0			86.93	85.94	4.82	3.34	5.2695×10^8	4.0117×10^8
-3.8			86.26	86.60	4.84	3.27	5.2685×10^8	4.0145×10^8
-3.6			79.80	80.57	4.66	3.66	5.2785×10^8	3.9984×10^8
-3.4			75.01	75.76	7.05	3.57	5.1461×10^8	4.0023×10^8
-3.2			72.03	72.67	8.54	3.42	5.0635×10^8	4.0085×10^8
-3.0			66.75	67.81	9.94	3.75	4.9863×10^8	3.9946×10^8
-2.8			36.83	31.38	12.25	1.58	4.8585×10^8	4.0849×10^8
-2.6			25.75	23.27	14.22	5.40	4.7493×10^8	3.9260×10^8

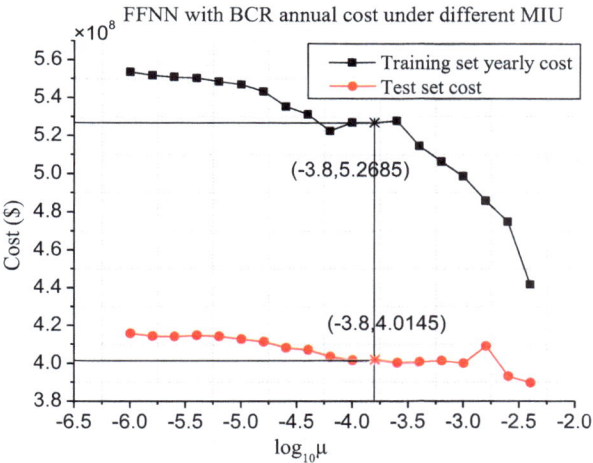

Fig. 1.43 Accuracy variation of FFNN with BCR under different MIU

In Fig. 1.44, a large value span of MIU is set for MIU boundary initially. Within this span, several sample values are selected for MIU experiments. By verifying the accuracy and cost of training sets and testing sets from all experiments, two MIU values are picked out: one is from the experiment whose accuracy is the positively closest to accuracy constraint (within constraint). The other one is the negatively closest to accuracy constraint (constraint broken). The optimal MIU should be located within these two values. If ending condition is met, the procedure will stop. Otherwise, the two picked values will be set as new MIU boundary and similar processes above will be repeated.

If experiments of all MIU values are all within constraint, a new value span with larger MIU values will be selected to repeat the processes. If experiments of all MIU values are all constraint broken, a new value span with smaller MIU values will be selected to repeat the processes. The procedures above can obtain a lower cost within accuracy constraint. In other words, the procedure is to draw a line with the value of minimum accuracy requirement in Fig. 1.42. There will be a cross point between this line and line of accuracy variation. The target MIU value is the horizontal ordinate of the cross point.

Conclusion and Future Work

Conclusion

Facing the inconformity between accuracy and benefit in DAM power scheduling submission, this study introduces a feed forward neural network with beneficial correlated regularization to improve LSEs' benefit with acceptable accuracy

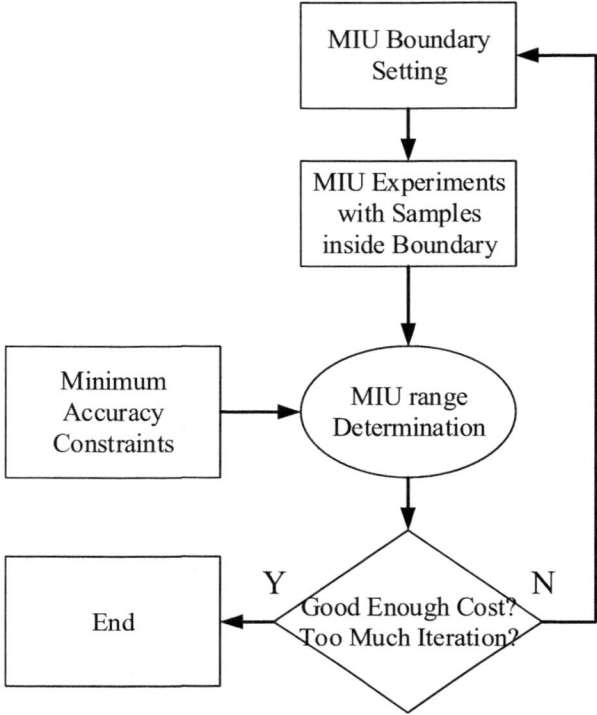

Fig. 1.44 MIU selection procedure

consideration. Results show that FFNN with BCR can reduce the power cost with acceptable accuracy.

In the numerical study, around 90% accuracy for testing set could be achieved. Practically, LSEs may construct a more relative feature space or a more details model framework to enhance the accuracy, including a more accurate environmental data, local high-influenced event indices, separated entire load to multiple consumer sets. In those cases, with more accuracy, decision maker may have a more cost reduction by using BCR in the prediction model.

One important feature of the proposed model is that price information of DAM and RTM is not required to be obtained ahead. It is only required in historical data set for training. Actually, BCR term with price information is only a regular factor inside the training optimization process. The reason of generalization effect on cost reduction without future price information is that PJM is a mature market as many other power market in the US. Behaviors of market participants converge to a comparatively stable regulations when facing many different kinds of market events. Market stability represents that price information at future predicted time can be obtained from historical price information via training.

Future Work

In FFNN with BCR, decision maker of LSEs decides its error acceptance line to ensure its accuracy level and selects the value of MIU to get the optimal cost reduction. But the price of DAM and RTM may change significantly if the permitted error is large. It means that the cost reduction effect may gradually lose effectiveness when acceptance error increases significantly. Though there is selection up to 20% cost reduction in Table 1.29, it is not suggested for over 100% error increasing. Facing this issue, one future work of this study is to expand the model effective range by considering the price–load elasticity of DAM and RTM with an economic model, so that the effect of price change can be considered.

1.3.1.6 Electricity Pricing Classification

In this section, cost-sensitive weighting and imbalance-reversed bagging for streaming imbalanced and concept drifting in electricity pricing classification will be presented.

Introduction

Electricity pricing plays a key role in determining short-term operating schedules and bidding strategies in competitive electricity markets [146]. Hence, many data-driven machine learning methods have been developed to predict short-term electricity market prices [147–150]. However, current methods tend to predict the exact value of prices while not all participants in the electricity market are interested in knowing the exact value of future values. For examples, demand-side market participants may only react when prices exceed certain thresholds considering the on/off nature of most electric loads [147]; some facilities only purchase electricity from the grid if the electricity price is below the marginal cost of operating the on-site electricity generation equipment [151]. In these types of applications, the exact value of prices is not primarily required and the price forecasting problem is turned into a price classification problem in which the task is to classify future prices into several classes of interest, for instances, whether the future price is higher than a threshold so that one should turn off most electric loads or whether the prices in a city will be higher than the other city so that a better schedule of electricity transmit between these two cities should be planned ahead of time.

Electricity pricing classification problem is not an easy task due to its streaming nature [152]. Data generated from the grid form a data stream, which introduces new challenges to traditional machine learning approaches, such as limited training and testing time, constraint of memory usage, and a single scanning of incoming samples [153]. More importantly, concept drift in streaming environment changes statistical characteristics of target concept over time which may lead to accuracy drop of classifiers being trained using past samples. Pattern classification problems in streaming environments become more complicated when both concept drift and class imbalance occur. In this work, electricity pricing classification refers to the prediction of the future electricity price to be higher or lower than the current time which is a classical two-class classification problem (high or low) in machine learning. In imbalanced classification problems, classifiers tend to classify most of samples into the majority class (In a two-class problem, a class containing more training samples than the other class is referred as the majority class while the other one is referred as the minority class) to gain a high overall accuracy and ignore the low accuracy in the minority class. This is not preferable when misclassifying a minority sample is much more expensive than misclassifying a majority sample. Current research studies focus on dealing with concept drift in streaming learning environments while very few efforts have been made to deal with the imbalanced streaming classification problems with concept drift.

Concept drift refers to the change of the joint probability distribution between inputs and true classifications in different time moments in a data streaming setting [154]. Classifiers trained using outdated samples would yield very poor generalization capability on samples in the future. Ensemble methods are often applied to relieve this problem because of their high performance and usefulness for streaming learning owing to the ease of being integrated with drift detection methods and dynamic updates [155].

Class imbalance problem is another major issue in data streaming environments. Class imbalance problems occur when the number of samples in at least one class is either much more or less than other classes. When class imbalance happens, classifiers trained using traditional methods (e.g., by the minimization of overall training error) yield poor generalization performance on the minority class. Therefore, proper techniques like data processing should be employed to deal with the class imbalance problem. Data processing is one of the key elements for the successful operation of complex systems such as smart grids [156, 157]. In the context of smart grid, classification with machine learning has been applied to fault cause identification [158, 159], future electricity market prices [147–150], electrical machines [160, 161], power quality disturbances classification [162], and cyberattacks detection [163, 164]. However, to date, the concepts of imbalanced streaming data and concept drifting in smart grid have rarely been studied. Seldom work has been done on imbalanced classification for power system problems. Authors in [165] claimed to be the first researches on investigating the outliers in electricity demand time series with imbalanced classification techniques. To assess power system short-term voltage stability, an oversampling technique and a cost-sensitive learning method are applied to deal with the predictions of the rarely-occur instability events [166].

Very few works focus on dealing with both the concept drift and class imbalance issues. Existing methods can be distinguished into two types. One is to retrain a new model using the most recent samples so that the trained classifier can react to the concept change fast, for example, the SERA (SElectively Recursive Approach) [167]. The SERA reserves all the minority samples seen so far, from which the most relevant ones are selected to combine with the most recent majority samples so that a preselected post-balance ratio is met. A classifier or ensemble is trained from this rebalanced data set. The other type is to dynamically update the model, for example, the Learn++.CDS (Concept Drift with SMOTE), Learn++.NIE (Nonstationary and Imbalanced Environments) [168] and the DWMIL (Dynamic Weighted Majority for Imbalance Learning) [169]. The CDS rebalances the most recent data chunk (A data chunk refers to a block of consecutive samples in between some time interval for the learning model to train or to predict) using the SMOTE (Synthetic Minority Oversampling TEchnique) [170] to tackle the class imbalance problem by generating new samples along a line connecting a minority sample and its nearest minority sample, while the NIE uses a bagging variation method to create several relatively balanced data set to train a classifier ensemble. Regarding the adaptation to the concept drift, both the CDS and the NIE apply a dynamic weight assignment scheme so that classifiers yielding high performance on the current data environment receive high weights. The major drawback of the Learn++ family is that all classifiers are maintained which increases the computational costs and lowers the prediction speed. To avoid this kind of problem, the DWMIL applies a time–decay function to its weight assignment scheme so that the weight of each classifier decreases automatically. When weights are lower than a threshold, corresponding classifiers are removed so that the number of classifiers maintained is much lower than the number of time moments.

The major concern of classifier training is their generalization abilities for future unseen samples in incoming data stream. However, current learning methods do not take generalization error of the classifiers into account when training classifiers. Therefore, we propose an incremental ensemble of ensembles learning with a Cost-sensitive Weighting and an Imbalance-reversed Bagging, i.e. CWIB, to deal with imbalanced classification problems in streaming environments with concept drifts, which significantly enhances the performances than the state-of-the-art methods in terms of accuracy, F1-measure, and G-Mean and ranks the first in terms of all performance metrics applied in this work. The CWIB relieves the class imbalance problem by applying an imbalance-reversed bagging method which builds a set of diversified base classifiers to form a component ensemble classifier. In comparison with methods building a single classifier with each data chunk and update weights of classifiers (e.g., CDS in the experiment), the proposed CWIB yields significantly better results in accuracy, F1-measure, and G-Mean value. This shows the effectiveness and satisfactory results of the CWIB using ensemble of classifiers. Ensemble of classifiers usually yields lower error rate in comparison with a single classifier [171]. Moreover, training an ensemble of classifiers using the current chunk looks to be very time-consuming, but these component classifiers are independent from each other and can be trained in parallel as suggested in [172]. In this way, time consumption will be roughly similar to that of training a single classier. Therefore, training an ensemble of classifier instead of a single classifier with each data chunk is a better choice. Then, component classifiers are fused together to form the final ensemble for the CWIB using the weighted sum method. To adapt to the concept changes across time, the weight of each component classifier is computed according to their cost-sensitive classification performances and stochastic sensitivities with respect to the current data chunk. Major contributions of this work are as follows:

1. An imbalance-reversed bagging (IRB) method is proposed to relieve the class imbalance issue in a data chunk. The IRB boosts the true positive rate while maintains a relatively low false-positive rate.
2. A new cost-sensitive stochastic sensitivity measure (ST-SM) is proposed to weight samples in different classes differently based on their ST-SM and a cost computed by the imbalance ratio.
3. A dynamic cost-sensitive weighting scheme based on the cost-sensitive ST-SM is applied to compute fusing weights of component classifiers. A larger weight is assigned to a component classifier yielding a good cost-sensitive performance on the current data chunk.
4. A fixed size of classifier ensemble is maintained, which is much smaller than the time moments and requires both less computational resources and less storage.

The section is structured as follows. The CWIB is proposed in section "Cost-Sensitive Weighting and an Imbalance-Reversed Bagging". Section "Experimental Studies with Electricity Price Classification" shows experimental results and discussion. Section "Conclusions and Future Work" gives the conclusion.

Cost-Sensitive Weighting and an Imbalance-Reversed Bagging

Algorithm 1.2 shows procedures for training the CWIB at time t. The CWIB training method consists of two components: one is to handle the class imbalance issue and the other one is to dynamically assign different weights to each component classifier for adaptation to changes in data. The overall procedure of the CWIB is as follows:

At time moment t, the current H^{t-1} is an ensemble of ensembles which consists of a set of component classifiers fused by a weighted sum while each component classifier consists of a set of base classifiers fused by a simple majority voting. When a new data chunk arrives, a new component classifier is trained using the IRB. The new component classifier h^t is expected to be more relevant to the current data environment, thus its weight is set to 1 (the largest weight). Existing component classifiers in H^{t-1} are then weighted according to their classification performance based on a cost-sensitive loss and their stochastic sensitivities on the current data chunk. Then, the H^{t-1} is combined with the h^t along with newly computed weights to from the H^t. If the number of component classifiers in H^t is larger than the preselected ensemble size, the worst-performing component classifier yielding the smallest weight is removed. The IRB and the dynamic weighing scheme are proposed in sections "Imbalance-Reversed Bagging" and "Dynamic Weight Assignment", respectively.

Algorithm 1.2 Training CWIB at Time Moment t

Input: t, current time moment; u, number of base classifiers in a component classifier; D^t, current data chunk; H^{t-1}, current classifier ensemble; M, maximum ensemble size.

Output: H^t, classifier ensemble at time t.

1: Apply the IRB on D^t to build the component classifier h^t and set its weight to 1

2: Compute the weight of each component classifier in H^{t-1} based on its cost-sensitive loss and stochastic sensitivity

3: Combine h^t and H^{t-1} with their newly computed weights to form H^t

4: If the number of component classifiers in H^t is larger than M then removes the component classifier yielding the smallest weight

Imbalance-Reversed Bagging

When a new data chunk arrives, u data sets are sampled from the original data chunk with replacement based on a probability distribution. The probability of a minority (majority) sample being sampled is equal to the number of samples in the majority (minority) class divided by the total number of samples in the current data chunk. Such that, the minority samples become the majority in the sampled data set. In this way, the class imbalance is reversed which forces the base classifier being built using this data set to bias to the minority class for improving its true positive rate. Then, a component classifier of the CWIB is built by fusing all u base classifiers using a simple majority voting.

However, in electricity pricing problems, the number of samples in the minority class may be larger than that of the majority class in some data chunks. In these anomaly cases, the probability of sampling will not be reversed as aforementioned to let the system to keep focus on the original minority class.

The random sampling with replacement from both classes creates diversified training datasets for base classifiers. By favoring the minority class in the IRB, each base classifier may yield a high false-positive rate. The bagging of diversified base classifiers maintains a low false-positive rate [173] to relieve this problem. By applying the IRB, the representation of the minority class is enhanced while a relatively low false-positive rate is maintained. The algorithm of the IRB is given in Algorithm 1.3.

Algorithm 1.3 Imbalance-Reversed Bagging

Inputs: u, number of base classifiers in a component classifier; D, current data chunk.

Outputs: A classifier ensemble h

1: Reverse the imbalance ratio of D if the number of minority samples is not larger than that of the majority class (anomaly case).

2: Randomly sample u data sets with replacement from D according to the imbalance ratio in Step 1.

3: Build u base classifiers using the u training datasets and form the classifier ensemble h with a simple majority vote fusion.

Dynamic Weight Assignment

The IRB proposed in the previous section builds a new component classifier whenever a new data chunk arrives. The adaptation to concept drifts in the nonstationary streaming data environment is achieved by a dynamic weighting of component classifiers according to their classification performances and stochastic sensitivities with respect to the current data chunk. The weight is ranged between [0, 1]. A larger weight is assigned to a component classifier if it yields a higher classification performance for the current data environment with smaller stochastic sensitivity with respect to small input perturbations. The final classifier ensemble of the CWIB is fused by the weighted sum method as follows:

$$H^t\left(x_b^t\right) = sign\left(\sum_{j=1}^{l^t} w_j^t h_j^t\left(x_b^t\right)\right) \tag{1.112}$$

where x_b^t, $h_j^t\left(x_b^t\right)$, w_j^t, and l^t denote the bth training sample, the predicted output of the jth component classifier given x_b^t, the weight of the jth component classifier and the number of component classifiers, at time t, respectively. For simplicity, the time t will be ignored in the following part of this section because all computations are finished within the same time moment. Therefore, Eq. (1.112) is rewritten as follows:

$$H\left(x_b\right) = sign\left(\sum_{j=1}^{l} w_j h_j x_b\right) \tag{1.113}$$

The classification performance of h_j (the jth component classifier) is evaluated by a cost-sensitive loss function. In class imbalance problems, misclassifying a minority sample is usually more costly than misclassifying a majority one. Therefore, a misclassification of a minority sample yields a larger penalty in the loss function. The logistic loss function is used in this work:

$$\varphi\left(h_j\left(x_b\right), y_b\right) = \log\left(1 + \exp\left(-y_b h_j x_b\right)\right) \tag{1.114}$$

where $y_b \in \{-1, +1\}$ and $h_j(x_b) \in \{-1, +1\}$ denote the true label and the predicted label of x_b, respectively. Then, the cost-sensitive loss function is defined as follows:

$$L_j = \sum_{b}^{m} C_b \varphi\left(h_j\left(x_b\right), y_b\right) \tag{1.115}$$

The misclassification cost (C_b) is equal to $\dfrac{N^-}{N^+}$ if x_b belongs to the minority class and 1 otherwise where N^-(N^+) denotes the number of majority samples (minority samples). Then, the classification weight ($w_{c|j}$) is inversely proportional to the cost-sensitive loss and written as follows:

$$w_{c|j} = \frac{\exp\left(-L_j\right)}{\sum_j \exp\left(-L_j\right)}, \quad j = 1, 2, \ldots, l^{t-1} \tag{1.116}$$

where l^{t-1} denotes the ensemble size of H at time moment $t-1$. The weight of the newly trained component classifier at t is equal to 1. Therefore, only weights for the l^{t-1} component classifiers in H^{t-1} need to be computed.

On the other hand, the sensitivity of a component classifier is evaluated by the cost-sensitive stochastic sensitivity measure (ST-SM). The ST-SM [174] has been widely applied in different applications, for instances neural network architecture selection [174], sample selection [175], MLPNN training [176], feature selection [177], steganalysis [178], and business intelligence [179]. The ST-SM of the jth component classifier is defined as the expectation of squared differences between outputs of training samples and samples located within a distance of Q in each dimension:

$$E_{S_Q}\left((\Delta y)^2\right) = \frac{1}{m}\sum_{b=1}^{m} E\left[\left(f\left(x_b + \Delta x\right) - f\left(x_b\right)\right)^2\right] \tag{1.117}$$

where $\Delta x \in [-Q, +Q]^n$ denotes the perturbation of the training sample and f is the real-valued outputs before thresholding to $\{-1, +1\}$ of a component classifier. Intuitively, the ST-SM measures the fluctuation of classifier outputs with respect to input perturbations, that is, it measures the stability of the classifier. Therefore, a classifier yielding a large ST-SM value is easily affected by small perturbations of inputs and more unstable. As a result, a smaller weight should be assigned to a classifier yielding a higher ST-SM value.

In this work, we propose the cost-sensitive ST-SM which is defined as follows:

$$S = \frac{1}{m}\sum_{b=1}^{m} C_b E\left[\left(f\left(x_b + \Delta x\right) - f\left(x_b\right)\right)^2\right] \tag{1.118}$$

A quasi-Monte Carlo-based method is adopted to calculate the cost-sensitive ST-SM of a classifier as in [176]. Specifically, Δx is generated via an n-dimensional Halton sequence [180] with each coordinate ranging from $[-Q, Q]$ using MatLab and 50 Halton points are used in the calculation of the expectation term in (1.118). According to experiments in [176], 50 Halton points yield only around 4% estimation error and the computational time is fast. Higher number of Halton points can be used for more accurate estimation but with higher computational costs.

It is difficult to automatically select the Q value theoretically. In implementations, $Q = 0.1$ is usually used which indicates a maximum of 10% of deviation from the training samples for data set with input features being normalized to [0, 1].

In the theory of the Localized Generalization Error Model [174], a good classifier should minimize both the classification error and the ST-SM. Therefore the proposed weighting scheme assigns larger weights to classifiers yielding smaller cost-sensitive ST-SM values. The sensitivity weight is inversely proportional to the cost-sensitive ST-SM and written as follows:

$$W_{s|j} = \frac{\exp(-S_j)}{\sum_j \exp(-S_j)}, \quad j = 1, 2, \ldots, l^{t-1} \tag{1.119}$$

Then, the fusion weight of the final ensemble of the CWIB is defined as the combination of the classification weight and the sensitivity weight as follows:

$$W_j = \eta W_{c|j} + (1 - \eta) W_{s|j}, \quad j = 1, 2, \ldots, l^{t-1} \tag{1.120}$$

where η is a trade-off coefficient between the classification performance and the stability of component classifiers. In our experiment, $\eta = 0.5$ is used to represent an equal importance of these two factors. The final decision of the ensemble is the weighted sum of outputs of all component classifiers:

$$H^t(x_b) = sign\left(\sum_j w_j \cdot h_j(x_b)\right) \tag{1.121}$$

Experimental Studies with Electricity Price Classification

As mentioned in [149], electricity price is a complex signal due to its characteristics of nonlinearity, time variant, and nonstationary behavior. More robust and accurate price classification and forecasting methods are still needed. As an example, for electricity price forecasting, authors in [149] proposed a complex electricity price forecasting technique based on feature selection and cascaded neuro-evolutionary algorithm (CNEA). The CNEA consists of cascaded forecasters, with each forecaster made up of an evolutionary algorithm and neural network. The adjustable parameters in the feature selection algorithm and the CNEA are fine-tuned with an iterative search procedure. However, the data segmentation for model training, i.e. optimal data size for training, was not well studied. To predict the day-ahead price, authors used a rule-of-thumb and the model was trained according to previous 50 days of data. The nature of the data, i.e. data imbalance or concept drift has not been considered prior training the model.

The work in [148] investigated several data mining approaches for electricity price classification. This includes correlation-based feature selection, multilayer perceptron, K-nearest neighbors, etc. Similar to forecasting problems, the data segmentation for classification problems has been arbitrary. For example, the authors used 20 historical days for the model training, with an argument as a fair comparison to the previous work in [147]. Evidently, it has been observed that previous research efforts have not considered electricity market price classification in data streams.

Table 1.31 Short descriptions on the methods used in the experiments

Methods	Short description
CWIB	When a new data chunk arrives, a new sub-ensemble is trained using the IRB to relieve the class imbalance problems. Each sub-ensemble is weighted according to their classification performance based on a cost-sensitive loss and their stochastic sensitivities on the current data chunk. The worst sub-ensemble is removed if a pre-selected ensemble size is reached.
CDS [161]	When a new data chunk arrives, a new classifier is trained using data rebalanced by the SMOTE. Each classifier is weighted based on a time-decay function and its performance on current data chunk.
NIE [161]	The differences between CDS and NIE are that NIE trains a sub-ensemble when a new data chunk arrives and NIE uses different error metrics to evaluate its sub-ensembles. By using different error metrics, the NIE can be distinguished into three variations, which are WRM (weighted recall measure), FM (F1-score measure), and GM (geometric-mean measure).
SERA [160]	When a new data chunk arrives, the SERA trains a new ensemble using the current data chunk and the most relevant historical minority samples. All minority samples seen so far are preserved and those with the smallest Mahalanobis distance from current minority samples are selected as part of the training samples so that a preselected post-balance ratio is met.
DWMIL [162]	The DWMIL trains a new sub-ensemble for each data chunk using UnderBagging and weights each sub-ensemble based on their performance to the current data chunk. The weights are reduced based on both a poor performance and/or the age of the sub-ensemble over time.

In this section, the CWIB is compared with other existing methods designed for imbalanced data streaming classification problems with concept drift. The electricity data set used in the experiment is introduced in section "Electricity Dataset". Section "Effects of Different Parameters" studies the effects of parameters to the CWIB. Section "Experimental Results" presents and discusses experimental results of the CWIB and other methods. The CWIB is compared with the following state-of-the-art methods in experiments: Learn++.CDS [168], Learn++.NIE [168], SERA [167], and DWMIL [169]. There are three variations of the Learn++.NIE using different error metrics: the weighted recall, the F1-measure, and the geometric mean. In our experiments, they are named as the WRM, the FM, and the GM, respectively. The default values for the parameters are used as suggested in the literatures. A short description on the methods used in the experiment is given in Table 1.31:

The numbers of component classifiers and base classifiers used in the CWIB are set to be 10 and 5, respectively. Larger numbers of component and/or base classifiers could be used to better adapt to the gradual drifts if higher computational costs are allowed. Radial Basis Function Neural Networks (RBFNN) with 10 hidden neurons are used as the base classifiers in all algorithms for fair comparisons. Neural networks have been successfully applied in future electricity forecasting [150, 181]. RBFNN is used here for its universal approximation capability [182] and its fast training speed compared with other types of neural networks, e.g. multilayer perceptrons. Ten independent runs are performed for all methods to reduce random

Table 1.32 Characteristics of the Elec2 data set

Dataset	Size of data chunk	# Features	# Time moments	Imbalance ratio
Elec2	328–329	5	82	0.27–1.63

effects. The AUC (area under curve), the F1-measure, the G-Mean (geometric mean), and the accuracy are used to compare performances of each method.

Electricity Dataset

The Electricity Pricing data set Elec2 [183] is used in our experiment to simulate the concept drifting and class imbalance environment, which originally contains 45,312 samples drawn from May 7, 1996 to December 5, 1998 with one sample for each half-hour from the electricity market in New South Wales, Australia. Samples with missing features have been removed so the remaining data set contains only 26,975 samples. This data set provides time and demand fluctuations in the price of electricity in New South Wales, Australia. The day, period, New South Wales electricity demand, Victoria electricity demand, and the scheduled electricity transfer between the two states are used as the input features to predict whether the price of New South Wales will be higher or lower than that of Victoria's in a 24-h period. Usually, a data chunk consists of 336 samples. However, samples with missing values are removed. Therefore, some data chunks may consist of fewer samples.

The concept drifts in this data set are natural and unavoidable because the electricity prices change with demand over different time periods. Moreover, the imbalance ratio between two classes changes over time and the majority and the minority classes may swap over time.

Table 1.32 shows the characteristics of the data set.

The imbalance ratio is defined as the ratio of the number of minority samples over that of majority samples. A special case of concept drift for this data set occurs when the imbalance ratio exceeds 1 because of the minority positive class becomes the majority class in some time moments. Figure 1.45 shows the imbalance ratios of the Elec2 data set over time, where the y-axis represents the imbalance ratio and the x-axis represents the time moment/time step of each data chunk arriving which contains roughly 1 week of data (since some data with missing values are removed). The effects of imbalance drift on the performance of learning models have been systematically analyzed in [184, 185], showing that without properly handling the drift of imbalance, changes in imbalance status would negatively affect the performance. In this work, we propose to apply the IRB to handle the class imbalance problem to avoid severe performance deterioration caused by the drift of class imbalance.

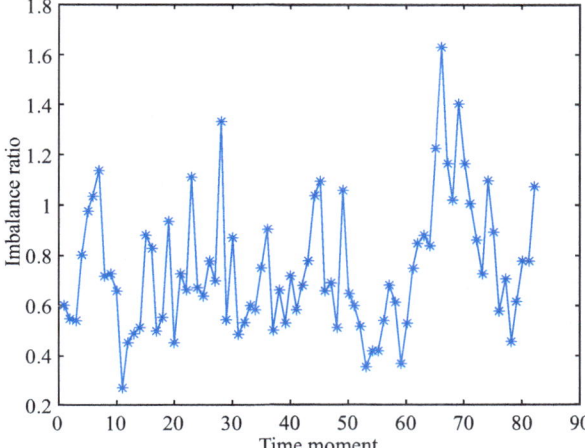

Fig. 1.45 Imbalance ratios of Elec2 data set over time

Effects of Different Parameters

The CWIB uses two parameters: the number of component classifiers and the number of base classifiers in a component classifier. Experiments are carried out to show different behaviors of the CWIB by using different sets of parameters.

Figures 1.46(a), (b) show the true positive rate (TPR) and false-positive rate (FPR) of both the minority and majority class by varying numbers of component classifiers and base classifiers, respectively. From Fig. 1.46(a), with the increment of the number of component classifiers, the true positive rate of the minority class increases while the false-positive rate of the minority class decreases. Both curves of the true positive rate and the false positive rate of the minority class tend to converge when the number of component classifiers is around 10. In contrast, the true positive rate of the majority class decreases and the false-positive rate of the minority class increases when the number of component classifiers increases. This is because the IRB reverses the class imbalance ratio and the representation of minority class is enhanced while the representation of majority class is diminished. The performance gained on the minority class is higher in comparison to the minor classification performance loss on the majority class. So, the overall performance of the CWIB is enhanced. Therefore, the number of component classifiers is set to 10 in our experiments to yield relatively high true positive rates and relatively low false-positive rates for both classes.

From Fig. 1.46(b), with the increment of the number of base classifiers, all four curves are quite stable and start to converge when the number of base classifiers is around 5. The number of base classifiers seems to have very minor effects on the performance of the CWIB. Hence it is set to 5 to maintain a low computational cost and achieve a high classification performance.

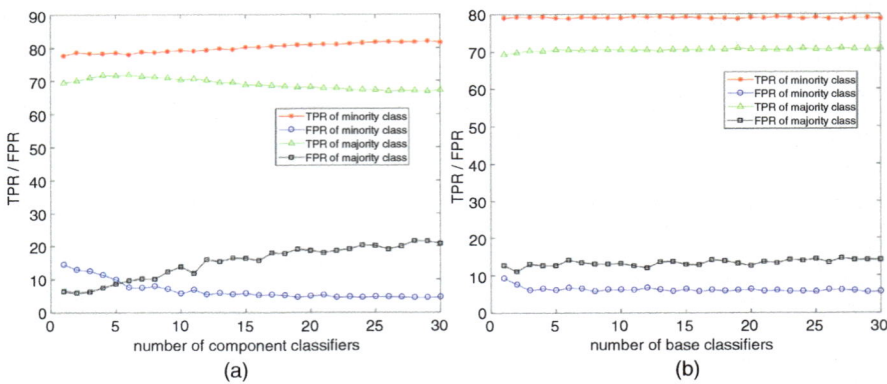

Fig. 1.46 Different behaviors of the CWIB by varying the parameters. (**a**) TPR and FPR of both classes by varying the number of component classifiers. (**b**) TPR and FPR of both classes by varying the number of base classifiers

Experimental Results

Figures 1.47(a)–(d) show average values of the four performance metrics of different methods over 10 independent runs over time, respectively. Table 1.33 shows the mean and the standard deviation values of different metrics for different methods over all data chunks. The bolded value of each column indicates the best result yielded for this metric and the symbol "*" indicates a statistically significant difference between the CWIB and the corresponding method by Student's t-test with 95% confidence. The number in the parenthesis is the rank of the method in terms of corresponding performance metric. The last column gives the average rank of each method over four metrics.

From Table 1.33, the CWIB yields the best average rank in terms of all metrics. The CWIB outperforms the FM, GM, WRM, and SERA significantly in terms of all metrics. In comparison to the CWIB, both the DWMIL and the CDS yield only a small degraded performance in AUC, but much worse performances in all F1-measure, G-Mean, and Accuracy (at least 2.74% differences). The high values of G-Mean and F1-measure yielded by the CWIB indicate that the combination of the IRB and the cost-sensitive weighting scheme in the CWIB enhances accuracies of both classes (i.e., true positive rate and true negative rate). The SERA yields a relatively high Accuracy (ranks the third) but very poor ranks of seventh in the AUC, the F1-measure, and the G-Mean. It may be due to the fact that decision boundaries created by the SERA are too biased to the majority class. This makes the SERA classify most samples as the majority class to achieve a high average accuracy but ignore the performance on the minority class. In contrast, the CWIB enhances the representation of the minority samples by applying the cost-sensitive weighting scheme, reversing the imbalance ratio, and at the same time a bagging method is

Table 1.33 Performance of different methods on the Elec2 data set

	Accuracy	AUC	F1-measure	G-Mean	Mean Rank
CWIB	**75.80 ± 0.52(1)**	**85.47 ± 0.36(1)**	**73.47 ± 0.44(1)**	**72.20 ± 0.37(1)**	**1**
DWMIL	73.06 ± 1.12*(2)	85.14 ± 0.37(3)	68.68 ± 1.30*(6)	64.74 ± 1.48*(6)	4.25
CDS	72.37 ± 0.54*(6)	85.40 ± 0.29(2)	70.00 ± 0.60*(3)	68.90 ± 0.68*(3)	3.50
FM	72.15 ± 0.51*(7)	83.30 ± 0.42*(5)	69.66 ± 0.49*(4)	68.41 ± 0.53*(4)	5
GM	72.84 ± 0.58*(4)	83.37 ± 0.45*(4)	70.37 ± 0.60*(2)	69.13 ± 0.59*(2)	3
WRM	72.41 ± 0.41*(5)	83.29 ± 0.41*(6)	69.47 ± 0.36*(5)	67.63 ± 0.44*(5)	5.25
SERA	72.98 ± 0.44*(3)	71.83 ± 0.75*(7)	65.53 ± 0.62*(7)	57.70 ± 1.73*(7)	6

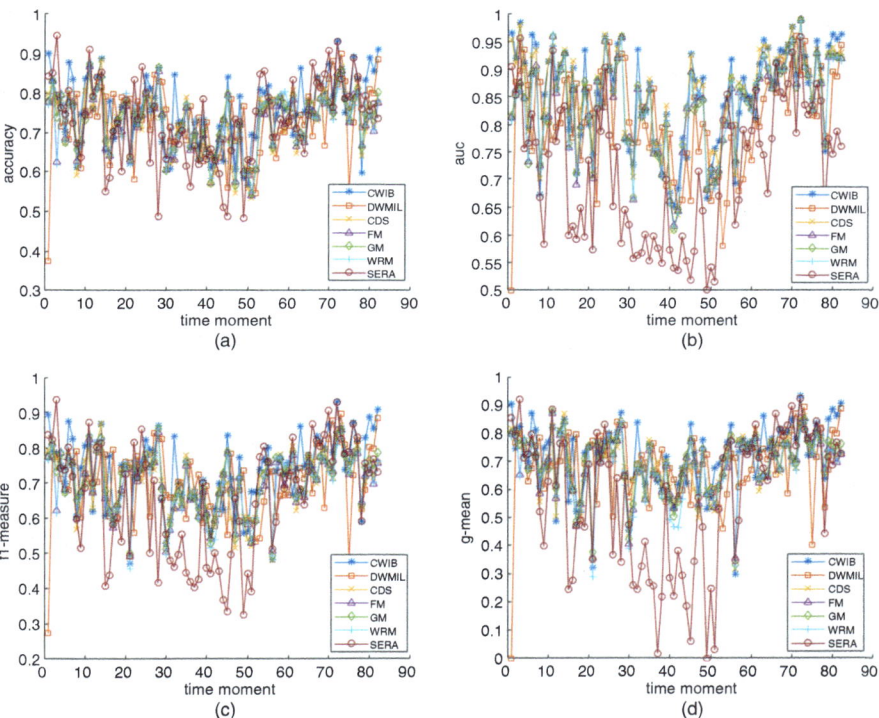

Fig. 1.47 Experimental results of different methods for the Elec2 data set. (**a**) Accuracy. (**b**) AUC. (**c**) F1-measure. (**d**) G-Mean

employed to maintain a low false-positive rate. Numerical results confirm the effectiveness of the proposed CWIB.

From Figures 1.47(a)–(d), performances of all methods fluctuate severely because of the type of concept drift is unknown and can be highly complicated. Moreover, the swapping between the minority and the majority classes further increases the difficulty of this learning task. The SERA yields the worst performance in all metrics and sometimes yields 0 value in terms of G-Mean because the SERA maintains too many outdated minority samples which consistently deteriorate its performance. The rest of methods perform similarly and fluctuates severely as time varies.

In summary, experimental results show that the CWIB is effective and yields statistically significantly better results in comparison to state-of-the-art methods. Moreover, the CWIB uses fewer storage and computational costs by using a small fixed size ensemble in comparison to the very large ensemble size (equal to t) used by the CDS, the WRM, the FM, and the GM for a large t and the variable ensemble size used by the DWMIL.

Conclusions and Future Work

Concept drifts occur commonly and are unavoidable in data streaming-based pattern classification problems, such as electricity price classification. The consideration of concept drift and imbalanced data for electricity price is a novelty of this work. In general, the problem is more complicated when numbers of samples in different classes are imbalanced. Therefore, the CWIB is proposed to deal with these two problems simultaneously. The CWIB dynamically weights component classifiers according to their classification performances and stochastic sensitivities with respect to input perturbations. New component classifier is trained using the Imbalance-Reversed Bagging (IRB) method to cope with the imbalance issue in a data chunk. The ensemble of the CWIB maintains a constant size by removing the component classifier yielding the smallest weight.

Experimental results show that the proposed CWIB yields better Accuracy, AUC, G-Mean, and F1-measure than state-of-the-art methods with statistical significance on an electricity pricing data set. This shows that the proposed method is useful to energy and power researches when the classification problems have a data set in a streaming form with class imbalance occurring in data chunks, e.g., prediction of outliers of electricity demand, fault diagnostic in power distribution system, and stochastic renewable energy generation, e.g., wind and solar. They are the future areas to be studied.

On the other hand, removing the component classifier yielding the smallest weight may not be the best method because it may reduce the diversity of the classifier ensemble. In our future works, we will research on the possibility of adding time as a component of the weight computation. The diversity between base classifiers and between component classifiers may also be added to the weight computation to enhance the overall diversity of the ensemble of ensembles of the CWIB.

In this work, the RBFNN is used as the base classifier. However, it may be more powerful to use multiple types of classifiers to create the classifier ensemble for the CWIB. The optimal combination and selection method will be one of an important future works.

1.3.1.7 Application of Big Data to Smart Energy

Introduction

The terms "big data" and "data analytics" can mean very different things, depending on the problems to be solved [186–189]. Many researchers and practitioners manage "big data" challenges by focusing on its high volumes of information rather than the information management. In general, big data is a term used to recognize the exponential growth, availability, and use of information in the data-rich environment of today and beyond. The term "big data" puts a focus on the issue of information volume. Many researchers and practitioners manage "big data" challenges by focusing on its high volumes of information rather than the information management and this can lead to short-sighted decisions that will restrict the information architecture in which decision makers and managers try to expand and upgrade to meet changing needs for the society, environment, and business. Too narrow a focus will force massive reinvestment to address other factors due to big data to reduce risks and increase costs. Data volume is growing annually at a minimum rate of 59% worldwide annually. Although volume is a significant challenge in managing big data, data information variety, and velocity must be focused on as well.

Volume: The increase in data volumes comes from traditional data types and new types of data. A very large size is a storage issue and too much data is also a massive analysis problem.

Variety: It is always a logical sequence to obtain data, derive information from it, and then make decisions. There are now many more varieties of information to analyze. The varieties include tabular data, documents, metering data, video, images, audio, financial transactions, and others.

Velocity: This involves streams of data, structured record creation, and availability for access and delivery. In some cases, real-time decision making is needed. Therefore, it is important to have the required velocity to produce the data and process it to meet the need.

Big Data technologies as a new generation of technologies and architectures designed to extract value economically from very large volumes of a wide variety of data by enabling high velocity capture, discovery, and/or analysis. The total amount of data has grown exponentially, it has been estimated that more data was produced between 2010 and 2012 than in all of preceding human history [190]. It is essential to make sense of big data and get patterns from it to help organizations to make better decisions. The big data in Smart Grid is generated from various sources, such as (1) power utilization habits of users, (2) phasor measurement data for situational awareness, (3) energy consumption data measured by the widespread smart meters, (4) energy market pricing and bidding data collected by automated revenue metering (ARM) systems, (5) management, control and maintenance data for devices and equipment in the power generation, transmission, and distribution networks acquired by intelligent electronic devices, (6) operational data for running utilities, such as financial data and (7) very large data sets, not directly obtained through the grid measurement but widely used in decision making, such as weather data, data from the National Lightning Detection Network (NLDN), and Geographic Information System (GIS) data.

The application of big data in power utilities is just started. By 2020, the number of installed smart meters in Europe will reach 240 million while North America will have 150 million smart meters in use, China is forecasted to install about 400 million smart meters by that date. Japan would deploy about 60 million smart electricity meters and South Korea would plan to deploy between 500,000 and 1.5 million smart meters per year in homes before 2020. With so many smart meters being deployed, utilities' data center will increase the amount of data by several TB per day. Research has shown that many of the current utilities have not fully explored the value of "big data." However, serious attention has been given to this field and some well-known institutions have updated their teaching and research programme to education students who have can take up the challenges of big data research and applications in the near future. Data analytics competitors are competing to bring a set of IT tools and capabilities that are largely new to the utility industry.

Related Techniques and Tools

Over the past few years, nearly all major companies, including IT giants like Oracle&IBM, grid giants such as General Electric, Siemens/eMeter, ABB/Ventyx&Schneider Electric/Telvent, and startups like AutoGrid, Opower, and C3. All have started their big data projects and are competing to bring a set of IT tools that are largely new to the utility industry. Scientists have developed a wide variety of techniques and technologies to capture, curate, analyze, and visualize Big Data. Big Data needs powerful techniques to efficiently process very large volume of data within limited time. The techniques involve a number of disciplines, including statistics, data mining, machine learning, signal processing, pattern recognition, optimization methods, and visualization method.

Techniques

Optimization methods have been applied to solve quantitative problems in a lot of fields, such as engineering, economics, physics, chemistry, and biology. In smart grid area, to tackle communication problems [191–194] and data processing problems [195], a variety of new scalable and distributed architectures or frameworks have been proposed. Data mining is a set of techniques to extract valuable information (patterns) from data, including clustering, classification, and regression. For meter data analytics, data mining methods have been adopted. The applied methods can be grouped in two classes, namely statistical methods to make estimation based on historical data, and AI methods to model risk and uncertainty [196]. Recently developed data mining methods include fuzzy wavelet neural network [197] and nonparametric estimation [198]. Customers are grouped according to their electricity consumption patterns and/or other characteristics, such as activities. Clustering methods are applied in generating residential load profiles [199] and to differentiate pattern variations due to seasonal and temporal impacts [200]. Fuzzy decision tree is used to classify power quality disturbances [201]. Data mining approaches have been used for price classification [148]. Machine learning is an important area in artificial intelligence. It is used to discover knowledge and make intelligent decisions automatically. Visualization method is used to create tables, images, and diagrams to understand data.

Tools

One of the most famous and powerful batch process-based Big Data tools is Apache Hadoop. It provides infrastructures and platforms for specific Big Data applications in business and commerce. For stream data applications, for example, electric power system operation, would require real-time response for data processing platforms such as Storm which is designed especially for real-time stream data analytics. For interactive analysis processing, the data are presented in an interactive environment. Users are directly connected to the computer and can interact with it in real time. Apache Drill is a distributed system for interactive analysis of Big Data. It has the capability to process petabytes of data and trillions of records in seconds.

Technologies

The ongoing or emerging technologies that are closely related to big data include cloud computing, Internet of things (IoT), granular computing, data center, and quantum computing [202].

Cloud Computing

Cloud computing delivers applications and services over the Internet. Cloud computing is closely related to big data. Big data puts stress on the storage capacity of a cloud system. The main objective of cloud computing is to use huge computing and storage resources to provide big data applications with fine-grained computing capacity [203]. The emergence of big data also accelerates the development of cloud computing. The distributed storage technology and the parallel computing capacity can effectively manage big data and improve the efficiency of acquisition and analysis of big data. CloudView is a framework for storage, processing, and analysis of massive machine maintenance data in a cloud computing environment, which is formulated using the Map/Reduce model and reaches real-time response. Cloud computing promotes transferring and sharing data.

Internet of Things (IoT)

The Internet of Things (IoT) is the network of physical objects or "things" embedded with electronics, software, sensors, and connectivity to enable it to achieve greater value and service by exchanging data with the manufacturer, operator, and/

or other connected devices. The internet of things represents an enormous amount of networking sensors embedded into various devices and machines in the real world. Such sensors deployed in different fields to collect various kinds of data, such as environmental data, geographical data, operation data, customer data in power utilities. Many devices including mobile equipment, home appliances could be used for data acquisition. IoT enables advanced applications for smart grid. The big data generated by IoT has special characteristics because of the different types of data collected, which have the characters of heterogeneity, variety, unstructured feature, noise, and high redundancy. It is expected that by 2030, the IoT data will be the most important part of big data, big data in IoT has three features that conform to the big data paradigm: (1) huge amount of data is generated from terminals; (2) data generated by IoT is usually semi-structured or unstructured; (3) the data processing capacity of IoT has fallen behind so it is extremely urgent to promote the development of IoT based on the introduction of big data technologies.

Granular Computing

In granular computing (GrC), granules such as classes, clusters, subsets, groups, and intervals are used to build computational models for complex applications with huge amounts of data, information, and knowledge. Granular computing can reduce the data size into different level of granularity. However, not all the Big Data applications can use the GrC techniques. It depends on the confidence and accuracy of results required. For example, power network sensor data needs to be processed and responded in time and with high accuracy for decision making.

Data Center

The data center is in charge of acquiring, managing, and organizing data, and leveraging the data values and functions. The emergence of big data brings development opportunities and great challenges to data centers. Organizations are experiencing rapid IT growth but their data centers are aging. International Data Corporate (IDC) puts the average age of a data center at 9 years old. Gartner, another research company says data centers older than 7 years are obsolete [204]. As the electric grid gets smarter, vast quantities of data arrive at utility companies, which promote the explosive growth of the infrastructure and related software of data center. The big data has more requirements on storage capacity and processing capacity, as well as network transmission capacity. Enterprises must take the development of data centers into consideration to improve the capacity of rapidly and effectively processing of big data under limited price/performance ratio.

Quantum Computing

Quantum computing studies quantum computers that make direct use of quantum-mechanical phenomena, such as superposition, to perform operations on data [205]. Quantum computers are different from digital computers which require data to be encoded into binary digits. Quantum computation uses quantum bits. The development of actual quantum computers is still in its infancy, but experiments have been carried out in which quantum computational operations were executed on a very small number of qubits. It is foreseen that large-scale quantum computers will be able to solve certain problems much more quickly than any classical computers that use even the best currently known algorithms such as integer factorization using Shor's algorithm. Small-scale quantum computers are existed, for example, D-Ware Systems Company developed their quantum computer, called "D-Wave one" with 128 qubits processor and "D-Wave two" with 512 qubits processor in 2011 and 2013, respectively.

Challenges and Opportunities

When handling Big Data problems, difficulties lie in data capture, transmission, processing, storage, searching, sharing, analysis, and visualization. Data are increasing at exponential rate, but the improvement of information processing methods is relatively slow. In many important Big Data applications, the state-of-the-art techniques and technologies cannot solve the real-life problems practically, especially for real-time analysis. Some of them are listed below:

Uncertainty

Utilities are uncertain about the costs and requirements for building a big data analytics infrastructure. There is also uncertainty about how big data analytics will fit into the current systems. There is a lot of data and people do not know how to manage, store it, and at what point does it become useless. Data curation needs to be considered. The data management through its lifecycle of interest and usefulness is to power utilities. Curation activities enable data discovery and retrieval, maintain quality, add value, and provide for reuse over time. The existing database management tools are unable to process Big Data that grow so large and complex. The size of Big Data keeps increasing exponentially, but current capability to work with is only in the relatively lower levels of petabytes, exabytes, and zettabytes of data. New framework for modeling uncertainty and predicting the change of the uncertainty is required.

Security

Most Big Data are stored in a distributed way, cloud data storage is often used. However, the network bandwidth capacity is the bottleneck in cloud and distributed systems, especially when the volume of communication is large. Smart meter installations have generated concerns about data privacy. Tremendous amounts of data about individuals, e.g., internet activity, energy usage, social interaction, are being collected and analyzed, which have the risk to cause damage to data provider. Significant security problems include data security protection, intellectual property protection, personal privacy protection, commercial secrets, network security, and financial information protection. Most developed and developing countries have already made related data protection laws to enhance the security.

Data Quality

As the size of data set is very large, sometimes in the region of several gigabytes or more and also the data origin is from many sources, real-world databases include inconsistent, incomplete, and noisy data. Therefore, a number of data preprocessing techniques, including data cleaning, integration, transformation, and reduction need to be applied to minimize noise, inconsistencies, and incompleteness in data. In many cases, the current techniques will be too slow to achieve a workable solution for real-life problems.

Data Analysis and Visualization

For real-time Big Data applications such as power system operation and protection, it is necessary to guarantee the time response requirement when the data volume is very large. Presently, it is a big challenge for stream processing. Big Data has encouraged the development of the hardware and software architectures such as the advancement in cloud computing which distributes multiple workloads into a large cluster of processors. In this direction, distributed computing is being developed at a very high speed. It is required to understand how Big Data works with real-time systems. However, it is not easy to link system state, measurements, and network topologies together with respect to time. It is particularly difficult to conduct data visualization because of the high dimension and size of Big Data since the current Big Data visualization tools mostly have poor performance in functionalities, scalability, and response time.

Data Explosion

Data from the UK Department of Energy and Climate Change [206] are used to describe appliance ownership patterns. These data enable energy efficiency ratings to be accounted for. However, smart metering will create an explosion in data availability. For example, in the UK, the 27 million domestic electricity consumers currently just have over 100 million data points per year collected quarterly or half-yearly for energy suppliers to record, store, and use in billing and other business operation. When smart metering fully deployed and operated at a 30-min sampling rate, energy suppliers will need to ingest, store, and process at least around 4500–9000 times more of the current data size, reaching 50 TB. To manage data sets in such a large volume, the main problem of using relational database management systems is its low scalability [207] and this requires a scalable solution that can grow for practical use.

Lack of Standards

To take advantage of these large new data sets, it is essential to gain access to data; develop the data management; and programming capabilities to work with large-scale data sets. New approaches to summarize, describe, and analyze the information contained in big data must be developed [208]. Integration of many different forms of data from systems like meter data management, outage management, customer management, billing platforms, and asset management is required. Standards for data description and communication are essential. These facilitate data reuse by making it easier to import, export, compare, combine, and understand data. Standards also eliminate the need for each data originators to develop unique descriptive practices [209]. The lack of worldwide industry standards around data from smart grids and meters could lead to concerns about sharing data with competitors, it also brings worry around data ownership, and concerns about data accuracy. A further worry is on data privacy and data protection.

Lack of Talents

The shortage of talent will be a significant constraint to obtain values from Big Data. Big Data is expected to rapidly become a key element of competition internationally. This kind of specialist is difficult to educate as it takes many years to train Big Data analysts that must have strong mathematical background and related professional knowledge. Specialized resource is a critical success factor for better data management. Utilities are concerned about the shortage of available data specialists and will need to undertake efforts to locate or develop such talents. To ensure a continued supply of skills in the future, in-house training is a likely means but a talent war is likely to get bigger.

Current Workers May Be Reluctant to Use New Technologies

Big Data Application Examples in Smart Grid Worldwide

Table 1.34 summarizes some potential application areas of big data in smart grid.

The key goals of U.S. utilities' big data efforts, mainly focused on converting the tens of billions of data points coming from the millions of smart meters deployed around the country and turning them into actionable information for the grid operations. For U.S. utilities most of the data analytics focus on improving grid reliability, outage response, and lowering the cost of distribution operations. European utilities can either be vertically integrated service providers for entire nations or contenders in deregulated and competitive energy markets. For example, Italy and the Scandinavian countries have largely completed their smart meter rollouts while the UK, France, Germany are just getting started. For European utilities, they

Table 1.34 Potential application areas of big data in smart grid

1	Prediction and analysis of economic situation and social impact
2	Development of scientific reasoning for decision making
3	Performance analysis for generation and storage systems
4	Load management with demand response and energy utilization & efficiency analytics
5	Consumer behavior analysis
6	Using AMI and intelligent electronic devices for state estimation
7	Pricing analytics and incentive implementation analysis
8	Grid infrastructures optimization
9	Demand and generation forecast under high uncertainties
10	Asset management
11	Service quality analytics

concentrate on integrating large-scale renewables and distributed generations and managing new loads such as plug-in electric vehicles.

EDF [210]

EDF has applied big data technology for weather forecasting for risk assessment and modeling the impacts of energy operations on the environment. As EDF begins its plan to roll out 35 million smart meters across France, it will incorporate this data into doing business. Smart meter data will be used to better estimate the condition of the grid, promote demand response, predict the lifetime of power lines, transformers, and other equipment. Demand response is an important part of managing grid assets in a world where new customer loads such as plug-in electric vehicles and the increasing share of power being generated by customers via distributed generations are altering supply-demand balances.

E.ON [211]

In September 2013, IBM announced that it has been selected by E.ON Metering to operate its Smart Metering IT infrastructure in a private cloud. The new platform will improve the deployment and management of smart meters, simplify the integration of renewables, and other innovative services, while also allowing E.ON to deliver personalized services that will put customers in better control of energy usage. Customers will have the ability to view their usage profiles for information about time-of-use-rates and changes in use patterns that can be compared with historical data. In addition, the platform's scalability and low start-up and operation costs will provide the flexibility for future growth. Real-time data evaluation in a smart grid is growing increasingly important for energy utilities due to the increased use of intermittent renewable energy.

USA EXELON [212]

In 2014, Baltimore Gas & Electric (BGE) started a project in the hope to save hundreds of millions of dollars by using C3's cloud-based data analytics to manage operations of the 2 million smart meters deployed and by tapping their data to discover and prevent energy theft and revenue losses. Trillions of data points are regressing and correlating against each other to extract values. BGE believes it can better detect, isolate, and reduce meter tampering and unbilled energy delivery. BGE parent company Exelon also plans to apply the same C3 platform for projects at its Chicago-based Commonwealth Edison and Philadelphia-based PECO utilities, which are deploying a collective 8 million smart meters between them. Integration of many different forms of data from systems like meter data management, outage management, customer management, billing platforms, and asset management is required. On the smart grid side, C3 is working with utilities including Southern California Edison, Northeast Utilities, and Entergy. It will be very difficult if not impossible to encode every rule for theft patterns because they change over time. Machine learning is used where you have to learn from the data without being able to code an algorithm. C3 is a grid data analytics company to cite machine learning as part of its suite of tools, this is a cutting-edge claim to apply techniques as yet untested fully in the utilities.

Pacific Gas and Electric Company [213]

Smart meters are providing utilities with unprecedented amounts of energy data. Utilities across the US, including Pacific Gas and Electric Company (PG&E) in California, are revolutionizing the use of such data to empower their customers. PG&E's groundbreaking work with its interval data analytics (IDA) program was developed to maximize the value of data captured by 9.4 million smart meters. PG&E is the largest US utility to install smart meters across its entire service territory. PG&E captures hourly or sub-hourly reads from each home. With more robust data, PG&E is also able to provide detailed advice to help customers better manage their energy use. PG&E's My Energy portal, which is available online to all customers, demonstrates the energy-saving benefits of personalized data and insights. Customers can examine their time-specific energy use, see how they compare within their neighborhood, understand how and why their consumption varies over time, set their usage goals, and, of course, discover more ways to save energy.

KEPCO [214]

In 2014, State-run electric utility Korea Electric Power Corp (KEPCO) launched two pilot projects on ways to use big data to improve demand management and risk forecasting. The first pilot set up an energy consulting business based on advanced metering infrastructure (AMI) data while the second established a risk forecasting system analyzing social networking service data. The objective of the first project is to help customers save electricity by providing comparable data on energy usage for similar business types based on AMI big data, while allowing KEPCO to manage demand and reduce brownouts. The second project aims to analyze a variety of business risks including blackouts, customer complaints, and climate change by merging information from social network service data, internet data, and complaints. KEPCO has stated it aims to train 300 big data professionals by 2016. To summarize, power utilities have diversity in the application of big data due to different, business structure, technology development, etc. Having more completed operational data, most of the companies put the demand side response and user service in a higher priority. It can be seen that in general, power utilities are cooperating with IT companies to develop big data applications.

Conclusions

Big data is still unclear for many power utilities. Smart grid operation and future energy management will be huge data intensive. There are many obstacles which affect the success of big data applications in smart grid. Presently experience in integrating big data with smart grid is limited. The utilities need to focus on turning the data that they collect into business intelligence to improve processes and customer experience. For example, the detected patterns of spatial and temporal electricity consumption can be used to help optimizing demand response management. This will lead to solutions that improve efficiency and enable alternative approaches to handle various aspects of the utility to drive company performance and minimize risks from new regulations and political interference under a low carbon economy.

1.3.2 Smart Water

With the United Nations reporting that 60% of the global population will call an urban area home by 2030, the key challenge for cities will be to manage the growth while transitioning to a sustainable reliable energy supply with very limited financial and spatial resources. This means that a city has to balance sustainable and stable supply of water, electricity, heat, and housing, while also integrating new sources of urban renewable energy. Cities will also need to update, renew, and expand their water and waste water networks and reorganize the city's transportation. These are huge challenges where advanced digital solutions can help to achieve the three goals: increase sustainability, resilience, and efficiency.

Proper water and energy efficiency often dictate a citizen's quality of life in any environment, the importance of a clean and sustainable water and energy supply is getting more and more attention across the globe to manage the growing population in cities. Not only to meet climate and safety goals, but especially to ensure the future economic competitiveness of a city. Therefore, control and monitoring of water networks will need to be improved to address water and waste water challenges. However, we can go beyond the basic technology. Digitalization helps in the coordination of the operation of all infrastructure assets in a city in order to minimize the operational costs associated with them.

Public authorities can bring smart city solutions to businesses and citizens by first defining a clear objective. For example, "What do we want to achieve in our city, how and why?" To define these objectives, public authorities should bring together key stakeholders including citizens, businesses, and companies as well as service and technology providers. Collaboratively, city-wide goals, a vision, strategy, and regulation can be defined with a robust road map for real innovation. Power and water utilities should be active players in these collaborative leadership teams, as they provide experience and expertise to improve operations both securely and sustainably.

Without a collaborative approach, only urgent needs will be addressed without any plan to better the lives of citizens in the long term. The installation of the foundation technologies like EV-Chargers, urban renewable generation, building automation, water metering, and leakage detection systems or even Free WiFi in public squares is very cost-intensive. However, to bring true value to citizens by combining and optimizing these technologies together, the backbone is the digital infrastructure. Only with smart connection and joint operational optimization with digital services, can a city free up financial resources that are otherwise tight, inefficiently and uncoordinatedly running all of these services in parallel. The money saved from digital integration can be re-invested to benefit citizens in the long term.

Water plays an essential role in the everyday lives of people, communities, and businesses. Without investment in water infrastructure, smart cities have limited room for development.

1.3.3   Smart Health

1.3.3.1   Introduction

The advent of information and communication technology (ICT), IoT, and wearable devices, has shed new lights on the healthcare sector in recent years. Reference [215] offers some insights about IoT and wireless sensor network (WSN) in healthcare context. Wireless sensors are used to monitor and collect healthcare-related information. The information may include vitals, mobility, location, food consumption, fluid consumption, and sleeping time, etc. These sensors could be deployed at home, office, hospital, or anywhere else. A wireless sensor network consists of a large number of sensors or sensor nodes. Each sensor node of a wireless sensor network is connected to at least one and possibly several other neighboring nodes and each sensor node is capable of collecting, processing, transmitting, and receiving information. Therefore, each node must have an antenna, a processor, source of energy, and some mechanism to be uniquely identified. Sensor networks, with big data analysis combined, can bring evolutionary innovations to the healthcare system, not only at personal level, but also at community level or hospitals. Figure 1.49 gives a simple WSN application scheme for healthcare.

Wireless sensor network technologies have the potential to change the way of living with many applications in entertainment, travel, retail, industry, medicine, care of the dependent people, and emergency management, and many other areas. Wireless sensors and sensor networks, pervasive computing, and artificial intelligence research together have built the interdisciplinary concept of ambient intelligence in order to overcome the challenges we face in everyday life [215]. One of the major challenges of the world for the last decades has been the continuous elderly population increase in the developed countries. It was forecasted that in the next 20 years, the 65-and-over population in the developed countries will be nearly 20% of the overall population [216]. Hence the need of delivering quality care to a rapidly growing population of elderly while reducing the healthcare costs is an important issue. One promising application in that area is the integration of sensing and consumer electronics technologies which would allow people to be constantly monitored. In-home pervasive networks may assist residents and their caregivers by providing continuous medical monitoring, memory enhancement, control of home appliances, medical data access, and emergency communication. Constant monitoring will increase early detection of emergency conditions and diseases for at-risk patients and also provide wide range of healthcare services for people with various degrees of cognitive and physical disabilities. Not only the elderly and chronically ill but also the families in which both parents have to work will derive benefit from these systems for delivering high-quality care services for their babies and little children. Researchers in computer, networking, and medical fields are working together in order to make the broad vision of smart healthcare possible. The importance of integrating large-scale wireless telecommunication technologies such as 3G, Wi-Fi Mesh, and WiMAX, with telemedicine has already been addressed by some researchers. Further improvements will be achieved by the coexistence of small-scale personal area technologies like radio frequency identification (RFID), Bluetooth, ZigBee, and wireless sensor networks, together with largescale wireless networks to provide context-aware applications. Besides providing pervasiveness with existing and relatively more mature wireless network technologies, the development of unobtrusive small sensor devices enabling not only accurate information but also reliable data delivery is of great importance. Moreover, the glue combining all these technologies is the application, which is the coordinator between the caregivers and the caretakers and between the sensor devices and all of the actors in the overall system cycle. Since the application is the core of the high-quality healthcare service concept, the need for intelligent, context-aware healthcare applications will be increased.

Given the importance of the subject, there are already several applications and prototypes on the subject. For example, some of them are devoted to continuous monitoring for cognitive disorders like Alzheimer's, Parkinson's, or similar cognitive diseases. Some focus on fall detection, posture detection, and location tracking and others make use of biological and environmental sensors to identify patients' health status. There is also significant research effort in developing tiny wireless sensor devices, preferably integrated into fabric or other substances, and be implanted in human body.

1.3.3.2   Smart Healthcare Deployment Scenarios

Smart Wearables

Medical professionals who would like to monitor their patients for a longer duration for data collection can prescribe use of a wearable device for a specific function such as monitoring heart/pulse rate, glucose level, physical activity, etc. The patients wear the prescribed devices for several hours/days. The devices with embedded sensors collect the data for a specified duration and the data are either retrieved from the device or transmitted to a medical professional. The data can be used for analysis, assessment, and/or for diagnosis. Having different components of data over a longer duration of time, medical professionals are equipped with the advantage to analyze and diagnose some of the conditions that occur infrequently or occur only under certain conditions.

Smart Homes

A smart collection of smart appliances in a smart home can make life a bit easier and improve many aspects of daily living. The aging population is rapidly growing. It was reported that by year 2050 the number of people aged 60 years and over is expected to reach 2 billion. This represents a growth from 12% of the world population in this category in 2015 to about 22% in 2050. To serve this segment of the population and to maintain or even improve their quality of life, smart home should become a dwelling that deploys emerging technologies including sensor networks and IoT to facilitate healthcare monitoring of its residents and facilitate their living independence. The smart home technologies cover many aspects, including smart appliances, assistive devices, smart controls, and integration of health monitoring with smart systems.

Smart Hospitals

Reference [217] discussed that by including smart devices to hospitals can save the operation cost, enhance the medical experience of patients, and reduce the labor intensity of medical staff. Here are a few examples.

1. Ward care: In the ward, the patient's real-time physiological sign like heart rate or the environmental information like the cleanliness can be collected by wearable devices or smart sensors. These data are then sent to the monitoring and supervision center by wireless communication. If the patient's physiological sign is abnormal, the paramedics can make the corresponding treatment in time.
2. Outpatient medical treatment: Outpatient doctors can get a comprehensive understanding of the patient's health based on physiological sign data collected

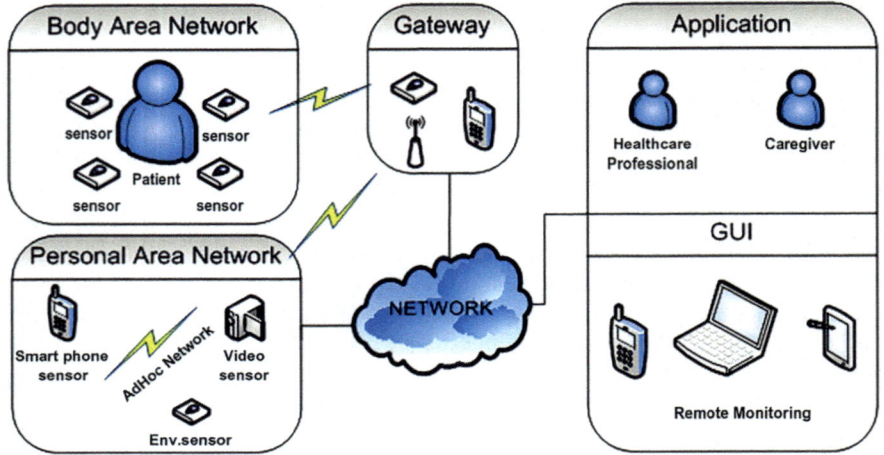

Fig. 1.48 A simple WSN application scheme for healthcare

by wearable devices, which can assist doctors to make accurate diagnosis, improve the efficiency of doctor's diagnosis, and save the patient's time.

3. Outdoor posture recognition: When the patient is outdoors, the patient's body and motion posture can be identified by posture sensors to identify whether a dangerous pose has occurred, which can also determine whether the abnormal physiological sign is a false alarm for normal situation.

4. Telemedicine monitoring: Some of the discharged patients need to be monitored at home. Wearable devices can monitor the patient's physiological sign remotely. When the patient's physical condition is abnormal, the device can notify the patient's family or the attending doctor in time to prevent accidental occurrence.

5. Other applications: We can realize intelligent meter reading by adding wireless communication module to the traditional electricity and water meters in the hospital. Some expensive medical equipment like gamma rays can be connected to the IoT system, then the equipment checking can be regularly completed. When some valuable medical item or medical waste is removal, the installed sensors can deliver the real-time location and status data to the cloud platform for effective monitoring.

1.3.3.3   Smart Healthcare Examples

Medicine Reminder

Patients can be alerted about taking their medications in a timely manner. Many elderly individuals forget to take their medicine regularly and on time. A smart medical device, like smart watch, can send alerts/reminders to the patients about the time to take a specific medicine. Apart from medicine taking notification, it can also assess if some of the medications should not be taken together and avoid serious medical errors. As an educational component, it can also generate a pattern of medicine intake based on prescription and the patient's habits. In case of some lapses, it can assess and share potential health risks. Such an information can possibly lead to more regular pattern of taking prescribed medicine as needed. NB-IoT smart watch proposed by China-Mobile is a great example for this application scenario [218].

Assistance for the Disabled

Residents with physical disabilities can make use of assistive devices that are connected to the home network. For instance, eye movement can be used to type commands, radio frequency identification (RFID) can be used in conjunction with sensor networks to open doors, or call for help, if needed, and provide identification information stored or appropriately implanted somewhere on a human body. These assistive technologies are priceless for individuals who are unable to communicate with objects and people, like healthcare providers, families, and friends otherwise.

Dementia

Older adults may experience cognitive decline but because they still retain a high degree of autonomy in their lives, this change may be too subtle to catch and treat. Nevertheless, patients in their early stage of dementia are prone to be spatially and temporally disoriented and forget important daily tasks [219]. With smart healthcare system, these changes translate into abnormal mobility patterns. The smart healthcare system analyzes these mobility patterns to communicate detected abnormalities to patients and care providers.

Depression

Another disease with subtle manifestations is depression. In some cases, symptoms are too faint for a person to note. With ICT-based psychiatry, changes in behavioral patterns such as lower activity levels, degrading sleep, decreased phone conversations, and even mobility patterns may point to a possible diagnosis of depression. Smart healthcare system intervenes when these changes are detected by recommending that the user contact a health care professional. ICT-based assistance can further extend from diagnosis to intervention by monitoring treatment compliance and medication effects.

Asthma

In Louisville, KY, USA, mobile ICT combined with citizen sensing helped the city to respond to asthma triggers and thus circumvent possible long-term chronic conditions for its residents. To identify where asthma triggers might be located throughout the region, sensor-enabled inhalers were distributed to asthma sufferers throughout Jefferson County. When the inhaler is used, the use is recorded on a smartphone app and the sensor monitors the nearby air for particulates that might be triggering the episode.

The sensor data spotlighted one particular road where inhaler use was three times as high as throughout the rest of the city. The city was able to respond to this information and planted trees that separate the congested road from residential neighborhoods. The result was a 60% decrease in particulate matter, addressing a major contributor.

Walkability

Urban planning can have huge impact on city residents' behavior and health. Specifically, the relationship between a community's "walkability" and a resident's behavioral routine and health profile is worth exploring. The impact of the built environment on lifestyle choices and resulting health is increasingly noticeable. In the smart healthcare at community level, information from smartphones to monitor daily routines in combination with neighborhood walkability for a population sample. Machine learning-based activity models label the captured sensor data with activity names and data mining techniques are used to analyze the relationship between walkability, activity, and demographics. To generate continuous monitoring information, participants wore GPS devices for 1–2 weeks while performing normal daily routines.

Based on the test and analysis, it is concluded that there does exist a relationship between the detected activities (including exercise and work) and body mass index (BMI) and that the correlations are not obtained by chance. The results of this study offer insights that city planners can use to generate urban analytics and modify city design to improve the health of city residents.

1.3.3.4   Smart Health Big Data Analysis

The greatest challenge of building a comprehensive healthcare system is handling the heterogeneous healthcare data captured from multiple sources. Reference [220] discussed the healthcare data classification, data-driven healthcare services, and the methods to manage the healthcare big data.

According to the data sources, data nodes can be divided into the following four groups.

1. Research data. Drug, medical equipment research and development institutions and other scientific research institutions have accumulated a large amount of research data, such as clinical trial data and high-throughput screening data. These data, including individual or clinical gene or protein data, can help identify the side effect and the new effect of a therapy.

2. Medical expense data. Medical treatments generate massive expense data, such as medical bills and medical insurance reimbursement. They are not the traditional healthcare data, but they can be used to analyze and estimate the medical cost, which is helpful for the government to formulate medical subsidy planning to alleviate patients' burden effectively. These data are generally stored in different databases of medical institutions, which are geographically dispersed and adopt unified data formats.

3. Clinical data. This kind of healthcare data is typical of medical data. These data could be clinical diagnosis such as electronic medical record (EMR), and medical images collected by medical service providers like hospitals, clinics. These data can be unified, managed, and opened to researchers with a necessary precondition for ensuring the privacy of the patient, to maximize the value of clinical medical data mining.

4. Individual activity and emotion data. This kind of data is not necessarily generated from the healthcare sector, but it is also relevant to personal health. For instance, individual retail consumption records reflect the individual's living patterns, habits, which can be used to assess individual's health status and make a personalized health plan. Furthermore, based on the physiological data collected by wearable devices, the health status of a user can be easily monitored and traced. The individual emotion data are can be extracted from the posts of their social media networks, as well as changes in behavioral patterns such as lower activity levels, degrading sleep, decreased phone conversations. These data can be used in mental health assessment and affective computing to reveal the patient's feelings and emotion states. Especially, for the recovering patients, a doctor may be able to adjust the treatment plan according to a patient's emotion indications. An emotion-aware healthcare service promotes the innovation of modern medical with humanistic treatment.

In terms of data management, distributed file storage (DFS) module and distributed parallel computing (DPC) module are supposed to support efficient management and analysis of heterogeneous data. DFS distributed file storage module uniformly manages multisource heterogeneous healthcare data. DPC distributed parallel computing module analyzes and processes data from DFS and ultimately discovers knowledge. DPC not only provides offline computation for massive unstructured data, but also supports real-time data analysis and query, and integrates various data mining and machine learning algorithms.

According to the technical complexity and commercial value, the data-oriented healthcare applications and services can be divided into the following four groups.

1. Statistics-based applications only provide basic statistics and report services. For example, an individual health status report is the representative application. In addition, drug misuse and outdate reports are available through the statistics of clinical trial data.
2. Monitoring-based applications are typically utilized to monitor individual vital signs. Through real-time analysis, a user's physiological changes can be immediately detected to avoid sudden diseases. Through offline analysis of historical data, the recovery procedure can be traced, which supports treatment optimization.
3. Knowledge-based applications are the most representative big data application. Supported by data mining and machine learning techniques, it is available to discover data correlation and dependence. Typical applications include chronic disease diagnosis, genetic disease analysis, treatment evaluation, side effect identification, and public health warning.
4. Prediction-based applications have the highest technical complexity and greatest commercial value. For example, individual eating habits can be deduced through retail records, and some potential health risks can be predicted, particularly diet-related diseases, such as obesity and high blood pressure. In addition, considering individual physiological features, individual treatment simulation is available to assess risk and make the optimal medical plan.

1.3.3.5   Issues and Challenges of NB-IoT Healthcare

Reference [217] introduced the background about NB-IoT based intelligent medical electronics and their limitations. The existing architecture cannot connect all types of devices in hospitals due to the limitation of wireless protocols. The promising NB-IoT technology well maybe bring formalized architecture to connect all intelligent things in smart hospitals and smart healthcare systems. Reference [221] argues that NB-IoT is extremely suitable for healthcare sector, because of its low power consumption and unharmful low power radiation. However, challenges still exist in the following.

Accuracy and Reliability of Data

Accurate and reliable data collection is a major challenge. Many of the current sensor techniques are beneath medical standards and inaccurate data are valueless in medical applications. The intrinsic sensor error and the external interference are two main contributors to the issue. External interference is difficult to be eradicated from sensing devices. For example, the change of the patient's posture may have significant effect on the measurement of blood pressure. In addition, the wearable devices have great limitations on the wearing position while bringing comfort and accuracy to patients. The medical devices' technical limitations, environmental factors, and network attacks altogether will possibly lead to data loss and faults of IoT devices. Faults and missing data will render incorrect diagnosis or false alarms, which could be fatal to the patients.

Security and Privacy

Large amount of valuable data generated by smart healthcare sector inevitably attracts attackers. Terminal devices and wireless communications are especially vulnerable to cyberattacks. On the one hand, the terminal devices are usually unable to run complex algorithms limited by the size and poor processing ability. On the other hand, the wireless communication protocols need to pass a large amount of data and open wireless channels for data transmission, which might lead to data interception and replay. In addition, medical data contain sensitive and private information about the patients, so security and privacy protection are a real challenge for NB-IoT smart healthcare. We need to improve the encryption mechanism of terminal devices, develop effective encryption algorithms to strengthen authentication, ensure the legal identity, and prevent illegal nodes to send, forge, and tamper information. In addition, there should be a complete data backup mechanism to restore data in time when unexpected situations occur.

Energy Consumption of Terminals

Terminal devices need to collect information with a relatively high frequency and transmit data to the cloud platform with wireless communication. However, it is a major challenge for IoT devices to work continuously for a long time due to its limited size, low battery capacity, and the inability to constantly replace the battery or charge. Therefore it is worth exploring new energy saving or energy self-generating technologies to make NB-IoT based smart healthcare devices more versatile and universal.

Ethics

The increased pervasiveness of smart healthcare systems e.g., wearable devices and autonomy of decision making raises serious moral–ethical concerns about the use of such technology [222]. Smart healthcare research primarily focuses on identifying lucrative business markets while social and moral considerations of the technology are ignored. We have to propose mechanisms that are able to implement the moral, ethical, legal, or cultural policies in smart healthcare context. According to ethical norms, only selective information should be shared with family members and doctor. For example, economic costs of a health condition like loan, etc. are to be disclosed only to a specific relative/guardian in case of emergency.

Table 1.35 Application and
data range and duration
requirements

Sensor	Data range	Duration
Heart rate	0–150 BPM	5 min
Respiratory	2–50 breaths/min	5 min
Blood pressure	10–400 mm Hg	30 min
Blood pH	6.8–7.8 pH units	30 min
Body temperature	24–44 °C	5 min
GPS position	0–180°	2 h
Motion sensor	–	2 h

1.3.3.6 Other Factors

Other main issues of NB-IoT healthcare applications include: (1) lack of robust real-time provisioning and other real-time applications due to the latency, (2) bandwidth insufficiency, (3) requiring additional doctors and nurses training for IoT practice [221]. Table 1.35 proposed in Refs. [217, 223] listed some requirements for NB-IoT in healthcare could only meet the data transmission rate requirement of certain healthcare applications, rather than all the situations.

Intravenous Infusion Monitoring System

Reference [217] proposed an architecture to connect intelligent things in smart hospitals based on NB-IoT and introduced edge computing to meet the low-latency requirement in certain medical process. In case study, they developed infusion monitoring system based on NB-IoT. It pointed out that the many intelligent devices in the hospital are numerous, and sensitive to power consumption, which fits NB-IoT characteristics.

It discussed the proposed architecture of smart devices in smart hospitals, which is comprised of sensing layer, base station layer, edge computing layer, and cloud computing layer. Sensing layer is where lots of terminal devices integrated with NB-IoT module collect and process data. Base station layer is where NB-IoT base stations are deployed, which need mechanisms of routing, congestion control, traffic scheduling, and security measures. Edge computing layer is especially worth mentioning. Traditionally, computing and storage of big data are carried out on the cloud platforms or cloud servers. However, the massive connections scheme in NB-IoT might make the centralized data processing mechanism a disaster. In addition, some healthcare applications have high requirements on latency control. Therefore, edge computing was introduced in the architecture to reduce latency and accomplish the real-time data processing. The edge server is relatively close to the terminals, the round-trip time of data is relatively short which greatly reduces the latency. Also, the edge server can increase the reliability of the NB-IoT based smart healthcare system and reduce energy consumption of the terminal devices.

Finally, in case study, the intravenous infusion monitoring system based on NB-IoT and infrared sensor is proposed. The injection monitoring system can count the drops and measure the remaining drug volume. It uses NB-IoT to transmit data from the monitoring terminal to the monitoring platform.

Fall Detection

Reference [224] is concerned about incorporating NB-IoT into fall detection system. An age-related risk is falling. Falling happens due to various reasons, including leg muscle degradation, loss of balance, and inability to walk. Falls lead to other adverse conditions, including fractures, closed head injuries, and even death. A fair amount of research already emerged, dedicated to falling detection and minimizing the harm that falling has on the elderly. Some researches tried to explain how to utilize sensors and microcontrollers to identify the occurrence of a fall, some tried to identify a fall by detecting and analyzing the vibration of the floor, or the sudden change of the electromagnetic wave in the room.

Some recent monitoring system with microcontroller unit receives inputs from the sensor and compare their values to the preset threshold and determine whether a fall occurs, enabling the system to have the ability to immediately recognize a fall. And then, the data are transmitted via advanced information service narrowband IoT network, and stored on the platform, which allows users to add or detract devices to the system. Once a fall occurs, the patient's location is automatically available without the need for voice contact by the patient. This NB-IoT based fall detection system can be used in hospitals or at the elderly's homes. This device will expedite the dispatch of a responder or a medical unit without the need for voice communication by the patient and this will result in earlier treatment and better patient outcomes.

Blood Glucose Measurement

A blood glucose measurement method was proposed with NB-IoT network. The traditional blood glucose monitoring scheme requires patients frequently collect and record blood glucose information manually. There are lots of disadvantages in such scheme. The blood glucose data collection mainly relies on manually recording or simply memory of patients, therefore, the recorded data are difficult to keep for a long time. Furthermore, doctors and healthcare providers could not obtain the patient's data promptly. In the traditional scheme, only when patients are in hospital monitoring blood glucose, doctors can get patients' data in time, but this adds time and cost burden to patients, and it is also easy to generate psychological pressure, which is not conducive to the patients' health condition. In the long run, data mining technology could not be used to analyze the collected blood glucose data and provide treatment suggestions based on data analysis since there is no decent dataset. To alleviate the blood glucose data collection and management problem, a blood glucose measurement system could incorporate with NB-IoT network. The blood glucose measurement module uses electrochemical detection method to measure blood glucose. When the patient uses the blood glucose meter to measure and collect blood glucose information, the microcontroller unit will calculate the blood glucose concentration value and then transmit the encapsulated information to the base station through the NB-IoT communication module, and then forward it to the Internet of things cloud platform. Doctors can use mobile phones or computers to have access to the relevant databases on the Internet of things cloud platform and view and manage the patient information.

Investment in Data Science

Nowadays, cancer could be considered as multiple subtypes of disease, so the need to generate evidence in each individual subtype means there are more questions need to be answered. This will demand high-quality data, shear of mistakes, and inconsistencies, which will be trusted by regulators such as the European Medicines Agency. Securing access to data from electronic health records is not that complicated. Making that data fit for use is the hard part. It is just over a decade since the American Recovery and Reinvestment Act spurred the widespread take-up of electronic patient records. Combining with AI, data offer hope of expanding the number of patients who can benefit from existing medicines or even unearthing entirely new drugs. Medical applications for data and AI have rapidly become attractive to investors. However, taking a drug from bench to bedside can cost $2.6 billion. According to the Tufts Center for the Study of Drug Development, this may take up to 14 years.

Digital data will be the fuel for the twenty-first century, but what will be the carbon footprint of the AI and machine learning revolution be? As the FT has reported, training up Google's Transformer model (which supports Google translate) could use as much as 626,155 lb of CO_2, which equates to about 315 New York–San Francisco return flights. If artificial intelligence is the twenty-first century equivalent as a disruptive technology, it is critical that we do not ignore the risks [225]. Digital transformation will build capacities, opportunities, and resilience for some organizations but increase vulnerability, exposure, and risks for others. We must seriously consider the economic, environmental, social, and normative risks of new technology.

1.3.4   Smart Mobility

Urbanization and growing population in most cities are causing more and more problems for the movement of the citizens in their city. Commuting or driving to work and home has become a hassle; congestion in the EU is often located in and around urban areas and costs nearly EUR 100 billion, or 1% of the EU's GDP, annually as estimated by the European Commission.

The mobility challenges are plenty and not limited just to traffic congestion. They are also about efficiently connecting (time, cost, effort) different neighborhoods with public means of transport, helping citizens and professionals at the last mile journey, giving access to the critical stations (train, airport, buses) with multiple means and from multiple regions, offering a variety of options to the people to move around (including bicycle), offering of parking slots, and many more. It is also about understanding how citizens move every day in order for city officials to plan accordingly the location of stations, bike routes, and traffic lights, as well as to optimize the schedule of each city activity without disturbing others.

By adopting IoT and AI-enabled solutions, cities can be improved and solve or at least reduce some of the main urban transportation issues. Here are some examples:

- Optimize availability of public parking slots through real-time parking sensors that can show to the drivers where the nearest parking is without going around blindly. Finding parking in less time can reduce both traffic jam and air pollution.
- Understand how and when people are moving in the city, from where to where and what is their profile. A city authority which can have this knowledge is able to take much better planning decisions based on data and facts. Some ways to achieve this are by analyzing the anonymous and aggregated mobile data from consumer phones. If these data are combined with other data generated by connected city furniture, then the insights are priceless. Smart city furniture could be connected lights, smart benches, and connected traffic lights, while other city assets could be connected bikes and buses, connected buses, and rubbish bins. The analysis of all these combined data can generate insights and automations that we could never think otherwise.
- Plan maintenance and improvements in the road and public transport network efficiently based on the collected data by the IoT enabled assets. For example, big halls on a street can be identified by the data generated from smart bikes/lights due to the shaking sensors. No need to send employees to check or ask citizens to report it (usually after accidents). At the same time, the schedule of when a local authority is appropriate to send the workers to cover the halls can be planned based on the available data from the sensors around that street, so traffic interruptions can be avoided.

Of course, there are even more IoT applications that can improve the mobility in the city. By improving mobility can improve also air quality. Based on European Commission statistics, urban mobility accounts for 40% of all CO_2 emissions of road transport and up to 70% of other pollutants from transport. The smart traffic systems throughout Amsterdam will be used as an illustration.

Amsterdam is the largest city in the Netherlands, with a population of 2.4 million. The city is also one of Europe's leading tourist destinations, attracting around 6 million people a year. Amsterdam's oldest quarter, the medieval center, is very small and has an incredibly complex infrastructure, with roads, tunnels, trams, metro, canals, and thousands of bicycles.

This creates one of the world's most challenging traffic management environments, which the office for Traffic and Public Space (Verkeer en Openbare Ruimte) meets through vision, action, and modern technology. This is typified by the new intelligent data communications network being installed to support the city's traffic control system, for which they have selected advanced Ethernet switching and routing technology from Westermo.

In 2015, the municipality of Amsterdam created its own team that was responsible for the development and operation of the data communication network that supports the intelligent traffic systems (ITS) in the city. Previously, this was managed by an external partner, but due to rising costs, and increasing performance and cybersecurity requirements, it was decided the best way forward was to take back full responsibility for the network. Some case studies are summarized below:

1.3.4.1   Traffic Light Control

There are several hundred traffic light systems throughout Amsterdam. These work autonomously, but can also be controlled centrally, which is one of the most critical tasks for the city's department for traffic and public space. In the event of traffic congestion, traffic control center operators can manage the flow of traffic and if necessary, reroute traffic to less crowded roads.

The traffic light control systems interconnect several traffic lights. The infrastructure connecting the traffic lights is a mix of existing copper cables and new fiber cables. However, in order to connect a string of traffic lights back to the control room, the city has been relying on leased lines. This solution is not only expensive, costing around EUR 2 million per year, but also does not provide the reliability required for a system of this magnitude. The savings made as a result of replacing the leased lines with the Westermo cellular routers is estimated to cover the cost of the network upgrade project within just 3 years.

1.3.4.2   Environmental Zone Enforcement
and Zero-Emission Transportation

An environmental zone has been established in the central part of Amsterdam with the aim of decreasing pollution from motor vehicles. Vehicles that are not environmentally friendly are prohibited to enter the "green zone" and automatic number plate recognition cameras have been installed to ensure that the restriction is followed by motorists. Approximately 80 control points have been established at the entrances to the city to monitor about three million cars every day. Between one and five cameras automatically read the vehicle registration numbers as they pass the control points. The photos are processed inside the camera, converted into simple text information, and sent to the control center through a secure encrypted VPN tunnel using a cellular router.

The City of Amsterdam plan to participate in the European C-ITS smart traffic project, which will allow real-time traffic optimization. This will mean that there will be a requirement for more bandwidth and lower latency so in time, the mobile connections will be replaced with a fiber optic network, using for example the Lynx and RedFox switches.

Turing to zero emission, there are a few major trends that shape the future of smart cities: renewable energy, e-mobility solutions, energy and waste management (using data analytics), etc. The question is how fast all of these will be scaled up.

Renewable energy sources are projected to account for more than one-quarter of global electricity production by 2020. The growth in electric vehicles (EVs) and hybrid electric vehicles (HEVs) is climbing, and by 2025, EVs and HEVs will account for an estimated 30% of all vehicle sales.

Today, about 17% of the world's buses are electric—425,000 in total. Ninety-nine percent of them are in China, which adds a London-sized electric bus fleet to its roads every 5 weeks. In the United States, a few cities have bought some electric buses, or at least run limited pilots, to test the concept out. California has even mandated that by 2029 all buses purchased by its mass transit agencies be zero emission.

1.3.4.3 Traffic Observation and Situation Assessment

The Amsterdam traffic is continuously monitored from the control center to help operators maintain the flow of traffic, reduce congestion, and minimize the risk of accidents. Operators make decisions based on the information provided by hundreds of cameras installed across the city. Many of the regular surveillance cameras are connected to the network via Westermo switches.

The real-time video feed from the cameras can also be viewed for traffic controlling purposes. These are connected to the control room using Westermo cellular routers, which provide secure encrypted VPN tunnels. When traffic congestion occurs, the traffic control managers are permitted to disable the environmental monitoring system and activate predefined scenarios that reroute the traffic to dissolve the congestion.

1.3.5 Smart Infrastructures

Sensing technologies are embedded in infrastructure and in which its devices interact with each other is defined as smart infrastructure. Sensors are connected to a communication backbone which allows real-time data acquisition and analysis. The information gathered is analyzed, interpreted, and delivered as reliable, robust, and meaningful information to infrastructure providers who can then make better decision making about the health and management of their assets.

In this sensing environment, smart infrastructure is able to respond in real time to the needs of the users. Self-aware infrastructure assets direct their own maintenance, leading to condition-based maintenance, reduced down time, and greater operational efficiency of the infrastructure overall.

Better information leads to an enhanced understanding of the behavior of infrastructure. The impact of this will lead to transformations in the approaches to design and construction as well as step changes in improved health and productivity, greater efficiency in design and performance, a low-carbon society, and sustainable urban planning and management.

The rollout of smart cities is well underway, however, to make a city truly smart all the individual elements need to work together, not just independently. Connecting these elements is the infrastructure which, like the IoT devices themselves, is full of sensors but with their own challenges to ensure they work successfully. These may be motion sensors, pollution sensors, parking sensor, or moisture sensors to name just a few, and all require safe, reliable, and energy-efficient power.

Imagine traveling into a smart city in your autonomous car. With connected devices controlling the car and city, you can sit back and relax. But behind the scenes, there is a lot going on to make this possible. Lots of sensors are working flat out in both the city and car to ensure everything runs smoothly. The car sensors ensure it can read reference points like roadworks, parking, and smart traffic lights. These, in turn, are full of sensors that can read the arrival of autonomous and non-autonomous cars. They gather data on vehicle movement and quantity to ensure smooth traffic flow through the city—safe and efficient movement being a key benefit of smart cities.

On public transport, sensors on the rail tracks monitor where the trains are at all times ensuring smoother running of the trains and more up-to-date information for the passengers. In addition, sensors enable remote condition monitoring of tracks and points collecting data that will flag problems and maintenance issues before they become costly to repair.

Lampposts in a smart city not only light the streets but also provide the opportunity for city management to monitor them. The EU program Sharing Cities is trialing smart city technology in various European cities. They state Europe's existing lighting network costs €3 billion a year to operate. Installing smart street lighting could reduce electricity costs to €900 million.

Sensors on lampposts can detect movement so they only light when required thus providing cost and energy savings. Sensors will also provide maintenance and fault detection data in advance so engineers will only need to visit a particular lamppost when it needs maintenance rather than on a scheduled health check routine.

Smart building technology is also on the rise with a predicted increase of approx. 30% a year. Buildings are becoming more complex with interconnected IoT systems offering energy and cost-efficient buildings. IoT infrastructures are being developed to offer many benefits, including optimizing room occupancy, turning lights on/off as needed, monitoring assets movement, and thus increasing security by knowing where people are located.

However, these automated systems will have more complex infrastructure requiring more communication technologies and wiring. Sensors will monitor the health of the building, providing predictive maintenance data. Especially in the case of retrofitting existing buildings, these can be in hard to reach locations creating their own set of challenges.

A smart city will also have elements outside the city. To ensure its population has safe, clean drinking water, wastewater, and water treatment plants will benefit from increased sensing and communication technology adoption. Sensors at these facilities will remotely monitor a range of equipment (such as water composition control testers), sending data back to a central control point. It is not feasible to have engineers at each location full time on the off chance there is a problem and the time and expense of them proactively driving round to the different locations is both costly and inefficient.

While it is clear that smart cities offer many benefits, a major challenge is making it happen to reap the full benefits. Just thinking about each individual IoT device does not work. We need to think about the infrastructure connecting the whole smart city and how it is installed and maintained. Key to this is powering all the sensors that will be required in the infrastructure. If the sensors required keep powering down and need constant maintenance the city will never be truly smart. City officials do not want to incur costs of frequently sending out engineers to change batteries powering the ever-increasing number of sensors. They need a form of power that is as smart and intelligent as the devices being powered.

Cabling power to the sensors is costly and often impractical. Installing batteries can also be difficult as sensors can be located in hostile conditions where they need to function despite high temperatures, dust, oil, and vibration. Smart lampposts need to be powered in an energy and cost-efficient way that can handle the intense heat from the bulbs. Batteries powering sensors on rail tracks need to function despite being located in dirty, hot environments covered in oil and dust. Autonomous vehicles will have so many sensors cabling isn't feasible as the weight of the cabling required will be too great. This is especially key as there are global environmental targets to lower the weight of cars in order to make them more efficient and environmentally friendly. Batteries need to be lightweight, able to work in high temperatures.

In addition, they need to power the sensors for a long time as frequently changing the batteries can be very expensive negating the cost benefits of a smart city. One of the main goals of intelligent cities and buildings is to be more cost, energy, and time efficient. A smart city is not smart if the elements that make it smart keep powering down.

Traditional batteries do not have the lifespan required and vibrations can cause dangerous leakages. Some batteries can handle extreme temperatures but these are often large and heavy. Fortunately, battery technology innovation has moved forward. Solid-state batteries are designed for powering wireless sensors in connected IoT devices. They offer:

- Long lifespan up to 10 years with minimal maintenance
- Efficient in hostile environments with extreme temperatures or humidity
- Nonflammable
- Scalable size from miniature to large scale
- Increased energy density

Smart cities need smart infrastructure to become truly smart and as invisible as possible. Infrastructure with "Fit and Forget" powering that is safe, reliable, and long lasting is vital to making this a reality. Then not only will all the connected devices work effectively and efficiently but will also produce accurate data that will be used to make the city even smarter. Solid-state batteries offer the power required to make this happen.

1.4 Smart Cities Examples Worldwide

The digital transformation of cities and regions is its early stages. Utility companies will play a critical and strategic role with their regional scale, infrastructure, capabilities, and presence. "Business as usual" is not an option. Utility companies need to develop a new vision and strategies. They will lead and thrive in the new era with new skills, business models, operational processes, and partnerships. The smart region journey starts now.

Cities play a significant role in global consumption, production, and pollution. To become sustainable, cities need to plan, innovate, and invest in their future. There are many aspects to discuss when it comes to sustainable urban transformation: governance, planning, innovation, consumption, etc.

Clean technology is one of the critical assets of a green economy. Renewable energy (especially solar and wind energy), e-mobility solutions, and IIoT (used in energy and waste management) have a vast potential to help fight climate change and create more resilient and sustainable cities.

Because we have crossed the price points where all these technologies make sense, a transformation in the next few years is going to happen very quickly. In my opinion, it is a matter of raising awareness, understanding, and experience. As these things develop and increase, clean technology grows exponentially, much faster compared to the industrial revolution.

C40 Cities published a report titled Cities leading the way: Seven climate action plans to deliver on the Paris Agreement. This showcases seven cities with climate action plans that put the city on a path to becoming emissions neutral by 2050 and more resilient to the impacts of climate change.

Cities included in the report are Barcelona, Copenhagen, London, New York City, Oslo, Paris, and Stockholm.

1.4.1 Barcelona Has Set a Zero-Energy Poverty Target by 2030

One of the biggest challenges Barcelona will face is a major increase in the vulnerable population impacted by climate change and energy poverty. The city has already started implementing a series of actions targeting the most vulnerable citizens. From 2019, facilities in the Barcelona Metropolitan Area and up to 20,000 of the city's residents will have access to sustainable energy supplied by Barcelona Energia, the public electricity distributor for the Barcelona Metropolitan Area.

1.4.2 Copenhagen Aims to Become the First Carbon-Neutral Capital by 2025

Despite a population growth of 16%, Copenhagen managed to reduce the annual CO_2 emissions by 38% compared to 2005 levels. As mentioned in the C40 Cities Report, "Most savings were achieved through increasing the share of green energy from biomass used in the city's combined heat and power plants and wind energy. Furthermore, the conversion of a power plant unit from coal to sustainable biomass is underway and is expected to be completed by the end of 2020." According to the same source:

- 20,000 street lamps have been replaced with LED lights, resulting in an energy saving of 57% compared to 2010
- Copenhagen introduced its first electric buses and
- The city plans to build more bicycle lanes

1.4.3 London Sets the Target for a Zero-Emission Transport Network by 2050

London bought 100 new electric double-deckers from China, which were delivered in July 2019. In 2020, London will see the introduction of the world's first hydrogen double-decker buses as part of London's push for zero-emission transportation. According to a report published in January, by the end of 2019, London will have 240 purely electric buses, less than 2.6% of its overall bus fleet. The city also plans for at least 300 rapid charge points to be installed by 2020, and it is also exploring the next generation of road user charging systems.

1.4.4 Oslo Aims to Cut City Emissions by 95% by 2030

According to C40 Cities Report, Oslo aims to reduce total city emissions by 36% by 2020, by 50% as soon as possible after 2020, and by 95% by 2030.

Since transport in the city accounts for more than 60% of total emissions, Oslo plans to make walking, cycling, and public transport more attractive, phasing out fossil fuels for public transportation and introducing road user payment systems.

Over the next 4 years, Oslo aims to reduce CO_2 from the energy and buildings sector, which accounts for 20% of total emissions. This will be achieved by phasing out fossil oil for heating through national and local support schemes. Almost 99% of energy sources at this moment consist of heat from the sewer system, recovered heat from waste, bioenergy, and electricity from hydropower.

Concerning waste management, 200,000 additional tonnes of CO_2 are expected to go unused by reusing, recycling, and sharing more and by applying carbon capture and storage technologies to Oslo's waste-to-energy plants. A pilot project has already demonstrated that 90% of CO_2 emissions can be captured.

1.4.5 Stockholm Plans to Achieve Net-Zero Emissions by 2040, Paris by 2050

By 2022, Stockholm intends to replace all fossil fuels with renewables. There is also a potential for 10% of the power used in the city to be generated from solar power produced in buildings and from bioenergy in combined heat and power plants. About 43% of emissions reductions must come from transport. By 2021, 70% of all food waste will be collected for conversion into biogas and automatically sorted in a plant using near-infrared technology.

To attain zero emissions at the local level, Paris' energy consumption will need to be halved, and 100% of the energy consumed will need to come from renewables by 2050, states the C40 Cities Report.

1.4.6 Others

New York is investing in electric vehicle infrastructure—a minimum of $10 million will be spent toward the installation of 50 fast-charging hubs across all five boroughs by 2020 (C40 Cities Report).

Santiago, the capital city of Chile, recently procured 200 electric buses, the largest electric bus fleet in Latin America, according to the Santiago Times.

In Nepal, the government has announced a decision to procure 300 pure electric buses for its capital of Kathmandu.

Singapore is considered as the world's smartest city with its innovative healthcare ecosystem, free public housing, and impressive transportation infrastructure. In 2019, several Singapore-based startups announced their plans to enhance the smart city systems with blockchain-powered solutions. In the future, Singapore citizens are expected to get digital wallets, secure payments, general insurance, and healthcare records handled by the distributed ledger systems.

Taipei City, the capital of Taiwan, has established the Taipei Smart City Project Management Office in 2016 to enhance smart transportation services, healthcare, public housing, education, and payment systems. Since 2018, Taipei City

Government signed a memorandum of cooperation with IOTA to create a blockchain-based sightseeing application to attract more tourists. Also, IOTA participates in the Taipei City Hall Waste Management project. The Tangle will be used for storing data collected by IoT sensors that will measure the fill levels of their waste bins in real time.

Austin, a city in Texas, is becoming a smart city pioneer in the state after the 2019 Texas Smart Cities Summit, sponsored by IOTA. The transportation department in the city of Austin has recently partnered with the IOTA Foundation. The goal of their collaboration is to use IoT and blockchain to improve the interoperability between various transportation systems.

The Smart City Long Beach projects aim to leverage advancements in technology, data management, and user-centered design in order to improve residents' quality-of-life and promote digital equity. Smart Long Beach will better prepare the city to utilize emerging technologies, which will be deployed responsibly to meet community-sourced needs. With a purpose to foster civic engagement and allow for improvements in service delivery to residents, the smart city project has developed the rightful guiding principles to support this effort. In 2019, the city of Long Beach was named a Top 10 Digital City in a survey conducted by the Center for Digital Government. This recognition refers to the efforts to build modern technology infrastructure and efficient foundational systems; protect public safety using technology; and improve public engagement through open data and enhanced payment systems.

Sweden's fifth largest city, Västerås, has embarked on an ambitious digitalization strategy to develop smart city solutions that will make it more attractive as a community for citizens and industry. The project is being led by Mälarenergi, who provided a broad range of essential services for the city's 150,000 residents and its businesses. The utility operates hydropower plants, the local power grid, a waste-to-energy plant, heating and cooling networks, water and wastewater treatment plants, a water distribution network, and a fiber-optic network.

In Trier Germany, they have developed a smart energy management system for the city's diverse range of generation sources—wind power, hydropower, solar photovoltaic, biomass, combined heat and power (both large-scale conventional and micro CHP), as well as for battery storage, heat pumps, electric vehicle chargers, and industrial loads.

Trier also connects the city to three other municipalities in France, Belgium, and Luxembourg, each of which operates its own power pool of diverse types of generation and storage. The solution will enable the compensation of fluctuations due to renewable energy by exchanging power with each other and using storage capacity intelligently. This, in turn, will maximize their use of renewables and minimize their dependency on the national grids.

Now, Trier is connecting the wastewater plant, the water network, on-site PV generation, and the CHP plant to reduce operating costs. This is another instance of how the coordination of utilities and services—in line with the city's vision, strategy, and targets—creates value for the municipality and its citizens.

1.4.7   Challenges

Urbanization is evolving with a variety of challenges that cities must address. Long Beach City faced the difficulty of improving energy distribution, streamline trash collection, traffic congestion, and air quality. Also, other, more sensitive facets of the community had to be addressed, including homelessness, mobility, climate change, government transparency, and operational efficiency. City faces challenges with infrastructure and its maintenance, city services such as street cleaning, the lack of mobility, etc.

Smart meters and smart appliances (which can respond to or initiate dynamic demand) need to be standardized. There is a concern on the use of suboptimal technologies and current technologies are changing too fast.

There is a need to determine what data should be communicated back to the utility or to the consumer and what immediate control actions would be necessary.

There is a lack of evidence to suggest that people will make firm decisions to save money and carbon emissions.

Training and informing customers about change of customers' behavior could incur significant costs. There is also a shortage of talented person to implement smart cities.

1.4.8   Some Practical Applications

In Los Angeles, where traffic has been a tremendous problem for decades, data from an array of magnetic road sensors and hundreds of cameras feed through a centralized computer system to control 4500 traffic signals citywide to help keep traffic moving. Completed in 2013, the $400-million system is credited with increasing travel speeds around Los Angeles by 16% and shortening delays at major intersections by 12%.

In San Francisco, SFpark uses wireless sensors to detect parking-space occupancy in metered spaces. Installed in 8200 on-street spaces in the pilot areas, the wireless sensors detect parking availability in real time. In 2013, 2 years after launching SFpark, San Francisco published a detailed report showing that the program reduced weekday greenhouse gas emissions by 25%. Traffic volume went down and drivers cut their search time nearly in half. By making it easier for people to pay for their parking and reducing loss due to broken parking meters, San Francisco also increased parking-related revenue by about $1.9 million.

London has begun tests on a smart parking project that allows drivers to quickly locate parking spaces and remove the need for lengthy searches for an open spot. This significantly alleviates urban traffic congestion, saves fuel and reduces harmful emissions [226].

In 2011, Autolib debuted an electric car-sharing program in Paris that has grown to over 3000 vehicles. The connected cars can be tracked via GPS, and drivers can use the car's dashboard to reserve parking spaces in advance, saving time and reducing the waste associated with long searches for parking spots.

Copenhagen uses sensors to monitor the city's bike traffic in real time, which provides valuable data on improving bike routes in the city. This is crucial, as more than 40% of the city's residents commute by bike each day.

To save water, the drought-plagued town of Fountain View, California implemented the FlexNetcommunication system, smart residential and commercial meters to cut water usage by 23% [227].

1.5 Conclusions

It can be seen that cities can benefit tremendously from technological advances that utilize the Internet of Things. It is also easy to see that as cities continue to grow and more devices get added to the infrastructure, the amount of data will be voluminous. In order to manage these demands, and to fully utilize the new technology, cities will need information management systems. The data alone will benefit nobody without a seamless system to analyze and aggregate the vast amount of information. An efficient messaging system will help cities take advantage of the new technology and improve city life for residents and businesses while reducing costs for everyone involved.

Acknowledgments   The permission given in using materials from the following papers is very much appreciated.

A. Y.F. Wang, L.P. Huang, M. Shahidehpour, L.L. Lai, H.L. Yuan, F.Y. Xu, Resilience constrained hourly unit commitment in electricity grids. IEEE Trans. Power Syst. **33**(5), 5604–5614 (2018)
B. Y.F. Wang, Z.H. Huang, M. Shahidehpour, L.L. Lai, Z.Q. Wang, Q.S. Zhu, Reconfgurable distribution network for managing transactive energy in a multi-microgrid system. IEEE Trans. Smart Grid. (2019). https://doi.org/10.1109/TSG.2019.2935565
C. Y. Wang, L. Huang, M. Shahidehpour, L.L. Lai, Y. Zhou, Impact of cascading and common cause outages on resilience-constrained economic operation of power systems in extreme conditions. IEEE Trans. Smart Grid. (2019). https://doi.org/10.1109/TSG.2019.2926241
D. H. Ruiwen, D. Jianhua, L.L. Lai, Reliability evaluation of communication constrained protection systems using stochastic-flow network models. IEEE Trans. Smart Grid **9**(3), 2371–2381 (2018)
E. F.Y. Xu, X. Cun, M. Yan, H. Yuan, Y. Wang, L.L. Lai, Power market load forecasting on neural network with beneficial correlated regularization. IEEE Trans. Ind. Inform. **14**(11), 5050–5059 (2018)
F. W.W.Y. Ng, J. Zhang, C.S. Lai, W. Pedrycz, L.L. Lai, X. Wang, Cost-sensitive weighting and imbalance-reversed bagging for streaming imbalanced and concept drifting in electricity pricing classification. IEEE Trans. Ind. Inform. **15**(3), 1588–1597 (2019)
G. C.S. Lai, L.L. Lai, Application of big data in smart grid, in *2015 IEEE International Conference on Systems, Man, and Cybernetics, Hong Kong, China*, Oct 2015, pp. 665–670

References

1. C.S. Lai, Y. Jia, Z. Dong, D. Wang, Y. Tao, Q.H. Lai, R.T.K. Wong, A.F. Zobaa, R. Wu, L.L. Lai, "A review of technical standards for smart cities", Clean Technologies 2(3), 290–310 (2020)
2. Z. Li, M. Shahidehpour, F. Aminifar, A. Alabdulwahab, Y. Al-Turki, Networked microgrids for enhancing the power system resilience. Proc. IEEE 105(7), 1289–1310 (2017)
3. Y. Wang, C. Chen, J. Wang, R. Baldick, Research on resilience of power systems under natural disasters—A review. IEEE Trans. Power Syst. 31(2), 1604–1613 (2016)
4. M. Panteli, P. Mancarella, D. Trakas, E. Kyriakides, N. Hatziargyriou, Metrics and quantification of operational and infrastructure resilience in power systems. IEEE Trans. Power Syst. 32(6), 4732–4742 (2017)
5. A. Gholami, T. Shekari, F. Aminifar, M. Shahidehpour, Microgrid scheduling with uncertainty: The quest for resilience. IEEE Trans. Smart Grid 7(6), 2849–2858 (2016)
6. M. Ouyang, L. DuenasOsorio, Multi-dimensional hurricane resilience assessment of electric power systems. Struct. Saf. 48, 15–24 (2014)
7. M. Panteli, D.N. Trakas, P. Mancarella, N.D. Hatziargyriou, Boosting the power grid resilience to extreme weather events using defensive islanding. IEEE Trans. Smart Grid 7(6), 2913–2922 (2016)
8. Y. Wang, C. Liu, M. Shahidehpour, C. Guo, Critical components for maintenance outage scheduling considering weather conditions and common mode outages in reconfigurable distribution systems. IEEE Trans. Smart Grid 7(6), 2807–2816 (2016)
9. C. Chen, J. Wang, F. Qiu, D. Zhao, Resilient distribution system by microgrids formation after natural disasters. IEEE Trans. Smart Grid 7(2), 958–966 (2016)
10. L. Wu, M. Shahidehpour, Y. Fu, Security-constrained generation and transmission outage scheduling with uncertainties. IEEE Trans. Power Syst. 25(3), 1674–1685 (2010)
11. A. Street, F. Oliveira, J.M. Arroyo, Contingency-constrained unit commitment with n-K security criterion: A robust optimization approach. IEEE Trans. Power Syst. 26(3), 1581–1590 (2011)
12. Q. Wang, J.P. Watson, Y. Guan, Two-stage robust optimization for N-k contingency-constrained unit commitment. IEEE Trans. Power Syst. 28(3), 2366–2375 (2013)
13. M. Shahidehpour, W. Tinney, Y. Fu, Impact of security on power system operation. Proc. IEEE 93(11), 2013–2025 (2005)
14. Y. Wang, Z. Li, M. Shahidehpour, L. Wu, C.X. Guo, B. Zhu, Stochastic co-optimization of midterm and short-term maintenance outage scheduling considering covariates in power systems. IEEE Trans. Power Syst. 31(6), 4795–4805 (2016)
15. L. Wu, M. Shahidehpour, T. Tao, Stochastic security-constrained unit commitment. IEEE Trans. Power Syst. 22(2), 800–811 (2007)
16. L. Wu, M. Shahidehpour, T. Li, Cost of reliability analysis based on stochastic unit commitment. IEEE Trans. Power Syst. 23(3), 1364–1374 (2008)
17. P. Xiong, P. Jirutitijaroen, A stochastic optimization formulation of unit commitment with reliability constraints. IEEE Trans. Smart Grid 4(4), 2200–2208 (2013)
18. L. Wu, M. Shahidehpour, T. Li, GENCO's risk-based maintenance outage scheduling. IEEE Trans. Power Syst. 23(1), 127–136 (2008)
19. R. Billinton, B. Karki, R. Karki, G. Ramakrishna, Unit commitment risk analysis of wind integrated power systems. IEEE Trans. Power Syst. 24(2), 930–939 (2009)
20. A. Arab, A. Khodaei, S.K. Khator, K. Ding, V.A. Emesih, Z. Han, Stochastic pre-hurricane restoration planning for electric power systems infrastructure. IEEE Trans. Smart Grid 6(2), 1046–1054 (2015)
21. G. Huang, J. Wang, C. Chen, J. Qi, C. Guo, Integration of preventive and emergency responses for power grid resilience enhancement. IEEE Trans. Power Syst. 32(6), 4451–4463 (2017)
22. C. Wang, Y. Hou, F. Qiu, S. Lei, K. Liu, Resilience enhancement with sequentially proactive operation strategies. IEEE Trans. Power Syst. 32(4), 2847–2857 (2017)

23. A. Gholami, F. Aminifar, M. Shahidehpour, Front lines against the darkness: Enhancing the resilience of the electricity grid through microgrid facilities. IEEE Electr. Mag. **4**(1), 18–24 (2016)
24. T. Gholami, F.A. Shekari, M. Shahidehpour, Microgrid scheduling with uncertainty: The quest for resilience. IEEE Trans. Smart Grid **7**(6), 2849–2858 (2016)
25. A. Gholami, T. Shekari, S. Grijalva, Proactive management of microgrids for resiliency enhancement: An adaptive robust approach. IEEE Trans. Sustain. Energy **10**(1), 470–480 (2019)
26. P. Bak, C. Tang, K. Wiesenfeld, Self-organized criticality. Phys. Rev. A **38**(1), 364 (1988)
27. B.A. Carreras, D.E. Newman, I. Dobson, A.B. Poole, Evidence for self-organized criticality in a time series of electric power system blackouts. IEEE Trans. Circuits Syst. I Reg. Papers **51**(9), 1733–1740 (2004)
28. I. Dobson, B.A. Carreras, V.E. Lynch, D.E. Newman, Complex systems analysis of series of blackouts: Cascading failure, critical points, and self-organization. Chaos **17**(2), 026103 (2007)
29. Y. Koç, M. Warnier, R.E. Kooij, F.M. Brazier, An entropy-based metric to quantify the robustness of power grids against cascading failures. Saf. Sci. **59**, 126–134 (2013)
30. B. Wang, H. Tang, C. Guo, Z. Xiu, Entropy optimization of scale-free networks' robustness to random failures. Physica A **362**(2), 591–596 (2006)
31. K. Anand, G. Bianconi, Entropy measures for networks: Toward an information theory of complex topologies. Phys. Rev. E **80**(4), 045102 (2009)
32. Z.J. Bao, Y.J. Cao, G.Z. Wang, Analysis of cascading failure in electric grid based on power flow entropy. Phys. Lett. A **273**(34), 3032–3040 (2009)
33. D.R. Cox, Regression models and life-tables. J. R. Stat. Soc. B **34**(2), 187–220 (1972)
34. M.P. Bhavaraju, R. Billinton, G.L. Landgren, M.F. McCoy, N.D. Reppen, Proposed terms for reporting and analyzing outages of electrical transmission and distribution facilities. IEEE Trans. Power Appl. Syst. **PAS-104**(2), 337–348 (1985)
35. M. Pereira, N. Balu, Composite generation/transmission reliability evaluation. Proc. IEEE **80**(4), 470–491 (1992)
36. M. Shahidehpour, H. Yamin, Z. Li, *Market Operations in Electric Power Systems* (Wiley, New York, 2002)
37. S. Gasmi, C.E. Love, W. Kahle, A general repair, proportional-hazards framework to model complex repairable systems. IEEE Trans. Rel. **52**(1), 26–32 (2003)
38. Z. Liu, Q. Wu, S. Huang, H. Zhao, Transactive energy: A review of state of the art and implementation, in *2017 IEEE Manchester PowerTech*, Manchester, 2017, pp. 1–6
39. The GridWise Architecture Council, GridWise transactive energy framework, The GridWise Architecture Council, Tech. Rep. PNNL-22946, 2015
40. C. Hertzog, Transactive Energy American Perspectives on Grid Transformations, 2013
41. W. Zhang, Y. Xu, Z. Dong, K.P. Wong, Robust security-constrained optimal power flow using multiple microgrids for corrective control under uncertainty. IEEE Trans. Ind. Inf. **13**(4), 1704–1713 (2016)
42. X. Liu, M. Shahidehpour, Z. Li, X. Liu, Y. Cao, Z. Bie, Microgirds for enhancing the power grid resilience in extreme conditions. IEEE Trans. Smart Grid **8**(2), 589–597 (2017)
43. S. Chanda, A.K. Srivastava, Defining and enabling resilience of electric distribution systems with multiple microgrids. IEEE Trans. Smart Grid **7**(6), 2859–2868 (2016)
44. H. Gao, Y. Chen, Y. Xu, C. Liu, Resilience-oriented critical load restoration using microgrids in distribution systems. IEEE Trans. Smart Grid **7**(6), 2837–2848 (2016)
45. H. Farzin, M. Fotuhi-Firuzabad, M. Moeini-Aghtaie, Enhancing power system resilience through hierarchical outage management in multi-microgrids. IEEE Trans. Smart Grid **7**(6), 2869–2879 (2016)
46. J. Chen, F.Q. Wang, D. Zhao, Resilient distribution system by microgrids formation after natural disasters. IEEE Trans. Smart Grid **7**(2), 958–966 (2016)

47. F. Lezama, J. Soares, P. Hernandez-Leal, M. Kaisers, T. Pinto, Z. Vale, Local energy markets: Paving the path towards fully transactive energy systems. IEEE Trans. Power Syst. **34**(5), 4081–4088 (2019)
48. N. Liu, X. Yu, C. Wang, C. Li, L. Ma, J. Lei, Energy-sharing model with price-based demand response for microgrids of peer-to-peer prosumers. IEEE Trans. Power Syst. **32**(5), 3569–3583 (2017)
49. J. Wu, X. Guan, Coordinated multi-microgrids optimal control algorithm for smart distribution management system. IEEE Trans. Smart Grid **4**(4), 2174–2181 (2013)
50. D. Gregoratti, J. Matamoros, Distributed energy trading: The multiple-microgrid case. IEEE Trans. Ind. Electron. **62**(4), 2551–2559 (2015)
51. Y. Liu, Y. Li, H.B. Gooi, Distributed robust energy management of a multi-microgrid system in the real-time energy market, in *IEEE PES General Meeting*, Atlanta, USA, 4–8 Aug 2019
52. D. Wang, X. Guan, J. Wu, P. Li, P. Zan, H. Xu, Integrated energy exchange scheduling for multimicrogrid system with electric vehicles. IEEE Trans. Smart Grid **7**(4), 1762–1774 (2016)
53. T. Morstyn, M. McCulloch, Multi-class energy management for peer-to-peer energy trading driven by prosumer preferences. IEEE Trans. Power Syst. **34**(5), 4005–4014 (2019)
54. H. Wang, J. Huang, Incentivizing energy trading for interconnected microgrids. IEEE Trans. Smart Grid **9**(4), 2647–2657 (2018)
55. A.M. Jadhav, N.R. Patne, J.M. Guerrero, A novel approach to neighborhood fair energy trading in a distribution network of multiple microgrid clusters. IEEE Trans. Ind. Electron. **66**(2), 1520–1531 (2019)
56. Z. Wang, B. Chen, J. Wang, M.M. Begovic, C. Chen, Coordinated energy management of networked microgrids in distribution systems. IEEE Trans. Smart Grid **6**(1), 45–53 (2015)
57. T. Lv, Q. Ai, Interactive energy management of networked microgrids-based active distribution system considering large-scale integration of renewable energy resources. Appl. Energy **163**, 408–422 (2016)
58. Z. Wang, B. Chen, J. Wang, J. Kim, Decentralized energy management system for networked microgrids in grid-connected and islanded modes. IEEE Trans. Smart Grid **7**(2), 1097–1105 (2016)
59. H. Gao, J. Liu, L. Wang, Z. Wei, Decentralized energy management for networked microgrids in future distribution systems. IEEE Trans. Power Syst. **33**(4), 3599–3610 (2018)
60. H.S.V.S.K. Nunna, D. Srinivasan, Multiagent-based transactive energy framework for distribution systems with smart microgrids. IEEE Trans. Ind. Inf. **13**(5), 2241–2250 (2017)
61. P. Kou, D. Liang, L. Gao, Distributed EMPC of multiple microgrids for coordinated stochastic energy management. Appl. Energy **185**, 939–952 (2017)
62. B. Kocuk, S. Dey, X.A. Sun, Strong SOCP relaxations for the optimal power flow problem. Oper. Res. **64**(6), 1177–1196 (2015)
63. R. Jabr, R. Singh, B. Pal, Minimum loss network reconfiguration using mixed-integer convex programming. IEEE Trans. Power Syst. **27**(2), 1106–1115 (2012)
64. Y. Wang, Z. Li, M. Shahidehpour, L. Wu, C. Guo, Critical components for maintenance outage scheduling considering weather conditions and common mode outages in reconfigurable distribution systems. IEEE Trans. Smart Grid **7**(6), 2807–2816 (2016)
65. A. Khodaei, Provisional microgrids. IEEE Trans. Smart Grid **6**(3), 1107–1115 (2015)
66. C. Nan, G. Sansavini, A quantitative method for assessing resilience of interdependent infrastructures. Reliab. Eng. Syst. Saf. **157**, 35–53 (2017)
67. M. Ouyang, L. Dueñas-Osorio, Multi-dimensional hurricane resilience assessment of electric power systems. Struct. Saf. **48**, 15–24 (2014)
68. Y. Jia, Z. Xu, L.L. Lai, K.P. Wong, Risk-based power system security analysis considering cascading outages. IEEE Trans. Ind. Informat. **12**(2), 872–882 (2016)
69. C.M. Rocco et al., Assessing the vulnerability of a power system through a multiple objective contingency screening approach. IEEE Trans. Reliab. **60**(2), 394–403 (2011)

70. T. Ding, C. Li, C. Yan, F. Li, Z. Bie, A bi-level optimization model for risk assessment and contingency ranking in transmission system reliability evaluation. IEEE Trans. Power Syst. **32**(5), 3803–3813 (2017)
71. X. Liu, M. Shahidehpour, Z. Li, X. Liu, Y. Cao, Z. Bie, Microgrids for enhancing the power grid resilience in extreme conditions. IEEE Trans. Smart Grid **8**(2), 589–597 (2017)
72. S. Ma, B. Chen, Z. Wang, Resilience enhancement strategy for distribution systems under extreme weather events. IEEE Trans. Smart Grid **9**(2), 1442–1451 (2018)
73. X. Wang, Z. Li, M. Shahidehpour, C. Jiang, Robust line hardening strategies for improving the resilience of distribution systems with variable renewable resources. IEEE Trans. Sustain. Energy **10**(1), 386–395 (2019)
74. L. Che, X. Liu, Z. Li, Screening hidden N-k line contingencies in smart grids using a multi-stage model. IEEE Trans. Smart Grid **10**(2), 1280–1289 (2019). https://doi.org/10.1109/TSG.2017.2762342
75. D.A. Tejada-Arango, P. Sánchez-Martın, A. Ramos, Security constrained unit commitment using line outage distribution factors. IEEE Trans. Power Syst. **33**(1), 329–337 (2018)
76. C. Shao, M. Shahidehpour, X. Wang, X. Wang, B. Wang, Integrated planning of electricity and natural gas transportation systems for enhancing the power grid resilience. IEEE Trans. Power Syst. **32**(6), 4418–4429 (2017)
77. D. Panteli, N. Trakas, P. Mancarella, N.D. Hatziargyriou, Power systems resilience assessment: Hardening and smart operational enhancement strategies. Proc. IEEE **105**(7), 1202–1213 (2017)
78. PACME Working Group, IEEE PES Reliability, Risk and Probability Applications Subcommittee, Effects of dependent and common mode outages on the reliability of bulk electric system—Part I: Basic concepts, in *2014 IEEE PES General Meeting—Conference & Exposition*, National Harbor, MD, 2014, pp. 1–5
79. R. Billinton, T.K.P. Medicherla, M.S. Sachdev, Application of common-cause outage models in composite system reliability evaluation. IEEE Power Eng. Rev. **PER-1**(7), 62–62 (1981)
80. R. Billinton, R.N. Allan, *Reliability Evaluation of Power Systems*, 2nd edn. (Plenum, New York, 1996)
81. Y.F. Wang, L.P. Huang, M. Shahidehpour, L.L. Lai, H.L. Yuan, F.Y. Xu, Resilience-constrained hourly unit commitment in electricity grids. IEEE Trans. Power Syst. **33**(5), 5604–5614 (2018)
82. W. Li, R. Billinton, Common cause outage models in power system reliability evaluation. IEEE Trans. Power Syst. **18**(2), 966–968 (2003)
83. J. Chen, J.S. Thorp, I. Dobson, Cascading dynamics and mitigation assessment in power system disturbances via a hidden failure model. Int. J. Electr. Power Energy Syst. **27**(4), 318–326 (2005)
84. J.S. Thorp, A.G. Phadke, S.H. Horowitz, Anatomy of power system disturbances: Importance sampling. Int. J. Electr. Power Energy Syst. **20**(2), 147–152 (1998)
85. B.A. Carreras, D.E. Newman, I. Dobson, North American blackout time series statistics and implications for blackout risk. IEEE Trans. Power Syst. **31**(6), 4406–4414 (2016)
86. J. Guo, Y. Fu, Z. Li, M. Shahidehpour, Direct calculation of line outage distribution factors. IEEE Trans. Power Syst. **24**(3), 1633–1634 (2009)
87. R. Billinton, M. Fotuhi-Firuzabad, T.S. Sidhu, Determination of the optimum routine test and self-checking intervals in protective relaying using a reliability model. IEEE Trans. Power Syst. **17**(3), 663–669 (2002)
88. A.H. Etemadi, M. Fotuhi-Firuzabad, Design and routine test optimization of modern protection systems with reliability and economic constraints. IEEE Trans. Power Deliv **27**(1), 271–278 (2012)
89. X. Liu, M. Shahidehpour, Y. Cao, Z. Li, W. Tian, Risk assessment in extreme events considering the reliability of protection systems. IEEE Trans. Smart Grid **6**(2), 1073–1081 (2015)
90. A. Bose, Smart transmission grid applications and their supporting infrastructure. IEEE Trans. Smart Grid **1**(1), 11–19 (2010)

91. H. Hajian-Hoseinabadi, Availability comparison of various power substation automation architectures. IEEE Trans. Power Delivery **28**(2), 566–574 (2013)
92. J. König, L. Nordström, M. Österlind, Reliability analysis of substation automation system functions using PRMs. IEEE Trans. Smart Grid **4**(1), 206–213 (2013)
93. D.M.E. Ingram, P. Schaub, R.R. Taylor, D.A. Campbell, Performance analysis of IEC 61850 sampled value process bus networks. IEEE Trans. Ind. Inform. **9**(3), 1445–1454 (2013)
94. L. Yang, P.A. Crossley, A. Wen, R. Chatfield, J. Wright, Design and performance testing of a multivendor IEC61850-9-2 process bus based protection scheme. IEEE Trans. Smart Grid **5**(3), 1159–1164 (2014)
95. K. Jiang, C. Singh, Reliability modeling of all-digital protection systems including impact of repair. IEEE Trans. Power Delivery **25**(2), 579–587 (2010)
96. H. Lei, C. Singh, A. Sprintson, Reliability modeling and analysis of IEC 61850 based substation protection systems. IEEE Trans. Smart Grid **5**(5), 2194–2202 (2014)
97. W.C. Yeh, A fast algorithm for searching all multi-state minimal cuts. IEEE Trans. Reliab. **57**(4), 581–588 (2008)
98. W.C. Yeh, A fast algorithm for quickest path reliability evaluations in multi-state flow networks. IEEE Trans. Reliab. **64**(4), 1175–1184 (2015)
99. S.G. Chen, Y.K. Lin, Search for all minimal paths in a general large flow network. IEEE Trans. Reliab. **61**(4), 949–956 (2012)
100. Y.K. Lin, D.-H. Huang, L.C.-L. Yeng, Reliability evaluation of a hybrid flow-shop with stochastic capacity within a time constraint. IEEE Trans. Reliab. **65**(2), 867–877 (2016)
101. M. Forghani-Elahabad, N. Mahdavi-Amiri, A new efficient approach to search for all multistate minimal cuts. IEEE Trans. Reliab. **63**(1), 154–166 (2014)
102. S. Zarezadeh, M. Asadi, Network reliability modeling under stochastic process of component failures. IEEE Trans. Reliab. **62**(4), 917–929 (2013)
103. C.C. Jane, Y.W. Laih, Computing multi-state two-terminal reliability through critical arc states that interrupt demand. IEEE Trans. Reliab. **59**(2), 338–345 (2010)
104. L.R. Ford, D.R. Fulkerson, Maximal flow through a network. Can. J. Math. **8**(3), 399–404 (1956)
105. J.E. Ramirez-Marquez, D.W. Coit, Composite importance measures for multi-state systems with multi-state components. IEEE Trans. Reliab. **54**(3), 517–529 (2005)
106. Working Group on Centralized Substation Protection and Control, IEEE Power System Relaying Committee, Advancements in centralized protection and control within a substation. IEEE Trans. Power Delivery **31**(4), 1945–1952 (2016)
107. C. Fan, Y. Ni, R. Dou, J. Shen, C. Gao, G. Huang, Analysis of network scheme for process layer in smart substation. Autom. Electr. Power Syst. **35**(18), 67–71 (2011) (in Chinese)
108. X. Dong, D. Wang, M. Zhao, B. Wang, S. Shi, A. Apostolov, Smart power substation development in China. CSEE J. Power Energy Syst. **2**(4), 1–5 (2016)
109. H. Wang, Z. Cai, Y. Zhang, X. Shao, Y. Li, Z. Zhu, Custom switching technology to improve reliability and real-time performance of information flow in smart substation. Electr. Power Autom. Equip. **34**(5), 156–162 (2014) (in Chinese)
110. J.C. Spall, Estimation via Markov chain Monte Carlo. IEEE Trans. Control Syst. **23**(2), 34–45 (2003)
111. J. Lin, F.H. Magnago, Introduction, in *Electricity Markets: Theories and Applications*, 1st edn., (Wiley-IEEE Press, 2017)
112. H. Algarvio, F. Lope, J. Sousa, J. Lagarto, Multi-agent electricity markets: Retailer portfolio optimization using Markowitz theory. Electr. Power Syst. Res. **148**, 282–294 (2017)
113. M.Y. Hassan, M.P. Abdullah, A.S. Arifin, F. Hussin, M.S. Majid, Electricity market models in restructured electricity supply industry, in *Power and Energy Conference*, 1–3 Dec 2008, pp. 1038–1042. IEEE: Piscataway, New Jersey, US
114. P. Zou, Q. Chen, Q. Xia, C. He, C. Kang, Incentive compatible pool-based electricity market design and implementation: A Bayesian mechanism design approach. Appl. Energy **158**, 508–518 (2015)

115. B. Cory, Power system restructuring and deregulation: Trading, performance and information technology. Power Eng. J. **16**, 22–22 (2002)
116. S. Dhanalakshmi, S. Kannan, K. Mahadevan, Market modes for deregulated environment—A review. Emerg. Trends Electr. Comput. Technol. **2011**, 82–87 (2011)
117. L. Mari, N. Nabona, A. Pages-Bernaus, Medium-term power planning in electricity markets with pool and bilateral contracts. Eur. J. Operat. Res. **260**, 432–443 (2017)
118. M.L. Song, L.B. Cui, Economic evaluation of Chinese electricity price marketization based on dynamic computational general equilibrium model. Comput. Ind. Eng. **101**, 614–628 (2016)
119. Y. Ni, J. Zhong, H. Liu, Deregulation of power systems in Asia: Special consideration in developing countries, in *Power Engineering Society General Meeting*, vol. 3, June 2005, pp. 2876–2881
120. L. Pingkuo, Z. Tan, How to develop distributed generation in China: In the context of the reformation of electric power system. Renew. Sust. Energ. Rev. **66**, 10–26 (2016)
121. M. Kohansal, H.M. Rad, Price-maker economic bidding in two-settlement pool-based markets: The case of time-shiftable loads. IEEE Trans. Power Syst. **31**(1), 695–705 (2016)
122. J.D. Lambert, *Creating Competitive Power Markets: The PJM Model* (PennWell Corporation, Tulsa, OK, 2001, Chap. 5), pp. 106–132
123. G. Mitchell, S. Bahadoorshngh, N. Ramsamooj, C. Sharma, A comparison of artificial neural networks and support vector machines for short-term load forecasting using various load types, in *IEEE Manchester PowerTech*, June 2017, pp. 1–4
124. M. Collotta, G. Pau, An innovative approach for forecasting of energy requirements to improve a smart home management system based on BLE. IEEE Trans. Green Commun. Netw. **1**(1), 112–120 (2017), Early Access
125. N.G. Paterakis, A. Tascikaraoglu, O. Erdinc, A.G. Bakirtzis, J.P.S. Catalao, Assessment of demand response driven load pattern elasticity using a combined approach for smart households. IEEE Trans. Ind. Inform. **12**(4), 1529–1539 (2017), Early Access
126. T. Pinto, H. Morais, T.M. Sousa, T. Sousa, Z. Vale, I. Praca, R. Faia, E.J.S. Pires, Adaptive portfolio optimization for multiple electricity markets participation. IEEE Trans. Neural Netw. Learning Syst. **27**(8), 1720–1733 (2016)
127. D. Saez, F. Avila, D. Olivares, C. Canizares, L. Marin, Fuzzy prediction interval models for forecasting renewable resources and loads in microgrids. IEEE Trans. Smart Grid **6**(2), 548–556 (2015)
128. D.M. Minhas, R.R. Khalid, G. Frey, Short term load forecasting using hybrid adaptive fuzzy neural system the performance evaluation, in *IEEE Int. Conf. PES*, June 2017, pp. 468–473
129. A. Khosravi, S. Nahavandi, Load forecasting using interval type-2 fuzzy logic systems: Optimal type reduction. IEEE Trans. Ind. Inform. **10**(2), 1055–1063 (2014)
130. M. Yang, Y. Lin, X.S. Han, Probabilistic wind generation forecast based on sparse Bayesian classification and Dempster–Shafer theory. IEEE Trans. Ind. Appl. **52**, 1–7 (2016)
131. S. Gupta, R. Kambli, S. Wagh, F. Kazi, Support-vector-machine based on proactive cascade prediction in smart grid using probabilistic framework. IEEE Trans. Ind. Inform. **62**, 2478–2486 (2015)
132. C.E. Borges, Y.K. Penya, I. Fernandez, Evaluating combined load forecasting in large power systems and smart grids. IEEE Trans. Ind. Inform. **9**(3), 1570–1577 (2017)
133. D.H. Vu, K.M. Muttaqi, A.P. Agalgaonkar, A. Bouzerdoum, Short-term electricity demand forecasting using autoregressive based time varying model incorporating representative data adjustment. Appl. Energy **205**, 790–801 (2017)
134. M. Chaouch, Clustering-based improvement of nonparametric functional time series forecasting: Application to intra-day household-level load curves. IEEE Trans. Smart Grid **5**(1), 411–419 (2014)
135. V. Thouvenot, A. Pichavant, Y. Goude, A. Antoniadis, J.-M. Poggi, Electricity forecasting using multi-stage estimators of nonlinear additive models. IEEE Trans. Power Syst. **31**(5), 3665–3673 (2016)

136. K.B. Sahay, M.M. Tripathi, Day ahead hourly load forecast of PJM electricity market and ISO new England market by using artificial neural network, in *IEEE Int. Conf. ISGT. PES 2014*, Feb 2014, pp. 1–5

137. S. Buhan, I. Cadirci, Multistage wind-electric power forecast by using a combination of advanced statistical methods. IEEE Trans. Ind. Inform. **11**(5), 1231–1242 (2015)

138. X. Xia, X. Rui, X. Bai, H. Wang, F. Jin, W. Yin, J. Dong, H. Lee, One-day-ahead load forecast using an adaptive approach, in *IEEE Int. Conf. SOLI*, Nov 2014, pp. 382–387

139. Y.-H. Hsiao, Household electricity demand forecast based on context information and user daily schedule analysis from meter data. IEEE Trans. Ind. Inform. **11**(1), 33–43 (2015)

140. PJM, Control center and data exchange requirements, http://www.pjm.com/library/manuals. aspx. Accessed 20 Aug 2017

141. PJM, PJM InSchedule user guide, http://www.pjm.com/library/manuals.aspx. Accessed 20 Aug 2017

142. PJM, Historical load data, http://www.pjm.com/markets-and-operations/ops-analysis/historical-load-data.aspx. Accessed 20 Aug 2017

143. PJM, Hourly real-time & day-ahead LMP, http://www.pjm.com/markets-and-operations/ energy/real-time/monthlylmp.aspx. Accessed 20 Aug 2017

144. National Oceanic and Atmospheric Administration, Local climatological data, https://www. ncdc.noaa.gov/cdo-web/datatools/lcd. Accessed 20 Aug 2017

145. M.A. Nielsen, *Neural Networks and Deep Learning* (Determination Press, 2015), pp. 83–87, ch. 3, sec. 3

146. A.J. Conejo, F.J. Nogales, J.M. Arroyo, Price-taker bidding strategy under price uncertainty. IEEE Trans. Power Syst. **17**(4), 1081–1088 (2002)

147. H. Zareipour, A. Janjani, H. Leung, A. Motamedi, A. Schellenberg, Classification of future electricity market prices. IEEE Trans. Power Syst. **26**(1), 165–173 (2011)

148. D. Huang, H. Zareipour, W.D. Rosehart, N. Amjady, Data mining for electricity price classification and the application to demand-side management. IEEE Trans. Smart Grid **3**(2), 808–817 (2012)

149. N. Amjady, F. Keynia, Day-ahead price forecasting of electricity markets by mutual information technique and cascaded neuro-evolutionary algorithm. IEEE Trans. Power Syst. **24**(1), 306–318 (2009)

150. M. Rafiei, T. Niknam, M.-H. Khooban, Probabilistic forecasting of hourly electricity price by generalization of ELM for usage in improved wavelet neural network. IEEE Trans. Ind. Inform. **13**(1), 71–79 (2017)

151. L.M. Saini, S.K. Aggarwal, A. Kumar, Parameter optimisation using genetic algorithm for support vector machine-based price-forecasting model in national electricity market. IET Gener. Transm. Distrib. **4**(1), 36–49 (2010)

152. J. Gama, I. Zliobaite, A. Bifet, M. Pechenizkiy, A. Bouchachia, A survey on concept drift adaptation. ACM Comput. Surv. **46**(4), 1–37 (2014)

153. D. Brzezinski, J. Stefanowski, Reacting to different types of concept drift: The accuracy updated ensemble algorithm. IEEE Trans. Neural Netw. Learning Syst. **25**(1), 81–94 (2014)

154. B. Krawczyk, L.L. Minku, J. Gama, J. Stefanowski, M. Wozniak, Ensemble learning for data stream analysis: A survey. Inform. Fusion **37**(C), 132–156 (2017)

155. H.M. Gomes, J.P. Barddal, F. Enembreck, A. Bifet, A survey on ensemble learning for data stream classification. ACM Comput. Surv. **50**(2), 23 (2017)

156. X. Dai, Z. Gao, From model, signal to knowledge: A data-driven perspective of fault detection and diagnosis. IEEE Trans. Ind. Inform. **9**(9), 2226–2238 (2013)

157. D. Alahakoon, X. Yu, Smart electricity meter data intelligence for future energy systems: A survey. IEEE Trans. Ind. Inform. **12**(1), 425–436 (2016)

158. L. Xu, M.-Y. Chow, L.S. Taylor, Power distribution fault cause identification with imbalanced data using the data mining-based fuzzy classification $ E $-algorithm. IEEE Trans. Power Syst. **22**(1), 164–171 (2007)

159. L. Xu, M.-Y. Chow, A classification approach for power distribution systems fault cause identification. IEEE Trans. Power Syst. **21**(1), 53–60 (2006)
160. M. Cococcioni, B. Lazzerini, S.L. Volpi, Robust diagnosis of rolling element bearings based on classification techniques. IEEE Trans. Ind. Inform. **9**(4), 2256–2263 (2013)
161. R. Razavi-Far, M. Farajzadeh-Zanjani, M. Saif, An integrated class-imbalance learning scheme for diagnosing bearing defects in induction motors. IEEE Trans. Ind. Inform. **13**(6), 2758–2769 (2017)
162. F.A.S. Borges, R.A.S. Fernandes, I.N. Silva, C.B.S. Silva, Feature extraction and power quality disturbances classification using smart meters signals. IEEE Trans. Ind. Inform. **12**(2), 824–833 (2016)
163. S. Pan, T. Morris, U. Adhikari, Classification of disturbances and cyber-attacks in power systems using heterogeneous time-synchronized data. IEEE Trans. Ind. Inform. **11**(3), 650–662 (2015)
164. D. Liang, J. Zhao, F. Luo, S.R. Weller, Z.Y. Dong, A review of false data injection attacks against modern power systems. IEEE Trans. Smart Grid **8**(4), 1630–1638 (2017)
165. F.J. Duque-Pintor, M.J. Fernández-Gómez, A. Troncoso, F. Martínez-Álvarez, A new methodology based on imbalanced classification for predicting outliers in electricity demand time series. Energies **9**(9), 752 (2016)
166. L. Zhu, C. Lu, Z.Y. Dong, C. Hong, Imbalance learning machine-based power system short-term voltage stability assessment. IEEE Trans. Ind. Inform. **13**(5), 2533–2543 (2017)
167. S. Chen, H. He, SERA: Selectively recursive approach towards nonstationary imbalanced stream data mining, in *Proceedings of International Joint Conference on Neural Networks*, 2009
168. G. Ditzler, R. Polikar, Incremental learning of concept drift from streaming imbalanced data. IEEE Trans. Knowl. Data Eng. **25**(10), 2283–2301 (2013)
169. Y. Lu, Y.M. Cheung, Y.Y. Tang, Dynamic weighted majority for incremental learning of imbalanced data streams with concept drift, in *Proceedings of 26th International Joint Conference on Artificial Intelligence*, 2017, pp. 2393–2399
170. N.V. Chawla, K.W. Bowyer, L.O. Hall, W.P. Kegelmeyer, SMOTE: Synthetic minority over-sampling technique. J. Artif. Intell. Res. **16**(1), 321–357 (2002)
171. J. Gao, W. Fan, J. Han, P.S. Yu, A general framework for mining concept-drifting data streams with skewed distributions, in *SIAM International Conference on Data Mining*, 2007
172. K. Wu, A. Edwards, W. Fan, K. Zhang, Classifying imbalanced data streams via dynamic feature group weighting with importance sampling, in *SIAM International Conference on Data Mining*, 2014
173. M.A. Tahir, J. Kittler, F. Yan, Inverse random under sampling for class imbalance problem and its application to multi-label classification. Pattern Recogn. **45**(10), 3738–3750 (2012)
174. D.S. Yeung, W.W.Y. Ng, D. Wang, E.C.C. Tsang, X.-Z. Wang, Localized generalization error model and its application to architecture selection for radial basis function neural network. IEEE Trans. Neural Netw. **18**(5), 1294–1305 (2007)
175. W.W.Y. Ng, J. Hu, D.S. Yeung, S. Yin, F. Roli, Diversified sensitivity-based undersampling for imbalance classification problems. IEEE Trans. Cybernet. **45**(11), 2402–2412 (2014)
176. D.S. Yeung, J.-C. Li, W.W.Y. Ng, P.P.K. Chan, MLPNN training via a multiobjective optimization of training error and stochastic sensitivity. IEEE Trans. Neural Netw. Learning Syst. **27**(5), 978–992 (2016)
177. W.W.Y. Ng, D.S. Yeung, M. Firth, E.C.C. Tsang, X.-Z. Wang, Feature selection using localized generalization error for supervised classification problems using RBFNN. Pattern Recogn. **41**(12), 3706–3719 (2008)
178. W.W.Y. Ng, Z.-M. He, D.S. Yeung, P.P.K. Chan, Steganalysis classifier training via minimizing sensitivity for different imaging sources. Inf. Sci. **281**, 211–224 (2014)
179. W.W.Y. Ng, X.-L. Liang, J. Li, D.S. Yeung, P.P.K. Chan, LG-trader: Stock trading decision support based on feature selection by weighted localized generalization error model. Neurocomputing **146**(1), 104–112 (2014)

180. L. Kocis, W.J. Whiten, Computational investigations of low discrepancy sequences. ACM Trans. Math. Softw. **23**(2), 266–294 (1997)

181. C. Wan, Z. Xu, Y. Wang, Z.Y. Dong, K.P. Wong, A hybrid approach for probabilistic forecasting of electricity price. IEEE Trans. Smart Grid **5**(1), 463–470 (2014)

182. J. Park, I.W. Sandberg, Universal approximation using radial-basis-function networks. Neural Comput. **3**(2), 246–257 (2014)

183. M. Harries, SPLICE-2 Comparative Evaluation: Electricity Pricing, Tech. Rep. 9905 (School of Computer Science and Engineering, Univ. New South Wales, Sydney, New South Wales, Australia, 1999)

184. S. Wang, L.L. Minku, X. Yao, A learning framework for online class imbalance learning, in *Computational Intelligence and Ensemble Learning*, 2013, pp. 36-45

185. S. Wang, L.L. Minku, X. Yao, A systematic study of online class imbalance learning with concept drift. IEEE Trans. Neural Netw. Learning Syst.. https://doi.org/10.1109/TNNLS.2017.2771290

186. L. Wiggins, Bringing big data up to the big leagues, *IBM Data Magazine*, 2013

187. J. Manyika, M. Chui, B. Brown, J. Bughin, R. Dobbs, C. Roxburgh, A.H. Byers, *Big Data: The Next Frontier for Innovation, Competition, and Productivity* (McKinsey Global Institute, New York, 2011) http://www.mckinsey.com/insights/business_technology/big_data_the_next_frontier_for_innovation. Accessed 1 Apr 2015

188. D. Laney, *3D Data Management: Controlling Data Volume, Velocity and Variety, Application Delivery Strategies* (Meta Group, Stamford, CT, 2001) http://blogs.gartner.com/doug-laney/files/2012/01/ad949-3D-Data-Management-Controlling-Data-Volume-Velocityand-Variety.pdf. Accessed 1 Apr 2015

189. Gartner Says Solving 'Big Data' Challenge Involves More Than Just Managing Volumes of Data, Gartner, Inc., June 2011, http://www.gartner.com/newsroom/id/1731916. Accessed 1 Apr 2015

190. M. Lynch, Data wars: Unlocking the information goldmine, Apr 2012, http://www.bbc.com/news/business-17682304. Accessed 1 Apr 2015

191. Y.-J. Kim, M. Thottan, V. Kolesnikov, W. Lee, A secure decentralized data-centric information infrastructure for smart grid. IEEE Commun. Mag. **48**(11), 58–65 (2010)

192. J. Zhou, R. Hu, Y. Qian, Scalable distributed communication architectures to support advanced metering infrastructure in smart grid. IEEE Trans. Parallel Distrib. Syst. **23**(9), 1632–1642 (2012)

193. X. Fang, S. Misra, G. Xue, D. Yang, Managing smart grid information in the cloud: Opportunities, model, and applications. IEEE Netw. **26**(4), 32–38 (2012)

194. Z. Fan, P. Kulkarni, S. Gormus, C. Efthymiou, G. Kalogridis, M. Sooriyabandara, Z. Zhu, S. Lambotharan, W.H. Chin, Smart grid communications: Overview of research challenges, solutions, and standardization activities. IEEE Commun. Surv. Tutorials **15**(1), 21–38 (2013)

195. M. Arenas-Martinez, S. Herrero-Lopez, A. Sanchez, J.Williams, P. Roth, P. Hofmann, A. Zeier, A comparative study of data storage and processing architectures for the smart grid, in *IEEE International Conference on Smart Grid Communications (SmartGridComm)*, Oct 2010, pp. 285–290

196. C. Borges, Y. Penya, I. Fernandez, Evaluating combined load forecasting in large power systems and smart grids. IEEE Trans. Ind. Inform. **9**(3), 1570–1577 (2013)

197. M. Amina, V. Kodogiannis, I. Petrounias, D. Tomtsis, A hybrid intelligent approach for the prediction of electricity consumption. Int. J. Electr. Power Energy Syst. **43**(1), 99–108 (2012)

198. N. Ding, Y. Besanger, F. Wurtz, G. Antoine, Individual nonparametric load estimation model for power distribution network planning. IEEE Trans. Ind. Inform. **9**(3), 1578–1587 (2013)

199. W. Labeeuw, G. Deconinck, Residential electrical load model based on mixture model clustering and markov models. IEEE Trans. Ind. Inform. **9**(3), 1561–1569 (2013)

200. A.M. Ferreira, C.A. Cavalcante, C.H. Fontes, J.E. Marambio, A new method for pattern recognition in load profiles to support decision making in the management of the electric sector. Int. J. Electr. Power Energy Syst. **53**, 824–831 (2013)

201. M. Biswal, P. Dash, Measurement and classification of simultaneous power signal patterns with an s-transform variant and fuzzy decision tree. IEEE Trans. Ind. Inform. **9**(4), 1819–1827 (2013)

202. C.P. Chen, C.-Y. Zhang, Data-intensive applications, challenges, techniques and technologies: A survey on big data. Inf. Sci. **275**, 314–347 (2014)

203. Intel, Big Data in the Cloud: Converging Technologies, Intel IT Center, Sept 2014

204. http://www.forbes.com/2010/03/12/cloud-computing-ibmtechnology-cio-network-data-centers.html. Accessed 10 Apr 2015

205. http://en.wikipedia.org/wiki/Quantum_computing. Accessed 6 Apr 2015

206. Department of Energy and Climate Change (DECC), Energy Consumption in the UK: Domestic Data Tables, UK, Tech. Rep. URN 12D/270, 2012

207. M. Ferguson, Architecting a big data platform for analytics, *IBM Data Magazine*, 2012

208. L. Einav, J. Levin, Economics in the age of big data. Science **346**(6210), 1243089 (2014)

209. C. Lynch, How do your data grow? Nature **455**, 28–29 (2008)

210. EDF's Big Data Vision for France. http://www.greentechmedia.com/articles/read/edfs-big-data-vision-for-france. Accessed 11 Apr 2015

211. https://www-03.ibm.com/press/us/en/pressrelease/41921.wss. Accessed 4 Apr 2015

212. http://www.greentechmedia.com/articles/read/c3-energy-unveils-first-big-test-of-smart-grid-data-analytics. Accessed 8 Apr 2015

213. A. Tinjum, PG&E is revolutionizing how utilities use data to empower their customers—One smart meter read at a time, 9 May 2014, http://blog.opower.com/2014/05/pgeis-revolutionizing-how-utilities-use-data-to-empower-theircustomers-one-smart-meter-read-at-a-time/. Accessed 10 Apr 2015

214. Metering International, KEPCO pilots big data projects for AMI and customer service systems, 14 July 2014, http://www.metering.com/kepco-pilots-big-data-projectsfor-ami-and-customer-service-systems/. Accessed 11 Apr 2015

215. M. Ilyas, Wireless sensor networks for smart healthcare, in *2018 1st International Conference on Computer Applications & Information Security (ICCAIS)*, Riyadh, 2018, pp. 1–5

216. H. Alemdar, C. Ersoy, Wireless sensor networks for healthcare: A survey. Comput. Netw. **54**, 2688–2710 (2010)

217. H. Zhang, J. Li, B. Wen, Y. Xun, J. Liu, Connecting intelligent things in smart hospitals using NB-IoT. IEEE Internet Things J. **5**(3), 1550–1560 (2018)

218. http://iot.10086.cn/product/read/id/814

219. D.J. Cook, G. Duncan, G. Sprint, R.L. Fritz, Using smart city technology to make healthcare smarter. Proc. IEEE **106**(4), 708–722 (2018)

220. Y. Zhang, M. Qiu, C. Tsai, M.M. Hassan, A. Alamri, Health-CPS: Healthcare cyber-physical system assisted by cloud and big data. IEEE Syst. J. **11**(1), 88–95 (2017)

221. S. Anand, S.K. Routray, Issues and challenges in healthcare narrowband IoT, in *2017 International Conference on Inventive Communication and Computational Technologies (ICICCT)*, Coimbatore, 2017, pp. 486–489

222. S. Sholla, R. Naaz, M.A. Chishti, Incorporating ethics in Internet of Things (IoT) enabled connected smart healthcare, in *2017 IEEE/ACM International Conference on Connected Health: Applications, Systems and Engineering Technologies (CHASE)*, Philadelphia, PA, 2017, pp. 262–263

223. H. Malik, M.M. Alam, Y. Le Moullec, A. Kuusik, NarrowBand-IoT performance analysis for healthcare applications. Proc. Comput. Sci. **130**, 1077–1083 (2018), ISSN: 1877-0509

224. W. Manatarinat, S. Poomrittigul, P. Tantatsanawong, Narrowband-Internet of Things (NB IoT) system for elderly healthcare services, in *2019 5th International Conference on Engineering, Applied Sciences and Technology (ICEAST)*, Luang Prabang, Laos, 2019, pp. 1–4

225. What will the carbon footprint be of the AI revolution? *Financial Times*, Monday, 27 Jan 2020

226. A. Meola, How Smart Cities & IoT Will Change Our Communities, 20 Dec 2016, BusinessInsider.com

227. Sensus.com, 2017

Chapter 2
Data Analytics for Solar Energy in Promoting Smart Cities

2.1 Solar Energy for Smart City

There is an increasing interest in installing solar photovoltaic (PV) systems combined with battery energy storage to provide backup power during electric grid outages; however, decision-makers are often unsure how to assign value to the lost power anticipated during an outage. As a result, the resilience benefit that a PV system with storage could provide is in general not accounted for when considering project cost-effectiveness.

Reference [1] explored the impact of resilience on the economics of PV and energy storage systems for commercial buildings. As storage costs decrease, and as outages occur more frequently, PV and storage are likely to play a larger role in planning, operation, design, and management investigation.

As severe weather events such as sand storm, low temperature, hurricanes, and heat waves become common, interest is increasing in resilient electric power systems. For a power system to be resilient, it must be capable of islanding and operating independently from the grid during outages. Installing additional devices, for example, transfer switches and critical load, these systems can act as self-sufficient microgrids, generating energy and powering critical loads until utility services are restored. Recent natural disasters such as Hurricanes Harvey, Irma, and Maria have reinforced the need for reliable power for essential services (namely, air conditioning, medical operation, and water pumps) and to keep critical businesses in place (such as gas stations and grocery stores).

Diesel generators are often viewed as the default solution for providing resilient energy, but they might not always be the most reliable or cost-effective solution. Reliance on traditional fuel reduces the resilience of an energy system because a disruption or contamination in the fuel supply can cause disasters.

While sustainability has not previously been a top priority in city planning, the current state of the environment is quickly changing that trend. Innovators and technologists are focusing on energy efficiency and their environmental impact more than ever before. Smart solution providers are providing high-tech infrastructure options that can help city governments save on energy costs even as they reduce their carbon output. This is an especially smart win-win situation because it allows municipalities to consolidate efforts to improve the quality of life and sustainability under one scheme.

Some countries incorporate solar energy, electric cars, and sensors, and mobile apps to improve public transportation, security, parking, lighting, and waste management. In addition to maximizing renewable energy sources like solar power, there is a focus on electric buses, green construction, smart grids, and rooftop farms in the city.

By adopting renewable energy to power smart cities is not without pain. Solar power is great as long as the sun is shining, but cloudy days can minimize the output from the solar panels. A single grid-tied home losing solar efficiency is not a huge problem, but when a city-wide infrastructure is tied to solar power production, being able to plan around periods of low efficiency is critical and essential.

Increased accuracy has the potential to save money and reduce reliance on coal and natural gas power plants that often have to pick up when clouds roll in.

With more solar power output in the future, a smart city is going to be more efficient, more connected, and more sustainable. Making cities smarter and greener will change the way to operate and help citizens maximize their potential as responsible, sustainable members of a global community.

Using solar power to charge on-site energy storage offers unique benefits that traditional diesel-fueled backup power systems do not have. As a result, solar technology combined with energy storage is increasingly being implemented in power system designs.

Unfortunately, although the benefit of having a resilient power system is clear when the electric grid goes down, putting a monetary value on additional resilience investments can be difficult. Each individual business or service provider might have widely varying values of resilience. Determining the expected utility cost savings and potential for revenue generation associated with an investment in a PV and battery energy storage system can be relatively straightforward; however, assigning a value to the improved resilience associated with a PV and storage system is much more challenging. When solar and energy storage technologies are configured to provide backup power, they create value by allowing businesses to stay open. When powering critical facilities such as hospitals and emergency shelters, resilient power systems might even prevent losses of life.

To quantify the effect of valuing resilience on PV and battery energy storage system design, researchers at the U.S. Department of Energy National Renewable Energy Laboratory (NREL) incorporated the avoided cost of a grid outage into the economics of determining cost-optimal system sizing for buildings in Anaheim, California [1]. For each of the building types analyzed, two scenarios were explored: one that places no value on resilience and one that values resilience in terms of dollars lost per hour of outage.

For each scenario, a solar and energy storage system is designed to maximize economic benefit during an assumed system lifetime of 20 years. The lifetime economic benefit is measured in terms of the net present value (NPV) of the system, which is the net difference between the benefits and the costs of the project. The project benefits include the bill savings delivered by the PV and storage systems during normal grid-connected operation as well as the additional benefit of surviving a grid outage. The project costs include the capital costs of installing PV and storage, system operating and maintenance expenses, and the cost of any outage period not survived.

A project with a negative NPV indicates that it would cost more to install and maintain the system than the savings realized throughout time. A system with a positive NPV indicates that it would be less expensive to build and operate the system than to continue normal operations without it. Systems costs, benefits, and optimal system sizes for each customer scenario was determined by balancing the cost of the system, the cost of electricity from the utility, and the cost of outages. For scenarios in which resilience is not valued by the customer, the cost associated with the outage is assumed to be zero (i.e., no assets were damaged, and no business was disrupted). When resilience is assigned a value, the cost of outages can be reduced by the ability of a resilient power system to survive some parts, or all, of anticipated grid disruptions.

When sizing solar system and storages, a number of variables must be considered, for example, the number of hours that a given PV and storage system can power critical loads which depend on several factors such as current electricity price, load profile, the average duration of outages, time of day when outages occur, time of year when outages occur, critical loads, other uses for storages, and the average cost of outages and amount of energy stored in a storage.

Turning to islanding a PV system is critical for resilience. PV panels on a rooftop that are grid-connected do not ensure that a building will have power during a grid outage. Any stand-alone PV and storage system require additional expenses that are more than the cost of a stand-alone system. These added costs depend on many factors. These might include additional hardware components, such as transfer switches and critical load panels; software components; and electrical design, and safety considerations. These must be factored when determining whether a resilient system is the most economical solution.

The costs to island can be highly variable and depend on a multitude of site-specific factors. The cost to island a system might add incremental expenses ranging from 10 to 50% of the stand-alone PV and storage system cost. The benefit of any avoided losses during grid outages must be balanced with these added costs of designing a system.

For a resilient power system to result in a net economic benefit for a customer, the cost to island must be no more than the added savings delivered by the system. Under current technology price assumptions, battery energy storage systems are often only cost-effective in locations that have relatively high utility demand charges or where there is a viable market for the grid services storage can provide. It demonstrates that even though a PV and storage system might not appear to be economical under traditional cost-benefit calculations, placing a value on the losses incurred from grid disruptions can make a PV and storage system a fiscally sound investment. In most cases, incorporating the value of resilience will increase the optimal sizing of both the PV and battery systems, but the added cost to make a system stand-alone must also be considered. Recent major weather events and widespread outages have raised awareness of and interest in the need for localized, resilient power systems as well as the limitations of current solutions such as the use of diesel generators. With technology costs declining and extended outages becoming increasingly common, more businesses and building owners are likely to consider the value of resilience and the viability of PV and storage to avoid outage-related losses.

2.2   Global Developments on PV Systems

Several countries are aiming to maximize their solar energy portfolios. Greenpeace states that it is possible to become 100% renewable by 2050 and therefore experiencing a very sharp increase in installations [2]. Figure 2.1 presents the trends for global penetration of solar PV systems from different literatures and case scenarios. Greenpeace has provided a forecast of global PV penetration under two scenarios, namely the "revolution scenario" and "reference scenario". It can be seen that there is a sharp linear increase in solar power capacity.

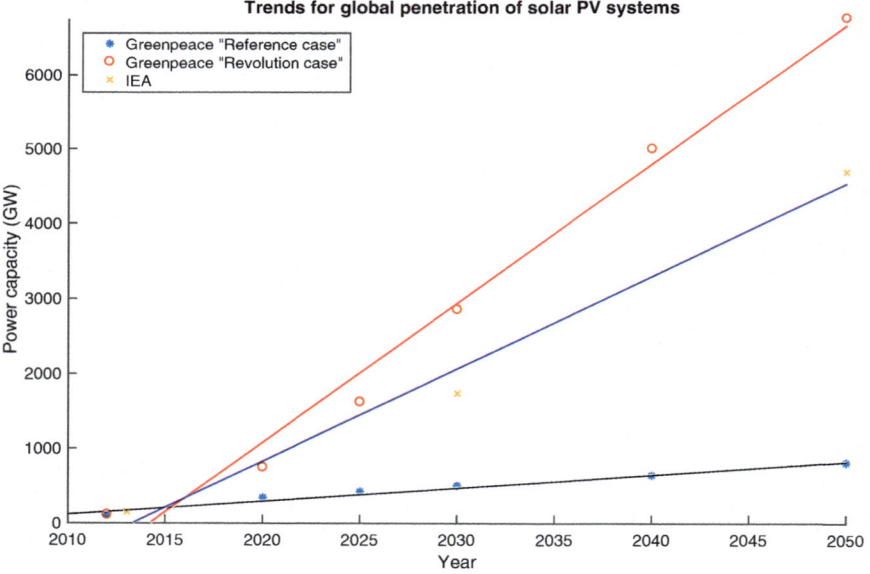

Fig. 2.1 Trends for global penetration of solar PV systems [3]

United States   Solar energy represents the largest renewable resource base in the United States, with the potential that considerably exceeds the total demand for electricity. The Energy Secretary has announced that up to $87 million will be made available to support the development of new solar energy technologies and the rapid deployment of available carbon-free solar energy systems [4]. Solar Grand Plan states that the country aims to meet 69% of the country's electricity demand by 2050 from PV while reducing CO_2 emission by 60% from 2005 levels; the PV contribution to this plan was assessed to be 250 GW by 2030 and 2900 GW by 2050 [5].

China   China has abundance of solar energy. Solar energy is currently mainly used in solar water heaters, solar stoves, and passive solar houses within the country. Solar power generation technologies are developing quickly in China. China produces approximately 18% of the PV products worldwide, as a result of more than 400 Chinese PV companies' production. Hainan's largest PV power plant is in operation since 2009. However, the amount of electricity generated with solar power within China is so far comparatively small. According to the plans unveiled by the National Development and Reform Commission (NDRC) in 2007, China's installed solar power capacity was originally planned to be 1.8 GW by 2020 [6]. By 2050, 2.7 TW of solar power will be installed with a total annual output of 9.66 trillion kWh, a contribution of 64% for China's total power generation [7].

India   In India, the Jawaharlal Nehru National Solar Mission has been initiated to promote the deployment of solar PV energy. It is expected that 20 GW of power will be produced by Solar PV by 2022 [8].

Australia   By the end of 2014, nearly 1.4 million Australian homes had photovoltaic (PV) systems on their roofs [9]. As a countrywide average, the installed cost of PV on household roofs declined from around AU$12/watt in 2008 to fewer than AU$2/watt in 2014. The decline in costs may be attributed to three main factors: an appreciation of the Australian dollar, large reductions in the price of solar panels, and greater competition among system installers in Australia.

Germany   Germany has the largest amount of installed PV capacity in the world [10, 11]. The country aims for PV penetration to reach 52 GW by 2020. The high PV installation rate during 2009–2012 has caused significant issues for some German distribution grids, with installed generation capacities exceeding the annual peak load on many occasions. Despite the expanding market for such PV battery systems, the market still lacks standards and approaches for comparing their performance and efficiency.

Germany's goal is to transform its electrical energy supply to one that is based on a renewable energy share of more than 80% by 2050.

Japan   The Japanese government began planning a completely new energy strategy following the nuclear power plant failures in Fukushima after the March 2011 earthquake [12]. The Innovative Energy and Environment Strategy was discussed in the

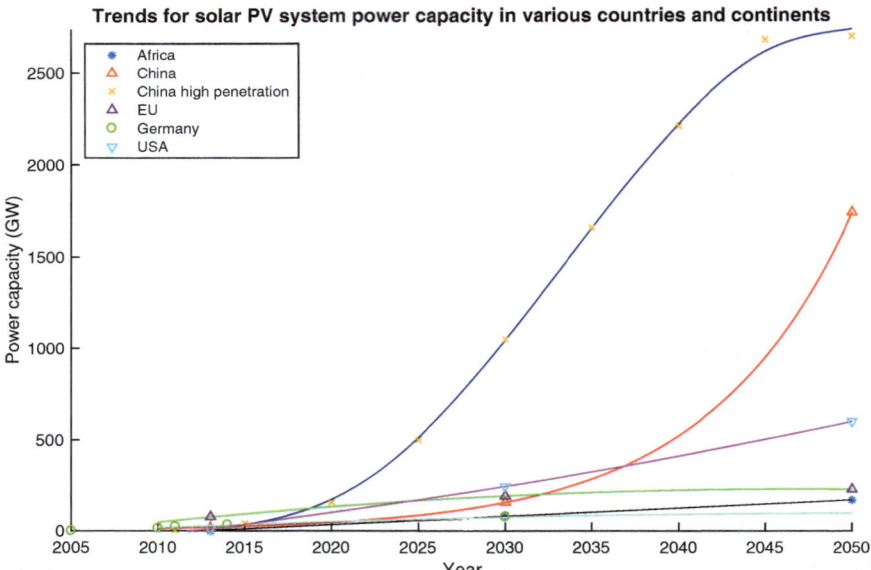

Fig. 2.2 Trend for solar PV system power capacity in various countries and continents [3]

cabinet on 14th September 2012. The main idea of the strategy is that all available efforts and resources are to be used to reduce the generation share of nuclear power, including the maximum deployment of all types of renewable energy. The FIT program went into effect on 1st July 2012 has brought about a significant change in the expansion of PV system installations in Japan. The national targets are to increase PV power generation to ten times its 2008 level, to 14 GW, and to 40 times the 2008 level, or an estimated 53 GW, by 2020 and 2030 respectively. In April 2009, the government formulated an economic stimulus measure named the J-Recovery Plan. By 2020, the country aims to increase 20 times the cumulative installed capacity as of 2009, to a level of 28 GW.

Figure 2.2 presents the future prediction for PV systems penetration of various countries in GW as reported by different organizations. It is expected that the share of PV will increase with time for many developed and developing countries.

Figure 2.3 presents the long-term average daily and annual solar irradiance received by the earth. It can be seen that Africa and Australia receive the highest proportion of solar irradiance compared to other continents. In general, the countries near the equator will benefit more from the solar resource. Subject to governmental policy, the global connection of PV and energy sources will be a feasible solution to improve asset management and fully utilize resources for countries with abundant solar irradiance. The term "Energy Internet" has been proposed to achieve flexible energy sharing for consumers in a residential distribution system with distributed renewable energy and distributed energy storage devices [13]. The development of the global energy internet and interconnection (GEI) is based on ultra high voltage AC/DC and smart grid technology [14].

Table 2.1 presents the top ten largest large-scale PV installation projects world-wide till 2016. It is noted that from 2011, the size of large-scale PV plants has significantly increased and reached 580 MW by the end of 2015.

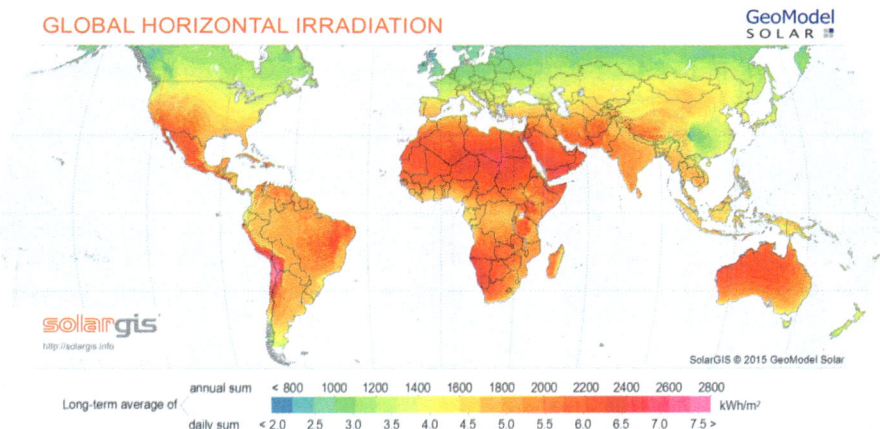

Fig. 2.3   The global long-term average of daily and annual solar irradiance in kWhm^{-2} [15]

Table 2.1   Top ten largest solar farms across the globe as of 2016

Location (State, Country)	Description	Size (MW)	Construction starting date	Completion date
USA, Rosamond, CA	Solar Star Projects [16]	579	2013	2015
USA, Riversize County, CA	Desert Sunlight Solar Farm [17]	550	2011	2015
USA, San Luis Obispo County, CA	Topaz Solar Farm [18]	550	2011	2014
China, Longyangxia Dam, Qinghai Province	Longyangxia Hydro-Solar PV Station [19]	480	2014	2015
India Charanka, Patan District	Charanka Park PV power plant [20]	345	2010	2016
France, Cestas, Bordeaux, Gironde	Centrale solaire de Cestas [21]	300	2014	2015
USA, Yuma County, AZ	Agua Caliente Solar Project [22]	290	2010	2014
USA, Boulder City, NV	Copper Mountain III Solar Facility [23]	250	2013	2015
USA, San Luis Obispo, CA	California Valley Solar Ranch [24]	250	2011	2013
USA, Lancaster, CA	Antelope Valley Solar [25]	242	2011	2014

2.3   Photovoltaic Cell Technology

A PV cell is an electrical device that converts the energy of light into electricity via the photovoltaic effect. The cells are the building blocks of PV modules, known as the solar panels. There are several challenges in photovoltaic cell technologies [26]. The crystalline-silicon photovoltaics heavily rely on an abundant amount of silicon and their production costs are relatively high. The thin-film solar cells can be produced more cheaply, but they use materials of limited availability. Cadmium telluride thin-film modules are the cheapest to produce, but there are concerns about the future availability of tellurium, and about the toxicity of cadmium used as a precursor to CdS and CdTe. Similarly, there are concerns about the availability of materials for Copper indium gallium (di)selenide (CIGS) technologies (i.e., gallium, indium), and its toxicity (i.e., cadmium, selenium). Some silicon technologies use potent greenhouse gases for reactor cleaning (e.g., NF3). Figure 2.4 presents the market share of various solar cell technologies. At present, Silicon-based solar cells are the most dominant with a total market share of over 90%.

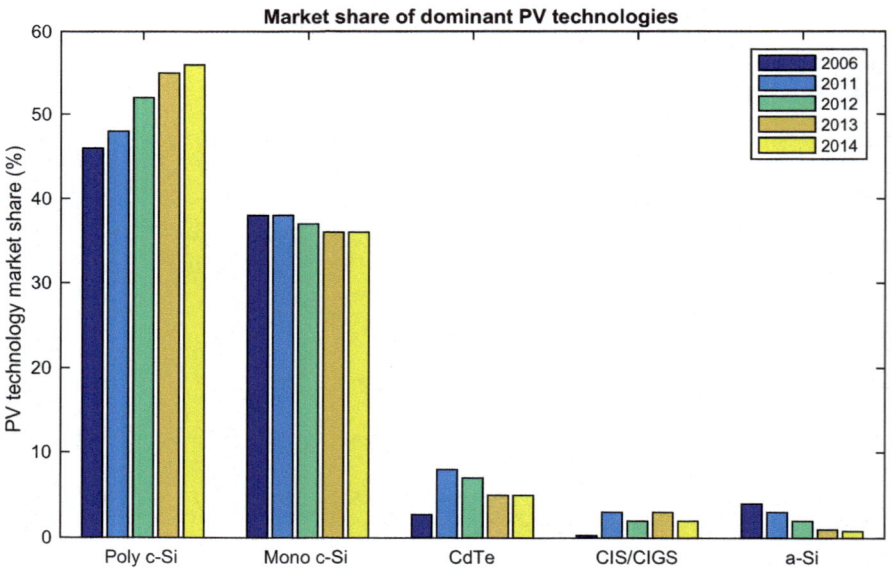

Fig. 2.4 Market share penetration of PV technologies in various years [27, 28]

In general, PV cell technology can be broadly classified into Fig. 2.4 four categories, namely the wafer-based, multi-junction, thin film, and the emerging PV cell technology. The wafer-based cells are made of crystalline silicon, the commercially predominant PV technology as shown in Fig. 2.4. The second-generation cells are thin-film solar cells. They are widely used in utility-scale PV power stations, building-integrated PV, or in small stand-alone PV power systems. The multi-junction solar cells are constructed with multiple p-n junctions from different semi-conductor materials. Since each material's p-n junction will produce an electric current in response to different wavelengths of light, the heterojunctions allow the absorbance of a broader range of wavelengths, improving the cell's sunlight to electrical energy conversion efficiency. As of 2016, the current highest solar cell efficiency is 46% from multi-junction technology [29].

Detailed literature on mature PV solar cell materials and technologies can be found in [30]. The emerging solar cell technologies include a number of thin-film technologies. Most of them have not yet been commercially available and are still in the research or development phase. Many use organic materials, often organometallic compounds as well as inorganic substances. Although their efficiencies had been low and the stability of the absorber material such as Perovskite was often too short for commercial applications, there is a lot of research invested into these technologies as they promise to achieve the goal of producing low cost, high-efficiency solar cells [31].

Table 2.2 presents an overview of current and emerging PV cell technologies. The following section gives a review of the emerging PV technologies that could be of significant interest.

Dye-sensitized cells   The dye-sensitized cell [32] is a type of thin-film solar cell. It is based on a semiconductor formed between an electrolyte and a photo-sensitized anode. The cell is simple to manufacture using conventional roll-printing techniques, is semitransparent and semiflexible which increases the use in applications, and most of the materials used are low cost (such as TiO_2). However, in practice, it has issues to eliminate the usage of a number of expensive materials, such as platinum (catalyst) and ruthenium (dye), and the liquid electrolyte is a serious issue in making a cell suitable for use under all weather conditions.

Perovskite cells   Perovskite cells first appeared in 2012 [33]. It was coined as a "meso-superstructured solar cell" and described as a low cost, solution-processable, based on a highly crystalline perovskite absorber with intense visible to near-infrared absorptivity cell. When it was first developed, it has a power conversion efficiency of 10.9% in a single-junction device under simulated full sunlight. It has an efficiency of 22.1% by 2016 [31].

This solar cell contains perovskite structured compound, most commonly a hybrid organic-inorganic lead or tin halide-based material, as the light-harvesting active layer. Perovskite materials such as methylammonium lead halides are simple to manufacture and inexpensive to create. One of the main advantages in comparison to previous cell technologies is it has a simplified device architecture, as it does not need the complex nanostructures [34, 35].

A major concern is the potential toxicology issue of lead, a crucial element for the light-harvesting active layer. Good semiconducting behavior could be achieved

Table 2.2 Categories of mature and emerging photovoltaic technologies [31]

Wafer-based	Multi-junction	Thin-film cell	Emerging technology
Crystalline silicon	GaAs and III-V single junction	Solar cells with multiple p-n junctions made of different semiconductor materials. This allows absorbance of a broader range of wavelengths	Absorb light 10–100 times more efficient than wafer-based technology. Also allow use of films of just few micron-meters thick	Has the potential to overcome Shockley-Queisser limit or are based on novel/advanced semiconductor technologies. Currently under research or development phase

			Multi-junction			Thin-film cell				Emerging technology			
Single crystalline silicon	Multi-crystalline silicon		Two-junction	Three-junction	Four-junction or more	Hydrogenated amorphous silicon (a-si:H)	Cadmium telluride (CdTe)	Copper indium gallium diselenide (CIGS)	Copper zinc tin sulfide (CZTS)	Dye-sensitized cells (DSSC)	Perovskite cells	Organic cells	Quantum dot cells

with organic–inorganic tin halide perovskites, but the instability of tin has proved to be an overwhelming challenge [36]. Another major challenge is the aspect of short-term and long-term stability due to the environment humidity. Under moisture environments, the water-solubility of the organic constituent of the absorber material causes the cell to rapidly degrade. A method to overcome this issue is to encapsulate the perovskite absorber with a composite of carbon nanotubes and an inert polymer matrix. However, no comprehensive encapsulation techniques and long-term studies are demonstrated for perovskite solar cells [37].

Organic cells/organic tandem cells Organic cells are built with organic electronics, a branch of electronics that deals with conductive organic polymers or small organic molecules. Polymer solar cell is an example of an organic solar cell. The reason for using organic cells in photovoltaics is the possibility of high throughput module manufacture by printing or coating from solution in continuous production [38]. Inexpensive manufacturing process together with the low quantities of organic semiconductors required could reduce the cost of modules to less than 1.1 $/watt. Therefore, it is possible to accelerate the process of photovoltaic electricity generation adoption. The features of flexibility, lightweight, and potential to tune the transparency and color of organic cells are reasons for the integration of PV into building components or other appliances. The hybrid tandem solar cell is presented in [39]. This solar cell is composed of an inexpensive and low temperature processed solar cell, such as an organic or dye-sensitized solar cell, that can be printed on top of one of a variety of more traditional inorganic solar cells. Organic solar cell can be added on top of a CIGS cell to improve its efficiency from 15.1 to 21.4%.

Quantum Dot cells (QDC) QDC [40] is a solar cell design that uses quantum dots as the absorbing photovoltaic material. QDC is highly recommended for the implementation of solar cells due to tunable bandgap, which could be achieved by changing the dots' size. The bandgap of the conventional bulk materials is fixed by the choice of a chemical element. Single junction implementations with lead sulfide (PbS) carbon quantum dots have bandgaps that can be in the range to far-infrared, these frequencies that are difficult to achieve with traditional solar cell technologies.

2.4 Clear Index Clustering

Solar photovoltaic distributed generation (PV-DG) systems are being integrated worldwide into distribution systems at a rapid rate [41]. Due to the intermittent nature of PV sources which are generally densely connected in a low-voltage distribution network. Voltage and power fluctuations on the grid must be considered. To study the fluctuations, statistical evaluation, and localized spectral analysis of the fluctuation power index should be further investigated [42]. As a result of the analytical monitoring costs, there are a limited number of studies on PV systems operation in remote areas. To reduce the costs, clustering results are needed for analyzing the performance and sizing of PV systems.

Given the statistical distribution of the solar irradiance, a large quantity of data can be characterized with only very few parameters. An example of the practical application of solar irradiance statistical modeling is provided in [43], for a case study in Tahifet, Algeria. It is learned that the installed PV system produces excess energy in October and energy storage is required in June and December.

Solar irradiance is characterized by short fluctuations mainly introduced by passing clouds. The analysis of these fluctuations with regard to solar energy applications should focus on the instantaneous clearness index (CI) [42]. CI can effectively characterize the attenuating impact of the atmosphere on solar insolation by specifying the proportion of extraterrestrial solar irradiance that reaches the surface of the earth. Performance analysis of the PV systems studied with the classification scheme of CI profiles provides useful insights [44, 45]. The ability of generalization of this technique allows the proposed method to be applied to other system configurations for evaluation purposes, such as sizing energy storage system [46]. In particular, it is shown that cloud-induced fluctuations in CI can be treated by statistical analysis.

This sub-section provides the grouping of daily CI profiles and to construct centroids with cluster analysis. Section 2.2 provides the literature review on the statistical analysis of PV and renewable energy sources. Section 2.3 presents the clear-sky solar model and real-life solar data collected for CI calculation purposes. The research problem and preliminary understanding will also be provided. Section 2.4 gives the clustering algorithms and distance metrics used for the clustering of daily CI profiles. Section 2.5 will present the clustering results for the four seasons with the five clustering techniques. To evaluate the usefulness of the clustering results for PV system planning, a case study based on sizing a stand-alone solar PV and storage system with anaerobic digestion biogas power plants is given in Section 2.6. Section 2.7 provides the conclusion and future work of the research.

Fractal analysis of daily solar irradiance measured with a time step of 10 min at Golden and Boulder located in Colorado is provided in [47], with the aim to perform the classification of daily solar irradiance. These results lead to three classes, namely clear-sky, partially covered sky, and overcast sky. The daily distributions of CI were classified by estimating a finite mixture of Dirichlet distribution in [48]. The results display four distinct classes of distributions corresponding to different types of days. However, in the two studies, the CI in different seasons or months has not been studied or given.

The use of models with CIs for any solar system applications, such as solar hydrogen production is appropriate and simple. This is due to the CI only needs the global solar irradiance data [49]. The knowledge of the statistical behavior of short-term variability of solar irradiance will provide a more accurate evaluation of the uncertainty in the long-term annual energy production of solar power plants [50]. CI can be used to train the Markov transitions matrix, in order to approximate the daily irradiance value with the Markov model [51]. Irradiance sequences can be generated via this method. Reference [52] uses CI to separate forecasting complexity into the prediction of solar geometry and the prediction of cloudiness and aerosol. The quadratic and cubic equations which are based on global solar irradiance data have the highest accuracy in predicting the diffuse fraction as a function of CI [53–55].

Wavelet analysis is applied to the daily CI profiles in [56], and which is decomposed into components to evaluate the endurance and magnitude of various fluctuations of the solar irradiance. The classification of typical meteorological days from global irradiance data is given in [57]. The classification was performed with aggregation Ward's method. It is learned that the recorded days are clustered in three, four, or five groups for monthly time step and three groups are classified for annual time-step. The authors relied on discriminant analyses to evaluate the number of clusters and this was achieved by visual inspection.

PV generations are commonly presented by Beta distribution [58]. This assumption has been widely used for system planning purposes. However, in reality, the underlying distribution may vary widely due to the hemisphere and climate of the location [59]. Reference [60] determined the parameters of the appropriate distribution that provide the best fit for CI. The global solar irradiance is thereafter predicted from CI using the inverse transformation of the cumulative distribution function. The proposed method is effective in predicting the monthly average global solar irradiance.

Pattern recognition and cluster analysis have been applied to other renewable sources. A statistical approach was proposed for the improvement of short-term wind electric power forecasts based on pattern recognition technique [61]. The predictions on wind speed and direction to identify patterns of the wind behavior at the location considered to obtain a stochastic distribution of the daily wind speed were studied in [62]. A statistical hybrid wind power forecast technique was proposed in [63], where weather events are clustered with respect to the most important weather forecast parameters.

2.4.1   Data Acquisition of Real-Life Solar Irradiance

The CI is developed with the solar irradiance data collected from the Skye Instruments SKS 1110 Pyranometer sensor [64, 65]. The cosine-corrected head, a sensor consists of a semiconductor diode, and a light filter system for the wavelength range 350–1100 nm was used to construct the pyranometer. Cosine-corrected head is required to avoid measurement errors when the sensor is not directly below the sun. The pyranometer can be used for energy balance studies, as the head is perfectly sealed and can be placed indefinitely in outdoor conditions. World Radiometric Reference [66] is used for the calibration of a sensor under open sky conditions.

The pyranometer sensor was placed on a perfectly flat surface in order for the top light-collecting surface to be exactly horizontal. Four years of solar irradiance data, from 2009 to 2012 were obtained in Johannesburg for this research. Johannesburg has a latitude of 26.21°S, longitude of 28.05°E and with an altitude of 1753 m. The data sampling rate is at 1 sample/30 min.

2.4.2   Clear-Sky Solar Irradiance Model

Under perfect atmospheric conditions, the earth will absorb the solar irradiance which is equal to the solar constant minus the amount absorbed by the atmosphere of the earth. The solar constant is at a value of 1367 Wm^{-2}. The global solar irradiance on a horizontal surface has two main components, namely the direct beam component and the diffuse sky irradiance.

The other factor in the attenuation of the atmosphere is a function of the concentrations of the various elements in the atmosphere [67]. Their impacts can be assessed by comparing the actual observed optical depth with the theoretical optical depth of a perfectly clean dry scattering Rayleigh atmosphere. The ratio of the two optical depths is known as the Air mass 2 Linke turbidity factor, T_{LK}. The clear-sky beam irradiance normal to the beam I_{model} at the surface is calculated as mentioned in [68, 69].

$$I_{model} = I_o \varepsilon \exp\left(-0.8662 T_{LK} m \delta_r (m)\right) \sin \gamma_s \tag{2.1}$$

$$\varepsilon = 1 + 0.0334 \cos\left(j' - 2.80°\right) \tag{2.2}$$

$$j' = \frac{J * 360}{365.25} \tag{2.3}$$

$$m = \left(p / p_o\right) / \left\{\sin \gamma_s + 0.50572\left(\gamma_s + 6.07995\right)^{-1.6364}\right\} \tag{2.4}$$

I_o is the solar constant, ε is the correction factor to mean solar distance, m is the optical air mass corrected for station height, γ_s is the solar altitude angle in degrees and δ_r is the Rayleigh optical depth, J is the Julian day and j is the Julian day angle. p/p_o is the pressure correction for station height and is calculated with Eq. (2.5) given below:

$$\frac{p}{p_o} = \exp\left(-\frac{z}{H_R}\right) \tag{2.5}$$

z is the site elevation above sea level in meter and H_R is a constant at 8400 m. δ_r is calculated as follows [70]:

$$\frac{1}{\delta_r(m)} = 6.6296 + 1.7513m - 0.1202m^2 + 0.0065m^3 - 0.00013m^4 \quad \text{if } m < 20 \ (2.6)$$

$$\frac{1}{\delta_r(m)} = 10.4 + 0.718m \quad \text{if } m >= 20 \tag{2.7}$$

The solar altitude angle is calculated as a function of time of day with Eq. (2.8) [68].

$$\gamma_s = \sin^{-1}\left(\sin\phi\sin\delta + \cos\phi\cos\delta\cos\omega\right) \tag{2.8}$$

$$\omega = 15(t - 12) \tag{2.9}$$

ϕ, δ, and ω are the latitude of the location, solar declination angle, and solar hour angle respectively. All are in degrees. t is the instantaneous time of the day in an hour with values between 0 and 23.

2.4.3   Real-Life Solar Irradiance Data Analysis

Solar irradiance data acquisition The SKS 1110 Pyranometer sensor developed by Skye Instruments [64, 65] was used to collect the solar irradiance data for the study. The sensor consists of a semiconductor diode, cosine-corrected head, and a light filter system for the wavelength range 350–1100 nm. Cosine-corrected head is built-in to eliminate measurement errors which may arise when the sun is not directly above the sensor, but at any angle within the hemisphere of measurement. The head is completely sealed and can be left indefinitely in exposed conditions in making it perfect for weather or energy balance studies. The sensor has been calibrated under open sky conditions against World Radiometric Reference [66]. The pyranometer sensor should be mounted perfectly leveled, so that its top light-collecting surface is exactly horizontal. The sensor is usually mounted in the same plane as the solar panel, in order to measure the radiation falling on its surface. Four years of solar irradiance data between 2009 and 2012 were collected in Johannesburg for the study. The sampling rate is at 1 sample/30 min.

Solar irradiance data analysis To examine the nature of the real-life irradiance data, the clear-sky model is used to provide comparisons. T_{LK} has been set to 5 to model the diffuse irradiance. A comparison of solar insolation data from different sources is summarized in Fig. 2.5. Further comparisons are made with the NASA data obtained in [71]. The maximum amount of insolation received is in December and the minimum amount is in June. The insolation is generally higher in Summer (Dec, Jan, Feb) season as compared to other seasons such as Spring (Sept, Oct,

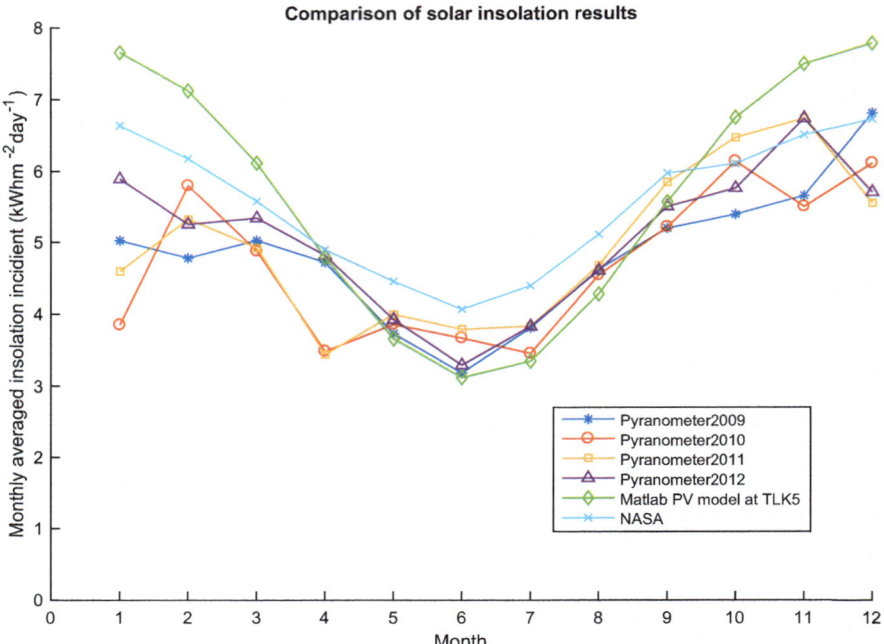

Fig. 2.5 Comparison of solar insolation data

Nov), Autumn (March, Apr, May), and Winter (June, Jul, Aug). NASA provides solar insolation for clear-sky conditions. The solar model and NASA data will have a higher monthly averaged insolation incident as compared to the real-life data. It can be seen that the three sources give a similar trend and this gives a good indication that the data is statistically accurate.

Clearness Index CI at instantaneous time t is expressed as a ratio between 0 and 1, where 1 signifies there is no loss in irradiance, i.e., all the insolation is of direct beam irradiance, and 0 means there is no irradiance due to a complete cloud cover. It is worth mentioning that CI can be undefined when no irradiance is available, such as before sunrise and after sunset. These conditions are not considered in this work as they are not applicable to the study. CI is calculated with Eq. (2.10).

$$\text{CI}(t) = \frac{I_{\text{pyranomter}}(t)}{I_{\text{model}}(t)} \qquad (2.10)$$

$I_{\text{pyranomter}}$ is the real-life solar irradiance and I_{model} is the clear-sky solar irradiance from the solar model. To calculate the solar model irradiance for CI, the TLK is set to 1 to remove the effect due to the clear-sky solar irradiance atmospheric absorption and scattering. These phenomena can be reflected in the CI, as it takes into

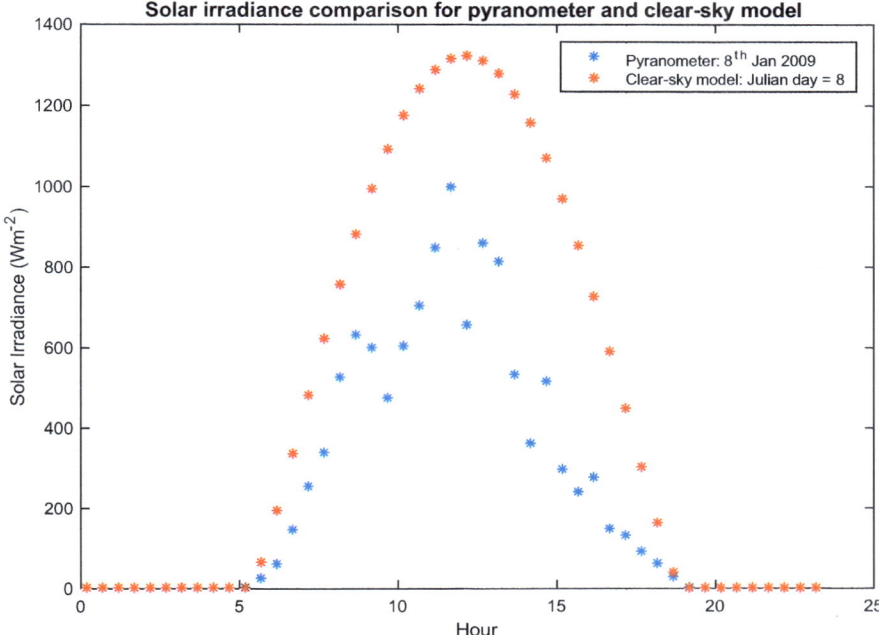

Fig. 2.6 Solar irradiance for clear-sky model and real-life (pyranometer) data

account the total irradiance reduction from the clear-sky irradiance. Fig. 2.6 presents the clear-sky and real-life solar irradiance for a typical day in January.

CI for four different seasons between 2009 and 2012 is shown in Fig. 2.7. Each color represents a CI profile for a day. Twenty profiles were plotted for each season due to the space limitation. It can be seen that in winter there are significantly more clear days, i.e., higher CI. In contrast, CIs in summer are mostly below 0.3. CI also displays the nature of uncertainty and the daily fluctuation.

2.4.4   Clustering Methods

This section describes the two clustering families.

Distribution-based clustering In distribution-based clustering, clusters can be defined as objects belonging most likely to the same distribution. A Gaussian Mixture Model (GMM) is a weighted sum of m components, i.e., the number of clusters. The Gaussian mixture densities for vectors x is given in Eq. (2.11) [72]:

$$g_{x;w,\theta}\left(x;,w;,\theta\right) = \sum_{i=1}^{m} w_i g_{x;\theta_i}\left(x;\theta_i\right) \tag{2.11}$$

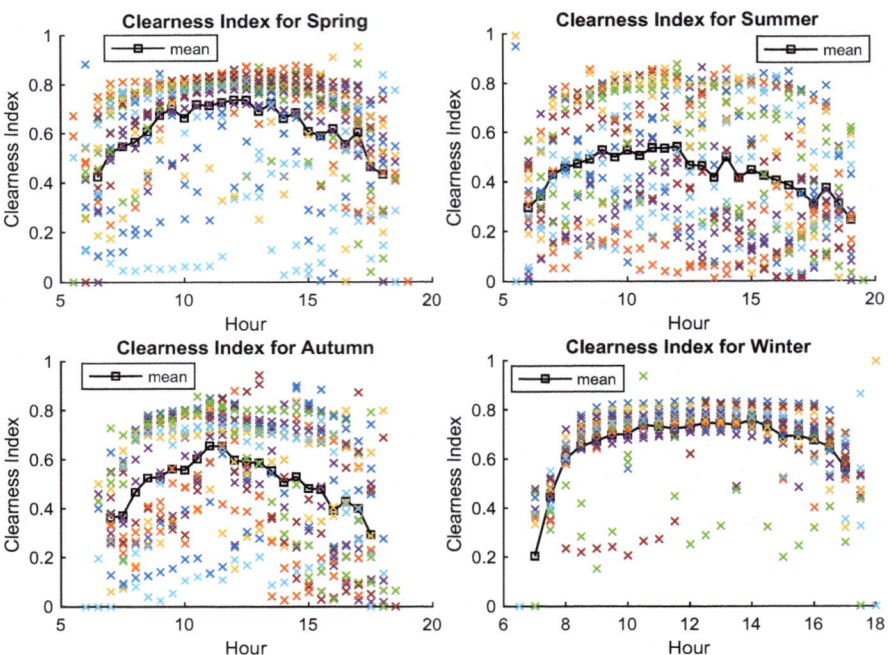

Fig. 2.7 Clearness index for the four seasons

w is the mixture weight with the constraints $w_i > 0$ and $\sum_{i=1}^{m} w_i = 1$. $g_{x;w_i,\theta_i}\left(x;,w_i;,\theta_i\right)$ is known as the component Gaussian densities. The parameter θ contains the component weights w_i, mean vectors, μ_i and the covariance matrices Σ_i. This is expressed with Eq. (2.12) [72]:

$$\theta = \left\{w_1, w_2, \ldots, w_m, \mu_1, \mu_2, \ldots, \mu_m, \Sigma_1, \Sigma_2, \ldots, \Sigma_m\right\} \tag{2.12}$$

For h_n to be the number of elements in the vector x_n, the log-likelihood function is given in Eq. (2.13) [72]:

$$L(\theta) = \sum_{n=1}^{h_n} \ln \sum_{i=1}^{m} w_i g_{x;\theta_i}\left(x_n;\theta_i\right) \tag{2.13}$$

The expectation-maximization (EM) algorithm aims to calculate the maximum likelihood estimation of the marginal likelihood in an iterative process. The process consists of two stages, the expectation step, and the maximization step.

1. Expectation step: Calculate the expected value of the log-likelihood function under the current estimate of the parameters $\theta^{(t)}$ at t iteration [72, 73].

$$B\left(\theta;\theta^{(t)}\right) = E_{X,\theta^{(t)}}\left[L\left(\theta\right)\right] \tag{2.14}$$

2. Maximization step: Find the parameter that maximizes the following quantity [72, 73]:

$$\theta^{(t+1)} = \underset{\theta}{\operatorname{argmax}}\, B\left(\theta;\theta^{(t)}\right) \tag{2.15}$$

Partition-based clustering K-Means clustering aims to classify the objects into the clusters with the nearest mean. It is an iterative algorithm and begins with choosing K initial cluster centers. The distances of all observations to each centroid are computed. The object is assigned to the cluster with the closest centroid. The new centroid locations are determined by calculating the average of the objects in each cluster. Given data with n vectors of equal lengths, $X = \{x_1, x_2, \ldots, x_n\}$, K-Means determines cluster centers for k clusters of vectors with equal lengths $V = \{v_1, v_2, \ldots, v_k\}$, by minimizing the objective function as given in Eq. (2.16) [74]:

$$\min_{V} \sum_{i=1}^{k}\sum_{j=1}^{n} D^2\left(x_j, v_i\right) \tag{2.16}$$

where D is the distance function, such as Euclidean distance (ED), Manhattan distance (MD), and dynamic time warping (DTW), etc.

The fuzzy C-Means (FCM) algorithm is an extended version of the K-Means algorithm by including the fuzzy-partition matrix. Each object can belong to more than one cluster. The iterative process is similar to K-Means. The objective function is given in Eq. (2.17) [74]:

$$\min_{V} \sum_{i=1}^{k}\sum_{j=1}^{n} \left(w_{ij}\right)^m D^2\left(x_j, v_i\right) \tag{2.17}$$

The fuzzifier m determines the level of cluster fuzziness, where $1 \le m \le \infty$. A large m results in smaller membership values. The centroid for FCM is calculated with Eq. (2.18) [74].

$$v_i = \frac{\sum_{j=1}^{n}\left(w_{ij}\right)^m x_j}{\sum_{j=1}^{n}\left(w_{ij}\right)^m}, i = 1, \ldots, k. \tag{2.18}$$

The fuzzy-partition matrix is given in Eq. (2.19) [74].

$$w_{ij} = \frac{1}{\sum_{h=1}^{k} \left(\frac{\mathrm{D}\left(x_j, v_i\right)}{\mathrm{D}\left(x_j, v_h\right)} \right)^{\frac{2}{m-1}}} \tag{2.19}$$

One of the crucial elements in partition-based cluster analysis is the function used to measure the similarity between time series. The distance metric can have a profound effect on the clustering of times series and their respective clusters.

Euclidean distance and Manhattan distance ED and MD are the most frequently used distance measures for data mining. Although both have been deployed in many time series application fields, the metric has many pitfalls for time series analysis. They can only be used for time series of equal length and are prone to noise and outliers, which are common in real-life temporal sequences especially when noise and uncertainty exist in the data source [75]. Another major issue with ED is that the metric is based on the comparison between data points at the same time interval. Time series regularly suffer transformations in the time axis although the series are in a similar shape, i.e., due to the perturbation to the solar irradiance and in the context of CI, there will be time discrepancies for sunrise and sunset in the clear-sky solar model and real-life solar irradiance data. Let x_i and v_j each be a d-dimensional vector, the ED and MD between the two vectors are presented in Eqs. (2.20) [74] and (2.21) [76], respectively.

$$ED = \sqrt{\sum_{f=1}^{d} \left(x_{if} - v_{jf} \right)^2} \tag{2.20}$$

$$MD = \sum_{f=1}^{d} \left| x_{if} - v_{jf} \right| \tag{2.21}$$

Dynamic time warping Dynamic time warping (DTW) has many features that may overcome the drawbacks of ED and MD. In essence, the objective of DTW is to find the optimal alignment between the two series by searching for the minimal path in a distance matrix that defines a mapping between them, whilst satisfying the moving restrictions during the searching process, i.e., only vertical, horizontal, and diagonal moves are allowed. This results in stretching and compressing of time series. The mapping for every pair of points in the series can be determined by distance metrics such as ED and Manhattan, etc. The distance metric used for FCM DTW is ED. The outputs of DTW are the cost matrix that denotes the cost values, i.e., the DTW distance between the two coordinates and the warping path. The main weakness of DTW is computational complexity. The algorithm for calculating the DTW for two-time series is given in [77].

The main challenge in applying DTW distance to partition-based clustering techniques is to calculate the average of a set of time series. To overcome this issue, DTW barycenter averaging (DBA) is used for DTW averaging. Unlike the traditional centroid calculation method where the mean is directly determined, DBA aims to minimize the sum of squared DTW distances between the centroid sequence and the set of sequences to be clustered. This is essentially achieved by performing two iterative procedures. The first stage is to perform DTW to the sequence to be clustered and the centroid to be refined. The associations between them are kept and will be stored in the vector or known as the association table. The second stage is to update each coordinate of the centroid with the barycenter of coordinates associated with it from the association table, by calculating the mean. The standard deviation of the centroid can also be calculated with the association table. The algorithm for DBA can be found in [78].

As previously explained, the cluster centers cannot be calculated with the traditional method in Eq. (2.18) when considering DTW as the distance function in Eq. (2.17). The cluster centers are calculated with DBA instead. To initialize the cluster centers in each cluster, the time series are assigned to the cluster having the maximum membership degree. The cluster centers will be refined in the iterative FCM optimization process. The partition matrix is calculated with Eq. (2.19), where D stands for the DTW distance.

Comparison of computational complexities   Let N, I, and d be the number of profiles, number of iterations, and dimension of profiles respectively. The K-Means computational complexity (CC) for ED or MD is approximated as $O(NKdI)$ [79]. For FCM with ED, the CC is approximated as $O(NK^2dI)$. The FCM suffers higher computation costs compared to K-Means, this is due to the need for updating the fuzzy-partition matrix in each iteration [80]. For GMM clustering, the CC is mainly associated with the EM algorithm. This is approximated as $O(INK(1 + d^2) + K)$ [75], which is higher than the partition-based clustering methods for high dimensional data [81, 82]. The covariance and mean matrix grow in the size of Kd^2 and Kd respectively. For FCM DTW, the differences in CC per iteration with respect to the standard FCM algorithm are the distance and centroid calculations. The CC of DTW is quadratic, d^2, unlike the linear computation costs for ED and MD [78]. Therefore, the total complexity for distance calculation is $O(Nd^2)$. The centroids are calculated with the DBA method and have a computation cost of $O(INd^2)$ [78]. The resulting FCM DTW computation cost is, therefore, $O\{Nd^2I(K^2Nd^2)\}$, which simplifies to $O(IN^2K^2d^4)$. FCM DTW has the highest CC compared to other clustering methods and future research should look for more efficient methods in using DTW for FCM clustering.

2.4.5   *Cluster Analysis and Discussions*

Cluster analysis aims to determine the smallest number of clusters for the daily CI profiles while minimizing the intra-cluster distance. To achieve this, it is required to minimize the total distance between each clustered profile with respect to their centroids. Five clustering methods described previously are studied and compared. These are FCM with Euclidean distance (FCM ED), FCM with Dynamic Time Warping (FCM DTW), K-Means with Euclidean distance (K-Means ED), K-Means with Manhattan distance (K-Means MD), and GMM. The cluster number to be evaluated is from 2 to 15. The total intra-cluster distance is calculated with Eq. (2.22) and is used as an indicator of merit.

$$D_{\text{Total}} = \sum_{i=2}^{k}\sum_{j=1}^{n} D^2\left(x_j, v_i\right) \qquad (2.22)$$

Due to the clustering algorithms contain random variables, 50 repeated tests were made and the minimum results are kept. The total intra-cluster distance with respect to the different number of clusters for the four seasons is provided in Fig. 2.8. Compared with different clustering methods, there is a huge difference in total intra-cluster distance for Winter and Summer cases. This can be explained by the level of uncertainty associated with the season. There is significantly more fluctuation in the Summer season, hence, the total intra-cluster distance will be increased. It is learned that FCM DTW has a significantly smaller total intra-cluster distance for all cases. FCM ED performs marginally better than K-Means ED in most cases, given the fact that it provides better performance in analyzing uncertainty, i.e., clusters with various sizes and shapes due to the fuzzifier used for calculation, where the profiles may belong to more than one cluster. It is worth mentioning that the intra-cluster distance for K-Means MD should not be compared with other intra-cluster distances, as ED and MD are different metrics.

ED is used to compute the intra-clusters distance for GMM. It can be seen that the total intra-cluster distance for GMM is marginally higher than K-Means methods. K-Means aims to minimize the intra-cluster distance with ED, while GMM aims at maximizing the maximum-likelihood via EM algorithm, which does not consider minimizing the distance.

Fuzzy decision-making One of the major challenges in using cluster analysis is to determine the optimal number of clusters. Traditionally, this is achieved by using the criteria such as Silhouette index, Dunn's index, and Calinski–Harabasz index. In essence, these indices aim to determine some form of relationship between the within-cluster cohesion and the cluster separation in order to evaluate the clusters' validity. The details of these indices and the distance suitability are provided in [76]. It is learned that the criteria are deemed ineffective for the data set with a significant amount of noise or uncertainty [77]. Also, the optimal number of clusters is problem dependent and the mentioned criteria rarely provide the same results. A technique in determining the optimal number of clusters is provided in this research, by first evaluating the number of clusters that provides the best trade-off for minimizing the total intra-cluster distance. Consequently, if the resultant centroids have similar characteristics, i.e., mean and standard deviation, the similar clusters will be grouped together with the new centroid calculated. Given no prior knowledge for the selection of candidate partitions in Pareto set, an un-weighted fuzzy logic decision-making strategy [83] is employed to yield the best trade-off solution.

It is assumed that the preferences for minimizing the number of clusters and total intra-cluster distance are unbiased. Fuzzy logic decision-making is formulated as follows. m is equal to 15 and stands for the number of non-dominated solutions and n is the number of objective functions. In this case, n is equal to 2 due to the objective is to minimize the number of centroids and the total intra-cluster distance. The fuzzy membership is defined below:

$$\mu_i(j) = \frac{f_i - f_i^{\min}}{f_i^{\max} - f_i^{\min}}, i = 1, 2 \tag{2.23}$$

f_i stands for the solutions in the ith objective function. The normalized membership for each solution is expressed as below:

$$\mu(j) = \frac{\sum_{i=1}^{n} \mu_i(j)}{\sum_{j=2}^{m} \sum_{i=1}^{n} \mu_i(j)} \tag{2.24}$$

The most satisfactory solution in this case is selected with the minimum fuzzy membership value [83]. The normalized fuzzy memberships for the four seasons, calculated with the intra-cluster distances given in Fig. 2.8 are presented in Fig. 2.9. The clustering methods show a similar trend for the normalized fuzzy membership with respect to cluster number. This explains that the optimal number of clusters is similar for the different clustering techniques.

Figure 2.10 shows the centroids for the Winter case with different clustering techniques. The centroids for FCM ED and K-Means ED in the Winter case show a similar shape. Recall the results in Fig. 2.8, the total distance for the two approaches are very similar. The number of representations for clear days are different. It is one centroid for GMM, two centroids for FCM ED and K-Means ED, and three centroids for K-Means MD and FCM DTW. A method is required to minimize the number of centroids with similar shapes and magnitude to reduce redundancy.

According to Fig. 2.11, the centroids in the Summer case show that the clustering performance of FCM ED, K-Means ED, K-Means MD, and GMM are similar. It is worth noting that the clustering problem for Summer cases is significantly more challenging than Winter cases. This is due to the fact that CI in Summer has more fluctuations.

Reduction of clusters with centroid evaluation The K-Means related clustering techniques consider solely the intra-cluster compactness, i.e., the distances between the objects and the centroids [84]. The inter-cluster separation, i.e., the distances between the centroids are not well considered during the clustering process.

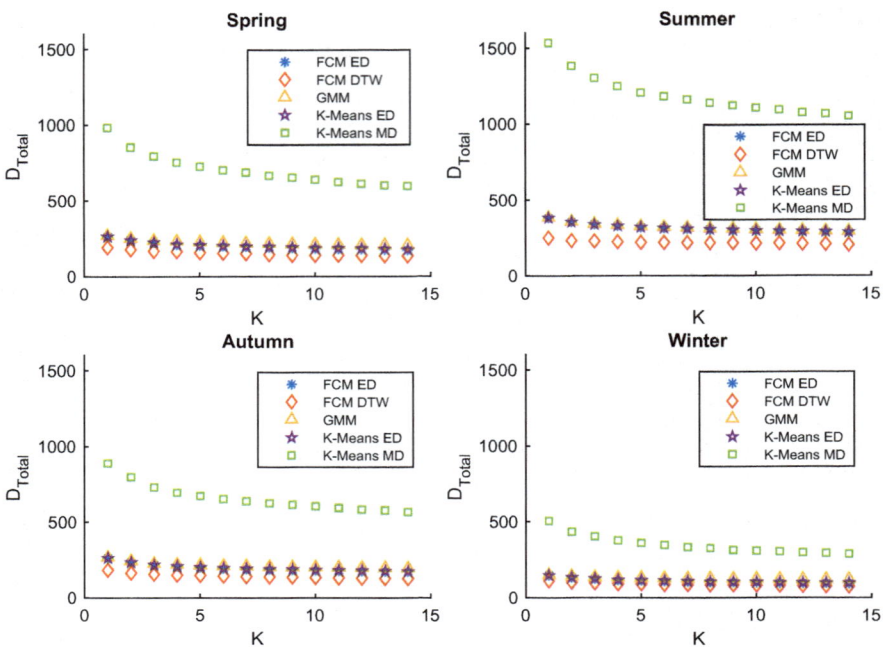

Fig. 2.8 Total intra-cluster distance for different number of clusters

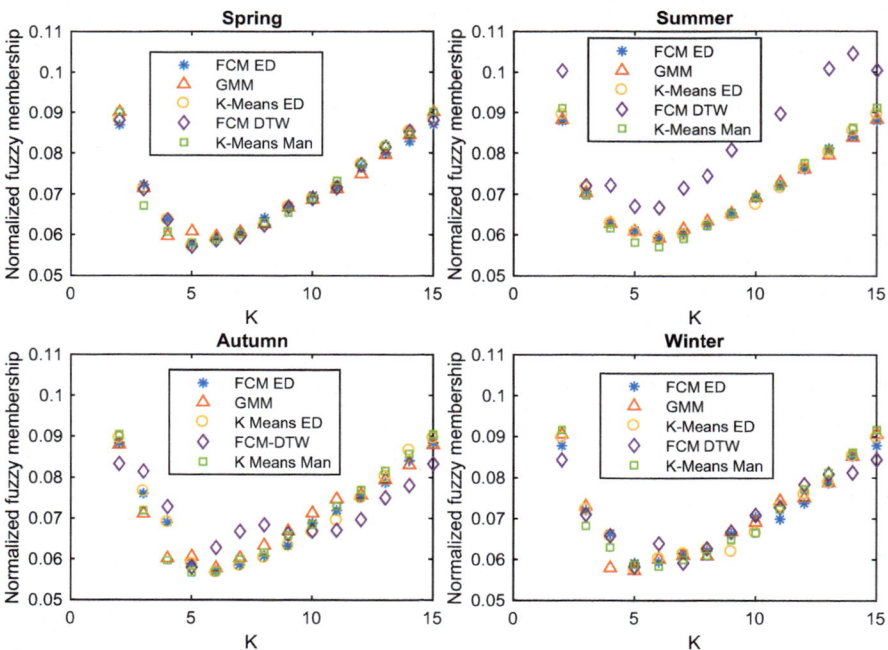

Fig. 2.9 Fuzzy decision-making for both case studies

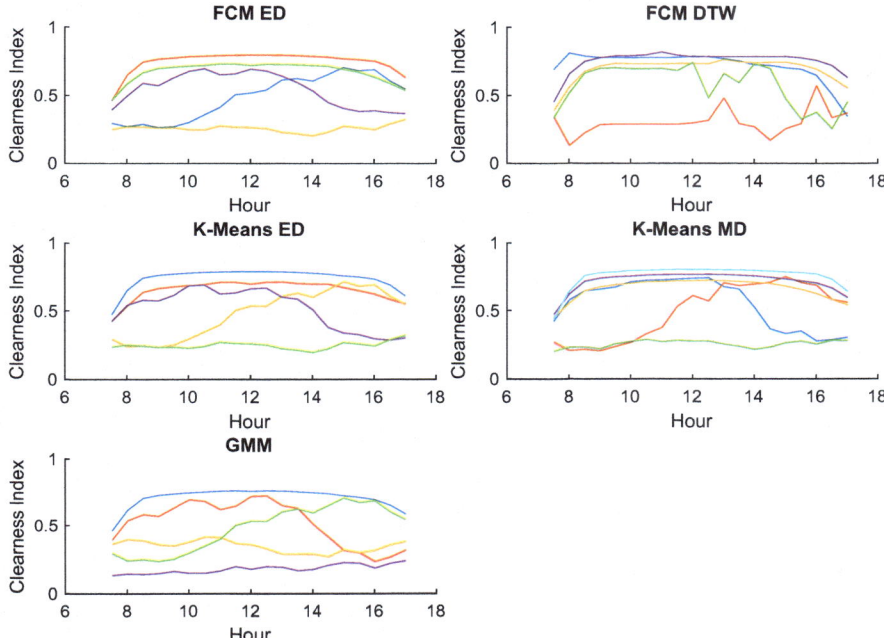

Fig. 2.10 Centroids for different clustering techniques in the Winter case

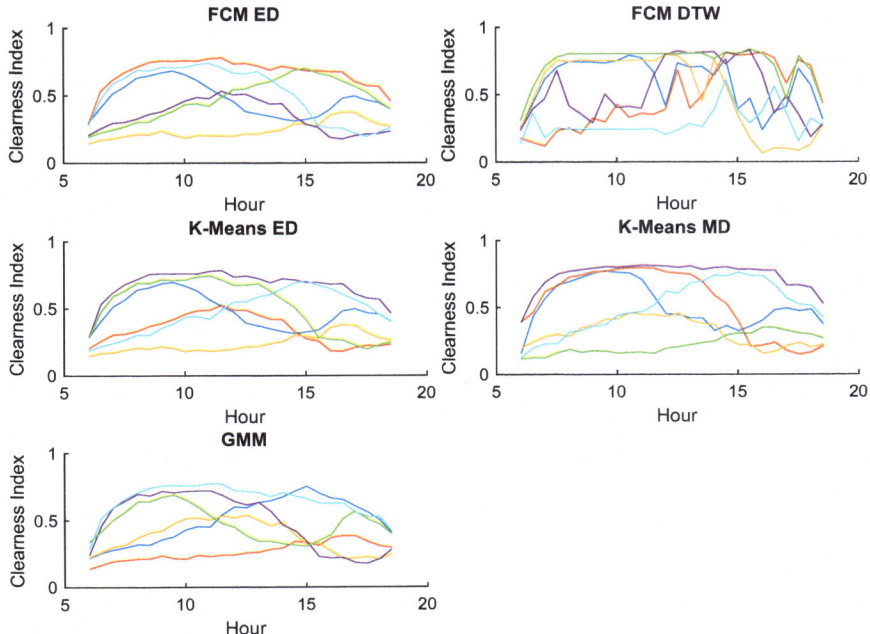

Fig. 2.11 Centroids for different clustering techniques in the Summer case

Table 2.3 Algorithm for reduction of redundant centroids

Input: $C = \{c_1, c_2, \ldots c_k\}$: the set of all centroids	
$card$: the cardinality of each cluster	
Output: C: the reduced set of centroids	
1.	Calculate the similarity matrix M
2.	$s = \{1, 2, \ldots k\}$
3.	Find the all-zero rows of M
	$s' \leftarrow$ the indices of all-zero rows
	$C = \{c_i \mid i \in s'\}$
4.	$s = s - s'$
5.	**while** s is non-empty

$$s' = \left\{ s_i \mid i = 1 \text{ or } M_{s_1, i} = 1 \right\}$$

$$C\_new = \frac{\sum_{i \in s'} c_i \cdot card_i}{\sum_{i \in s'} card_i}$$

$C = C' \; C\_new$	
$s = s - s'$	
end	

Redundant or similar clusters and centroids can be eliminated by evaluating the centroid's standard deviation and mean. In the first step, a mean similarity matrix and standard deviation similarity matrix, both with size K by K are constructed to determine if the clusters are similar. If the element in the matrix falls below a predefined threshold, in this case, 0.1 for both variables, then the element will be set to 1. This signifies that the two clusters have similar characteristics. The corresponding elements in the two matrices with binary numbers will be multiplied together to give the similarity matrix, M. Once the similar clusters are grouped together, the new centroid is calculated by calculating the average of the centroids by considering the weights with the number of profiles in the original cluster. The procedure is presented in Table 2.3.

Figure 2.12 shows that the centroids for FCM DTW Winter case can be reduced from the original five centroids as shown in Fig. 2.10 to three centroids. It also presents the daily CI profiles with their respective centroids for FCM DTW in the Winter case. The percentage day covered for clear days in Cluster 3 is 78%, which is the highest compared to other seasons. The black, blue, and red lines in Figs. 2.12 and 2.13 give the centroids, the centroids with plus one and minus one standard deviation respectively, calculated with the association table in DBA.

Figure 2.13 presents the daily CI profiles with respect to their centroids for FCM DTW Summer case. The clear days are presented in Cluster 5, which takes into 19.10% of the days of the season. Cluster 3 shows that the perturbation takes place

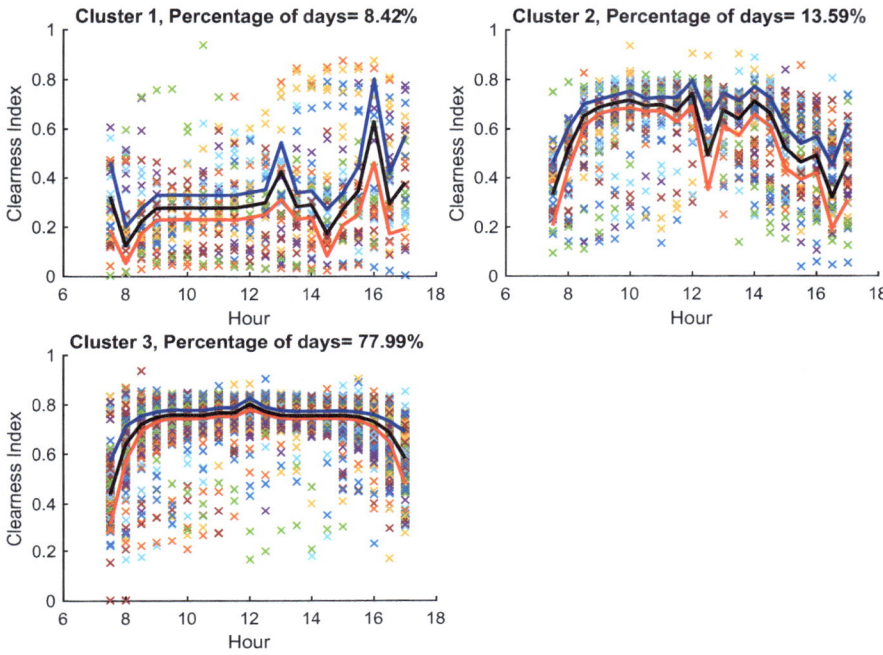

Fig. 2.12 CI Profiles in respective centroids for FCM DTW Winter case

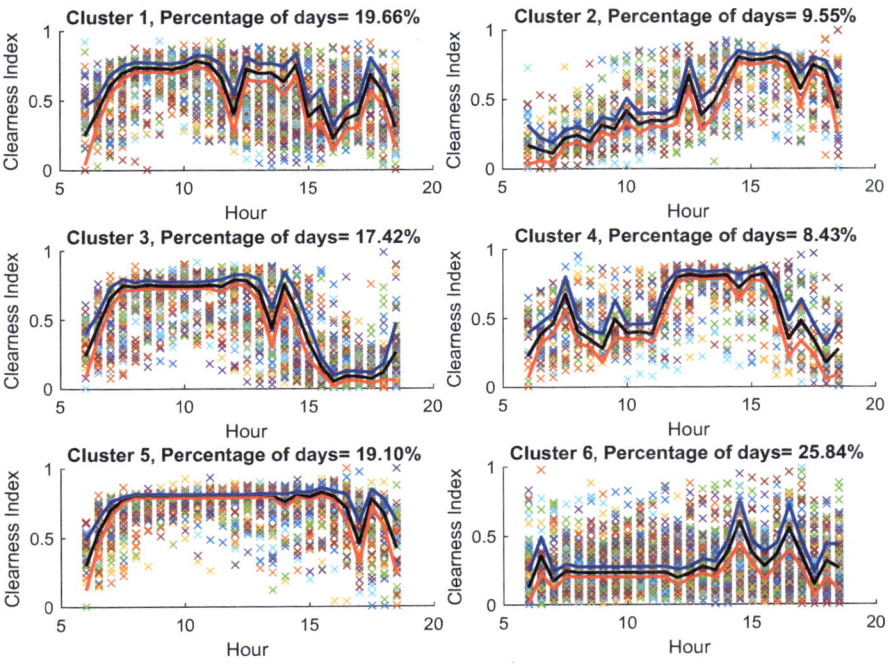

Fig. 2.13 CI Profiles in respective centroids for FCM DTW Summer case

Table 2.4 Optimal number of clusters

	Spring	Summer	Autumn	Winter
FCM ED	5	6	6	4
FCM DTW	4	6	5	3
GMM	6	6	6	5
K-Means ED	5	6	6	4
K-Means MD	5	6	5	4

during the late afternoon. Cluster 6 presents the CI profile where the CI is generally low for the whole day. The clustering results display an interesting pattern and could be understood and quantified. The optimal number of clusters for the four seasons with the five clustering techniques is provided in Table 2.4.

2.4.6   Case Study: Sizing of Stand-Alone PV and Storage System with Anaerobic Digestion Biogas Power Plants

In contrast with using the actual real-life daily solar irradiance profiles for system sizing in [85], this sub-section uses the daily solar irradiance profiles constructed from the cluster centroids with CIs. The daily clear-sky solar irradiance profile for Autumn, Spring, Winter, and Summer are calculated with the equinoxes (20th March and 23rd Sept.), the Winter solstice (21st June) and the Summer solstice (21st Dec.) respectively. The equation for the calculation of constructed solar irradiance is given in Eq. (2.25):

$$I_{\text{construct}}\left(t\right) = \text{CI}\left(t\right) * I_{\text{model}}\left(t\right) \tag{2.25}$$

The constructed solar irradiance profiles for FCM DTW Winter and Summer cases are presented in Figs. 2.14 and 2.15 respectively. To consider the dispersion of the clustered data, the plus one and minus one standard deviation of the centroids are included for sizing purposes.

Sizing of solar panels The required PV solar panel areas to meet the energy deficit of the solar PV hybrid energy system are determined with Particle Swarm Optimization with Interior Point Method [85]. Figure 2.16 presents the panel area results with the irradiance profiles developed from five different clustering techniques. The population of results in a form of boxplot for PV panel sizing for the four seasons is represented in Fig. 2.16 and the PV farm power capacities are given in Fig. 2.17. The calculation of PV capacity from the panel area can be referred to [85]. In Fig. 2.17, it can be realized that FCM DTW and GMM need a required PV capacity of 5 MW, whereas FCM ED and K-Means ED need a 4 MW PV capacity, and finally K-Means MD needs a PV capacity of 3 MW. The energy balance of generation and demand are highly related to the shape and arbitrariness of the solar

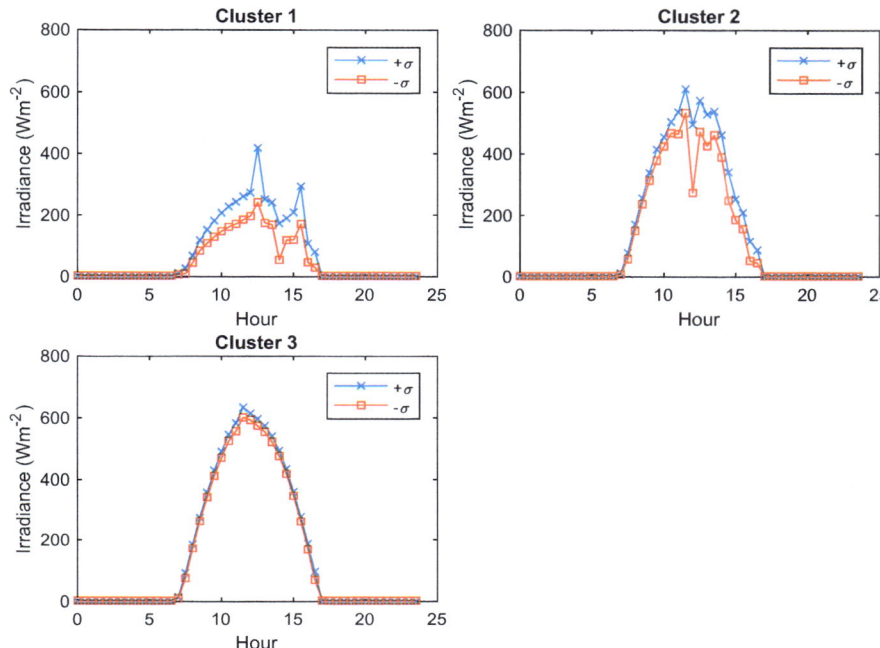

Fig. 2.14 Constructed irradiance from centroids in FCM DTW Winter case

irradiance profile. These are better captured by GMM and FCM DTW clustering methods, which are reflected in the PV panel sizing results in Fig. 2.16 and the centroids in Figs. 2.10 and 2.11.

Sizing of storage Turning to the sizing of storage, the aim is to determine the maximum energy deficit of the system with the centroid profiles. For a 5 MW solar farm, the maximum energy deficit with GMM results is 4.14 MWh and occurs in Summer. The maximum energy deficit with FCM DTW results is 3.49 MWh and also occurs in Summer, at Cluster 6 in Fig. 2.15 with the minus one standard deviation. To understand the implications of the energy deficit results, Fig. 2.18 shows the energy deficit computed with 4 years of real-life solar irradiance data.

The real-life solar irradiance profile study shows that the energy deficit can reach 5 MWh. It is also worth mentioning that the total number of clusters for FCM DTW is less than GMM. The total number of sizing cases, i.e., the total count is 1457. The number of additional cases where 4.14 MWh meets the energy demand in contrast to 3.49 MWh is 16. At 3.49 MWh, 1436 cases of energy deficit are covered. The difference in energy storage capacity between FCM DTW and GMM is $(4.14 - 3.49)/3.49 = 18.62\%$. By increasing the storage capacity from 3.49 to 4.14 MWh, the additional cases of energy deficit covered is $16/1436 = 1.11\%$. This concludes that the system can meet an additional 1.11% cases of energy deficit with

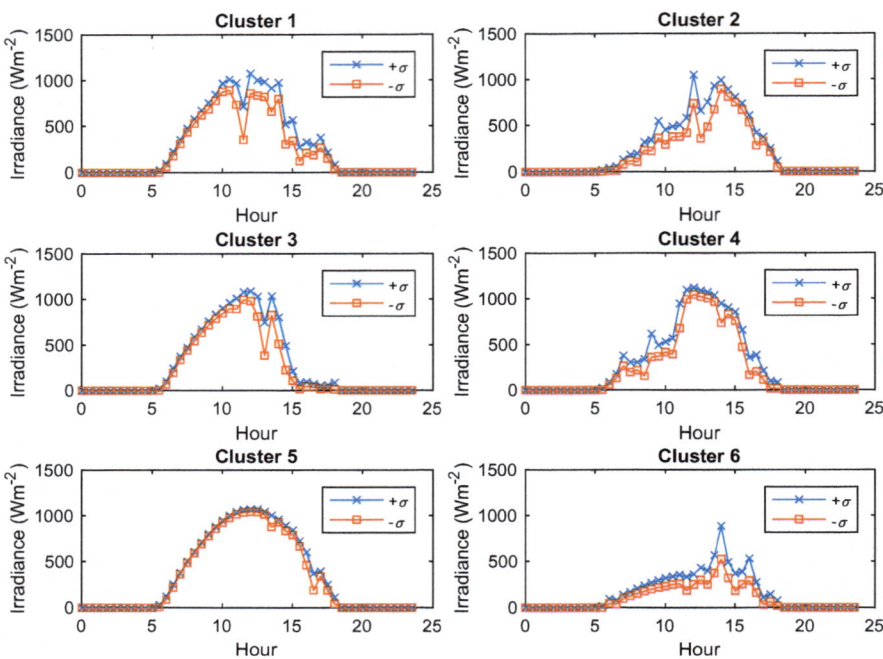

Fig. 2.15 Constructed irradiance from centroids in FCM DTW Summer case

an additional 0.65 MWh storage capacity. This may not be an economical solution and the issue with energy deficit can be overcome by optimal scheduling or demand-side management, which will be future work.

2.4.7   Conclusion and Future Work

This sub-section presents feature extraction for daily clearness index profiles with five different cluster analysis techniques. An optimal sizing case study for a PV system with energy storage and anaerobic digestion biogas power plants is used to compare the clustering results for PV system planning. As different to the 1457 daily irradiance profiles used in [85] for the system sizing, the data set can be represented with 36 and 46 profiles, with Fuzzy C-Means (FCM) dynamic time warping (DTW) and Gaussian mixture model clustering respectively. It is worth mentioning that the optimal number of clusters is problem dependent and may vary depending on the application.

For future work, it is possible to include an extra-temporal constraint in DTW, by limiting the number of vertical or horizontal steps that the path can take consecutively. This adjustment avoids the matching of points that are very far from each other in time and, in addition, it reduces the computation cost. The fuzzifier

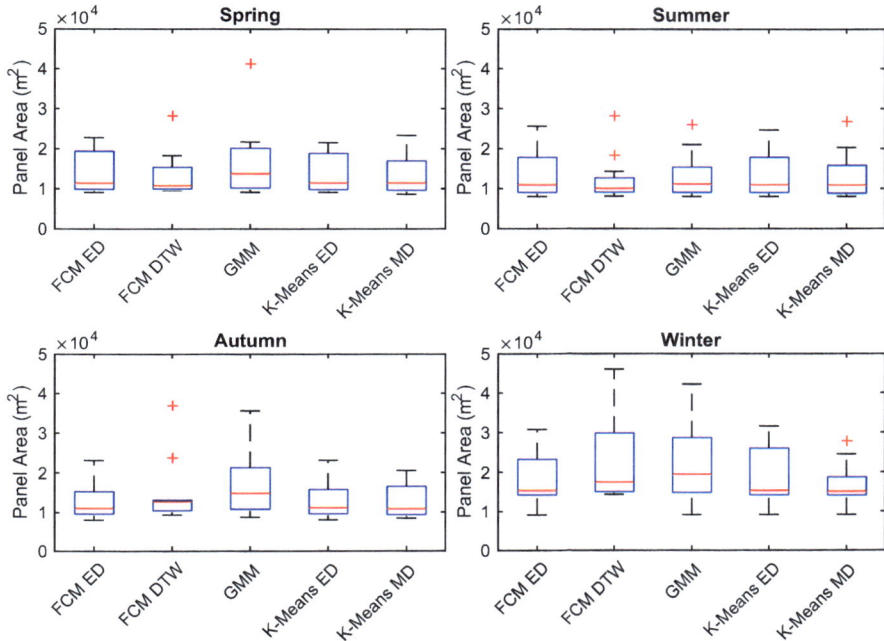

Fig. 2.16   Optimization results for PV panel sizing.

parameter for FCM can be further explored. The centroids can be used for other planning and operation purposes for PV systems, such as optimal placement of phasor measure unit and evaluation of scheduling algorithms.

2.5   Robust Correlation Framework

Correlation analysis is one of the fundamental mathematical tools for identifying dependence between classes. However, the accuracy of the analysis could be jeopardized due to variance error in the data set. This sub-section provides a mathematical analysis of the impact of imbalanced data concerning Pearson Product Moment Correlation (PPMC) analysis. To alleviate this issue, the novel framework Robust Correlation Analysis Framework (RCAF) is proposed to improve the correlation analysis accuracy. A review of the issues due to imbalanced data and data uncertainty in machine learning is given. The proposed framework is tested with an in-depth analysis of real-life solar irradiance and weather condition data from Johannesburg, South Africa. Additionally, comparisons of correlation analysis with prominent sampling techniques, i.e., Synthetic Minority Over-Sampling Technique (SMOTE), and Adaptive Synthetic (ADASYN) sampling techniques are conducted. Finally, K-Means and Wards Agglomerative hierarchical clustering is performed to

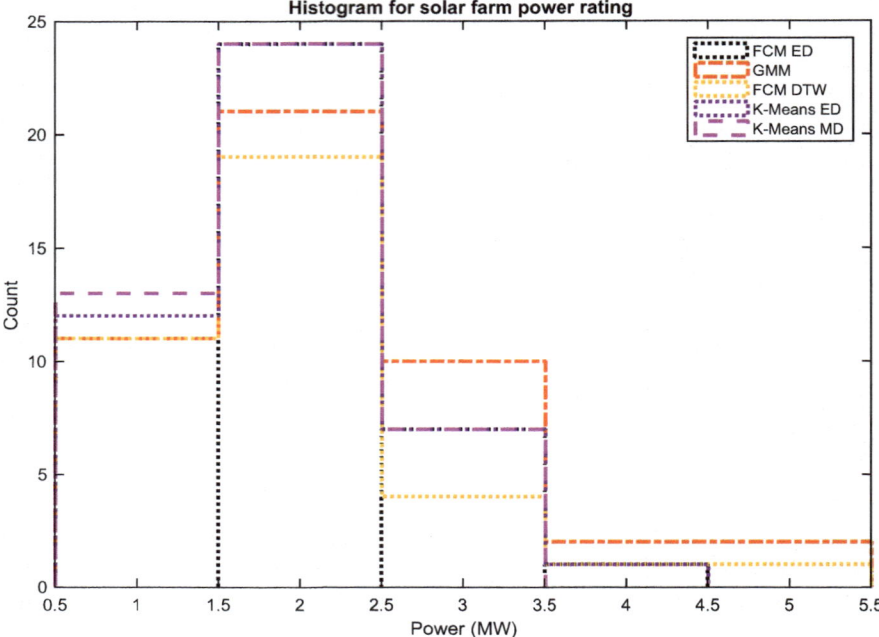

Fig. 2.17 PV farm rated capacity with different clustering techniques

study the correlation results. Compared to the traditional PPMC, RCAF can reduce the standard deviation of the correlation coefficient under imbalanced data in the range of 32.5–93.02%.

With the exponential increase in the amount of data introduced by an increasing number of physical devices, the large-scale advent of incomplete and uncertain data is inevitable, such as those from smart grids [86, 87]. For sparse data, the number of data points is inadequate for making a reliable judgment. This has been an issue for the successful delivery of megaprojects [88]. In machine learning and data mining applications, redundant data can seriously deteriorate the reliability of models trained from the data.

Data uncertainty is a phenomenon in which each data point is not deterministic but subject to some error distributions and randomness. This is introduced by noise and can be attributed to inaccurate data readings and collections. For example, data produced from GPS equipment are of uncertain nature. The data precision is constrained by the technical limitations of the GPS device. Hence, there is a need to include the mean value and variance in the sampling location to indicate the expected error. The major opportunities and challenges for learning from imbalanced data are also highlighted in [89]. The number of publications on imbalanced learning has increased by 20 times from 1997 to 2007. Imbalanced data can be classified into two categories, namely, intrinsic and extrinsic imbalanced. Intrinsic imbalance is due to the nature of the data space, whereas extrinsic imbalance is not. Given a data set

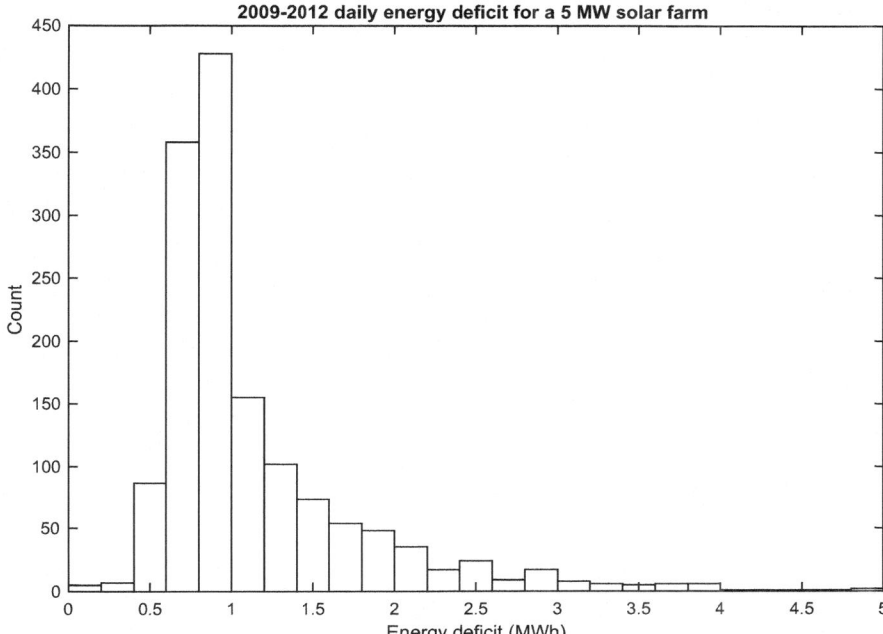

Fig. 2.18 Histogram for system energy deficit for 4 years of daily case study

sampled from a continuous data stream of balanced data with respect to a specific period; if the transmission has irregular disturbances that do not allow the data to be transmitted during this period, the missing data in the data set will result in an extrinsic imbalanced situation obtained from balanced data space. An example of intrinsic imbalanced could be due to the difference in the number of samples of different weather conditions, i.e., in general, the "Clear" weather condition has the most occurrences throughout the year, whereas "Snow" may only have a few occurrences.

There is a growth of interest in class imbalanced problems recently due to the classification difficulty caused by the imbalanced class distributions [90, 91]. To solve this problem, several ensemble methods have been proposed to handle such imbalances. Class imbalances degrade the performance of the derived classifier and the effectiveness of selections to enhance classifier performance [92].

This sub-section proposes and validates a new framework for the impact of imbalanced data on correlation analysis. The impact of imbalanced data is described using a mathematical formulation. Additionally, RCAF is proposed for correlation analysis with the aim of reducing the negative effects due to an imbalanced ratio. This will be investigated with a theoretical and real-life case study.

Section 2.5.1 provides a literature review on the imbalanced data problem, followed by the correlation analysis of imbalanced data. Section 2.5.2 provides an overview of the critical features and the impacts on correlation analysis. Simulations will be conducted to support the findings. Section 2.5.3 proposes a new framework for correlation analysis. Section 2.6 provides a real-life case study, based on solar irradiance and weather conditions, to evaluate the new framework. Different imbalanced data sampling techniques will be used to compare the correlation analysis performance. Cluster analysis of weather conditions will be given to understand the implications of the correlation results.

2.5.1 Correlation Analysis for Imbalanced Data Problems

Imbalanced data refers to unequal variable sampling values in a data set. For example, 90% of sampling data can be in the majority class, with only 10% of the sampling data in the minority class. Therefore, the imbalanced ratio is 9:1. Imbalanced data appears in many research areas. As mentioned in [93], when TV recommender systems perform well, the number of interactions for users to express positive feedback is anticipated to be greater than the number of negative interactions on the recommended content. This is known as class imbalanced. The misclassification of the unwanted content can be recognized by TV viewers easily, therefore, system performance could decrease.

Commonly, modifying imbalanced data sets to provide a balanced distribution is carried out using sampling methods [90, 94, 95]. From a broader perspective, over-sampling and under-sampling techniques seem to be functionally equivalent, since they both can provide the same proportion of balance by changing the size of the original data set. In practice, each technique introduces challenges that can affect learning. The major issue with under-sampling is straightforward, classifiers will miss important information with respect to the majority class, by removing examples from the majority class [96]. The issues regarding over-sampling are less straightforward. Since over-sampling adds replicated data to the original data set, multiple instances of certain samples become "tied," resulting in overfitting. As proposed in [97], one solution to the over-sampling problem is to add a small amount of random noise to the predictor so the replicates are not duplicated, which can minimize overfitting. This jittering adds undesirable noise to the data set but the negative impact of imbalanced data sets has been shown to be reduced. Under-sampling is a favoured technique for class-imbalanced problems; it is very efficient since only a subset of the majority class is used. The main problem with this technique is that many majority class examples are ignored.

Class imbalanced learning is employed to resolve supervised learning problems in which some classes have significantly more samples than others [91]. The study of multiclass imbalanced problems and the dynamic sampling method (DyS) for multilayer perceptron is provided in [98]. The authors claim that the DyS method could outperform the pre-sample methods and active learning methods for most data sets. However, a theoretical foundation is necessary to explain the reason a simple method such as DyS could perform so well in practice.

Support Vector Machine (SVM) is a popular machine learning technique that works effectively with balanced data sets [99, 100]. However, with imbalanced data sets, suboptimal classification models are produced with SVMs. Currently, most research efforts in imbalanced learning focus on specific algorithms and/or case studies. Many researchers use machine learning methods such as support vector machines [99], cluster analysis [101], decision tree learning [97, 102], neural networks [103–105], etc., with a mixture of over-sampling and under-sampling techniques to overcome the imbalanced data problems [95, 106]. A novel machine learning approach to assess the quality of sensor data using an ensemble classification framework is presented in [107], in which a cluster-oriented sampling approach is used to overcome the imbalance issue.

The issues of class imbalanced learning methods and how they can benefit software defect prediction are given in [108]. Different categories of class imbalanced learning techniques, including resampling, threshold moving, and ensemble algorithms, have been studied for this purpose. Medical data are typically composed of "normal" samples with only a small proportion of "abnormal" cases, which leads to class imbalanced problems [94]. Constructing a learning model with all the data in class imbalanced problems will normally result in a learning bias toward the majority class.

Imbalanced data can influence the feature selection results. As mentioned in [109], traditional feature selection techniques assume the testing and training data sets follow the same data distribution. This may decrease the performance of the classifier for the application of adversarial attacks in cybersecurity. For real-life applications, the distribution of different data sets and variables may be significantly different and should be thoroughly studied. Feature selection based on methods such as feature similarity measure [110], harmony search [111, 112], hybrid genetic algorithms [113], dependency margin [114], cluster analysis [115] has been developed. The methods have contributed to the quality enhancement of feature selection. However, the fundamental issues of the uncertainty and imbalanced ratio in data sets have not been studied.

Many correlation analyses have been conducted on imbalanced data sets. For example, community question answering (CQA) is a platform for information seeking and sharing. In CQA websites, participants can ask and answer questions. Feedback can be provided in the manner of voting or commenting. Reference [116] proposed an early detection method for high-quality CQA questions/answers. Questions of significant importance that would be widely recognized by the participants can be identified. Additionally, helpful answers that would attain a large amount of positive feedback from participants can be discovered. The correlation of questions and answers was performed with the Pearson R correlation to test the dependency of the voting score. The classification accuracy with imbalanced data, i.e., the ratio between the number of data for positive and negative feedbacks has not been addressed.

Gamma coefficient is a well-known rank correlation measure that is frequently used to quantify the strength of dependency between two variables in the ordinal scale [117]. To increase the robustness of this measure in data with noise, Ruiz and

Hüllermeier [117] studied the generalization of the gamma coefficient based on fuzzy order relations. The fuzzy gamma is advantageous in the presence of noisy data. However, the authors did not consider the imbalanced data issue for correlation analysis.

In clinical studies, the linear correlation coefficient is frequently used to quantify the dependency between two variables, e.g., weight and height. The correlation can indicate if a strong dependency exists. However, in practice, clinical data consists of a latent variable with the addition of an inevitable measurement error component, which affects the reproducibility of the test. The correlation will be less than one even if the underlying physical variables are perfectly correlated. Francis et al. [118] studied the reduction in correlation due to limited reproducibility. The implications of experimental design and interpretation were also discussed. It is confirmed that with large measurement errors, the measured correlation for perfectly correlated variables cannot be equal to one but must be less than one [118]. Francis et al. [118] described a method which allows this effect to be quantified once the reproducibility of the individual measurements is known. However, the sub-section has not resolved the correlation inaccuracy problem and only provides an indication of the effect of noise on the correlation in an imbalanced data set. The sub-section concludes that the designers of experiments can relieve the problem of attenuation of correlation in two ways. First, the random component of the error should be minimized, with the aim of improving reproducibility. Technical advances may allow this to occur, but relying on them is not always practical. Random measurement error can also be attenuated statistically but this requires care and logical judgement. Note that some variance errors in the data are inevitable, such as solar irradiance where unexpected phenomenon such as birds flying cannot be avoided.

2.5.2   Impact of Imbalanced Ratio and Uncertainty on Correlation Analysis

Classes exist in various machine learning models and can be in the form of dichotomous variables. The features can be represented by binary classification, i.e., 0 or 1. For example, different weather conditions for solar irradiance prediction can be classified (0 for "Clear" and 1 for "Rain").

In statistical analysis, a dependency is defined as the degree of the statistical relationship between two sets of data or variables. Dependency can be calculated and represented by correlation analysis. The most commonly used formula is parametric and known as the Pearson Product Moment Correlation (PPMC) coefficient. By definition, the PPMC coefficient has a range from the perfect negative correlation of negative 1.0 to the perfect positive correlation of positive 1.0, with 0 representing no correlation [110].

The following problem is used to describe this research issue.

Assumption: Given two variables X and Y, where $X = \{x_a, x_b\}$, $Y \in \mathbb{R}_0^+$. *In the obtained sampling data set, the number of samples in* x_b *is* n_a *and the number of samples in* x_b *is* n_b, *with* $n_a + n_b = N$. *The noise, i.e., sampling error, occurs in Y. The relationship between each value of* $Y(y_i)$ *and each value of* $X(x_i)$ *is* $y_i = f(x_i) + \mathrm{Err}_i$, $i = \{a, b\}$. *Each noise* Err_i *follows a certain distribution K with the mean error* μ_{me}. *The square of noise error* Err_i^2 *follows the distribution L with mean square error* μ_{mse}.

Figure 2.19 presents the PPMC correlation with a variable, i.e., weather being dichotomous. The regression line depicts a negative correlation between Clearness Index (CI) and the two weather conditions. This means the weather transition from "Clear" to "Mostly Cloudy" will reduce the number of solar resources received.

The PPMC coefficient is given in Eq. (2.26):

$$
\begin{cases}
\rho_{XY} = \dfrac{A - B}{C \times D} \\[2ex]
A = \left(n_a + n_b\right) \displaystyle\sum_{i=1}^{n_a+n_b} x_i y_i \\[2ex]
B = \displaystyle\sum_{i=1}^{n_a+n_b} x_i \cdot \sum_{i=1}^{n_a+n_b} y_i \\[2ex]
C = \sqrt{\left(n_a + n_b\right) \displaystyle\sum_{i=1}^{n_a+n_b} x_i^2 - \left(\sum_{i=1}^{n_a+n_b} x_i\right)^2} \\[2ex]
D = \sqrt{\left(n_a + n_b\right) \displaystyle\sum_{i=1}^{n_a+n_b} y_i^2 - \left(\sum_{i=1}^{n_a+n_b} y_i\right)^2}
\end{cases}
\tag{2.26}
$$

For C to become zero, possible factors include $n_a + n_b = 0$ and all x are zero. Based on Fig. 2.19, if there is no data, i.e., $n_a + n_b$ and the sample size is zero, it is impossible to conduct the correlation. All x equal to zero signifies there is no value in the variable. Similarly, for D to become zero, possible factors include $n_a + n_b = 0$ and all y are zero. The average value of the sampling set is equal to the expectation of the distribution. Equation (2.27) depicts this relationship while Eqs. (2.28) and (2.29) are true.

$$
\begin{cases}
\mu_{me} = \dfrac{\displaystyle\sum_{i=1}^{N} \mathrm{Err}_i}{N} \\[3ex]
\mu_{mse} = \dfrac{\displaystyle\sum_{i=1}^{N} \mathrm{Err}_i^2}{N}
\end{cases}
\tag{2.27}
$$

$$\frac{\sum_{i=1}^{n_a} \mathrm{Err}_i}{n_a} = \frac{\sum_{i=1}^{n_b} \mathrm{Err}_i}{n_b} \tag{2.28}$$

$$\frac{\sum_{i=1}^{n_a} \mathrm{Err}_i^2}{n_a} = \frac{\sum_{i=1}^{n_b} \mathrm{Err}_i^2}{n_b} \tag{2.29}$$

By considering $y_i = f(x_i) + \mathrm{Err}_i$ in Eq. (2.26), further expressions are presented in Eq. (2.30).

$$\begin{cases} A - B = n_a n_b \left(x_a - x_b \right) \left[f\left(x_a \right) - f\left(x_b \right) \right] \\ C = \sqrt{n_a n_b \left(x_a - x_b \right)^2} \\ D = \sqrt{n_a n_b \left[f\left(x_a \right) - f\left(x_b \right) \right]^2 + \left(n_a + n_b \right)^2 \cdot \left(\mu_{mse} - \mu_{me}^2 \right)} \end{cases} \tag{2.30}$$

By considering $n_b = \alpha * n_a$, where α is the number ratio between the value x_a and value x_b, Eq. (2.30) can be transformed into Eq. (2.31).

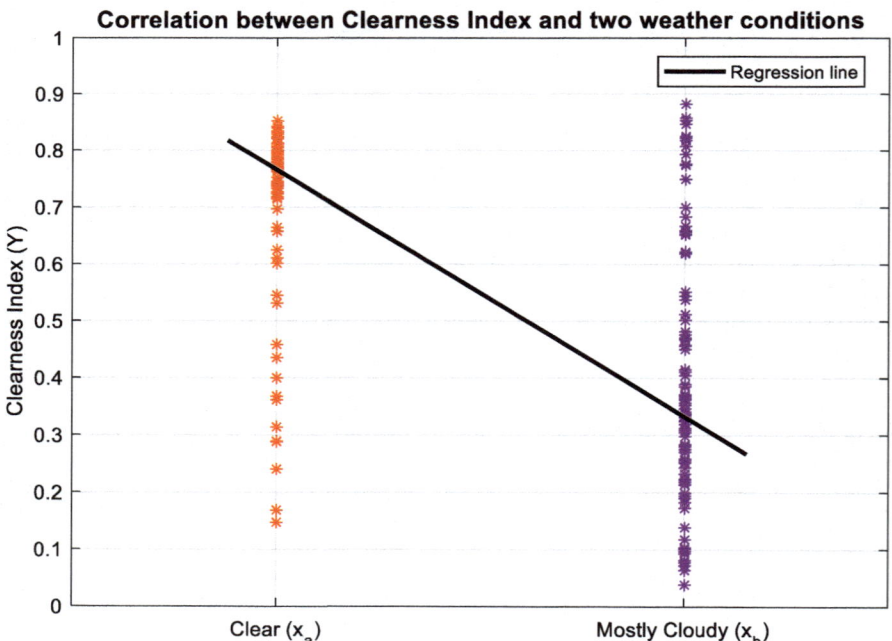

Fig. 2.19 Correlation analysis with a dichotomous variable

$$
\begin{cases}
\rho_{XY} = \dfrac{A-B}{C \times D} \\[4mm]
= \dfrac{x_a - x_b}{\left| x_a - x_b \right|} \cdot \dfrac{f(x_a) - f(x_b)}{\left| f(x_a) - f(x_b) \right|} \cdot R
\end{cases}
\tag{2.31}
$$

$$
R = \dfrac{1}{\sqrt{1 + \dfrac{\mu_{\text{mse}} - \mu_{\text{me}}^2}{\left[f(x_a) - f(x_b) \right]^2} \cdot \left(\dfrac{1}{\alpha} + \alpha + 2 \right)}}
$$

If $x_a \neq x_b$ and $f(x_a) \neq f(x_b)$, the type of correlation can be expressed by Eq. (2.32).

$$
\rho_{XY} =
\begin{cases}
R, \left(x_a < x_b, f(x_a) < f(x_b) \right) \\[2mm]
-R, \left(x_a < x_b, f(x_a) > f(x_b) \right)
\end{cases}
\tag{2.32}
$$

Equation (2.31) shows the correlation may not be $+1/-1$ given there is an increasing/decreasing linear relationship between X and Y. It is also related to the Momentum Ratio R. For the case $f(x_a) = f(x_b)$, based on Fig. 2.19, this means the "actual" (excluding error variance) CI for "Clear" is the same as the actual CI for "Mostly Cloudy." Since the variance of Y is zero, the denominator is zero which makes the correlation coefficient undefined.

The imbalanced ratio in the data set is presented by α in Eq. (2.32). Equation (2.33) extracts the section of R in Eq. (2.32) as given below:

$$
\text{coe}_\alpha = \frac{1}{\alpha} + \alpha + 2
\tag{2.33}
$$

In Eq. (2.33), the minimum point occurs at $\alpha = 1$. This indicates R is maximized if the sampling data set contains an equal number of x_a and x_b. In this section, two functions are employed to study the imbalanced data sets and the correctness of Eq. (2.32). Equation (2.34) introduces the two functions. The error of each sampling point is assumed to follow a standard normal distribution $N(0, 1)$. The first function in Eq. (2.34) establishes a negative relationship while the second function establishes a positive relationship. The correlation can be computed using two methods. Method 1 uses the derived Eq. (2.32) and Method 2 uses the conventional Eq. (2.26).

$$
\begin{pmatrix} x_a = 1 \\ x_b = 2 \end{pmatrix}
\begin{cases}
\text{fun}_1 : y = \sin\left(\dfrac{\pi}{2} x \right) + \text{Err} \\[3mm]
\text{fun}_2 : y = \ln(x) + \text{Err}
\end{cases}
\tag{2.34}
$$

Figure 2.20 shows the simulation results for the two functions in Equation (2.34). n_b is fixed at 100 and a sensitivity analysis is conducted for n_a from 1 to 3000. For Function 2, the correlation absolute value increases from 1 to 100 and decreases

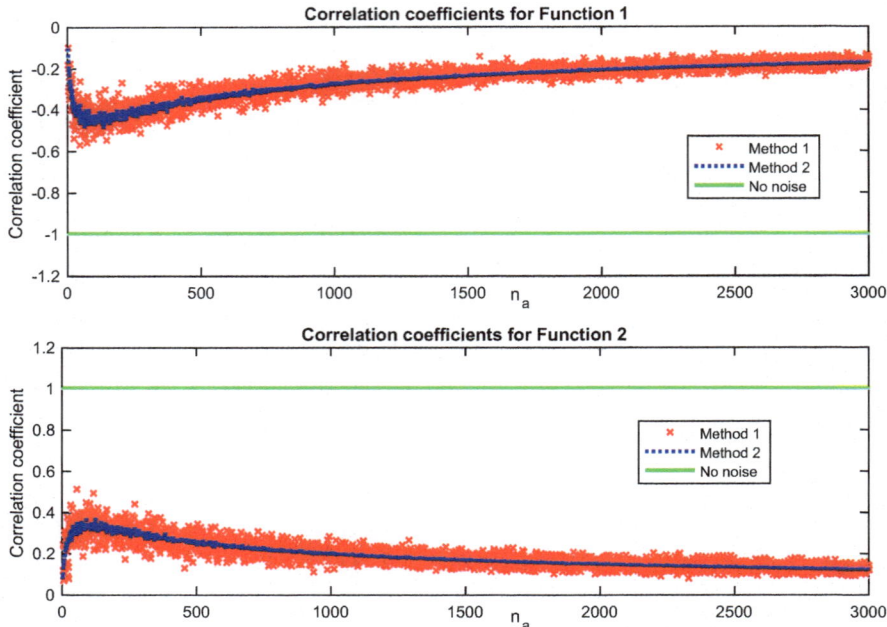

Fig. 2.20 Correlation for the two functions with imbalanced data set

from 100 to 3000. This shows that Method 1 and Method 2 produce similar results. The simulations in Fig. 2.20 have proved that Eq. (2.32) is valid. The maximum absolute value of the correlation occurs at $n_a = n_b = 100$, where $\alpha = 1$.

Figure 2.20 indicates that although variables X and Y have a confirmed dependence, the correlation may be distorted by imbalanced data. The reason the correlations obtained from Method 1 have more fluctuations than Method 2 is due to the assumption made with Eq. (2.27). A general recognition of correlation with high dependency is usually between 0.7 and 1.0, neutral dependency is between 0.3 and 0.7, and low dependency is between 0 and 0.3. However, for Function 2 in Eq. (2.34), the correlation reaches 0.12 when n_a is 3000 ($\alpha = 30$), which is far from the maximum value of 0.37. This may misinterpret the correlation from "neutral dependency" to "low dependency." The optimal correlation can be realized when the data sets have equal sizes.

The contribution of noise to the correlation is presented by Eq. (2.35). Noise represents an unconsidered impact that can cause deviation from the actual value of a variable, which contributes to variance error. It can be recognized as the inaccuracy of the measured data.

$$\text{coe}_{\text{noise}} = \mu_{\text{mse}} - \mu_{\text{me}}^2 \tag{2.35}$$

As shown in Eq. (2.32), correlation may be distorted by the imbalanced ratio, with an exceptional condition that coe_{noise} in Eq. (2.35) is equal to zero. If all noise is rejected by a perfect sensor, Eq. (2.32) indicates the correlation will not be influenced by an imbalanced ratio and the resultant Momentum Ratio becomes 1. A simulation is conducted with Eq. (2.34) without noise. The correlation results without noise are presented in Fig. 2.20. The correlations of the two functions in Eq. (2.34) are shown to be perfectly correlated, i.e., 1 (or −1) when noise does not exist. As n_a increases, the no-noise correlations maintain a value of 1 (or −1). This phenomenon indicates the imbalanced ratio does not influence correlation when noise is removed. Noise is one of the key factors that affect correlation with respect to the imbalanced ratio.

The contribution of the output difference to correlation is presented by Eq. (2.36).

$$coe_{out\_diff} = \frac{1}{\left[f(x_a) - f(x_b)\right]^2} \qquad (2.36)$$

In Eq. (2.34), coe_{out_diff} decreases and R in Eq. (2.32) increases if the difference between $f(x_a)$ and $f(x_b)$ increases. This indicates that R can be controlled by the output difference. A larger output difference can counteract the effect of an imbalanced ratio. Similar to Eq. (2.32), for the case $f(x_a) = f(x_b)$, the correlation coefficient is undefined when the variance of Y is zero.

$$\begin{pmatrix} x_a = 1 \\ x_b = 2 \end{pmatrix} \begin{cases} fun_1 : y = \beta.\sin\left(\frac{\pi}{2}x\right) + Err, \beta = \{1,3,6,9\} \\ fun_2 : y = \beta.\ln(x) + Err, \beta = \{1,3,6,9\} \end{cases} \qquad (2.37)$$

Figure 2.21 presents the simulation results for Eq. (2.37). Note that $[f(x_a) - f(x_b)]^2$ increases as β increases. In addition, the correlation at the same imbalanced ratio is closer to a strong correlation (1 or −1) with an increased β. This indicates that a larger output difference may increase R and counteract the impact of imbalance.

2.5.3   Robust Correlation Analysis Framework

This sub-section introduces a novel correlation analysis framework to alleviate the negative impact of imbalanced data with noise in the correlation analysis. Figure 2.22 presents the structure of the framework. In Fig. 2.22, X has two values (x_a, x_b) in the sampling data set. The number of data points in x_a and x_b are n_a and n_b, respectively. Each x value and its corresponding y value construct a data pair (x, y). The correlation analysis framework consists of the following two main steps:

- **Step 1: Creating groups of balanced data sets:** The first step is to determine which variable X has the largest amount of data. For example, x_a is selected if

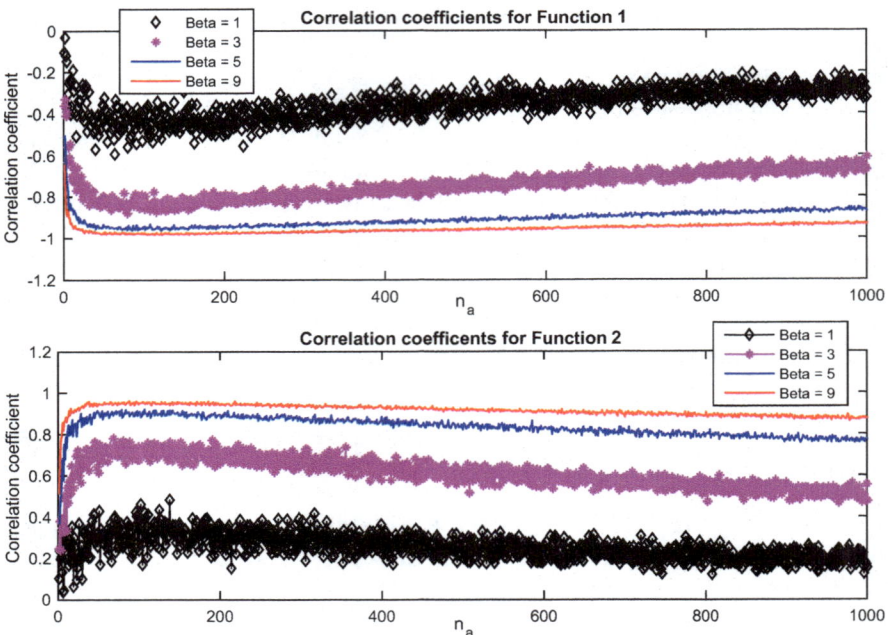

Fig. 2.21 Correlation on specified function with imbalanced data set

$n_a > n_b$, then, select n_b amount of x_a and combine them into pairs with x_b. In this data set, the number of data points in x_a and x_b is equal to n_b. The procedure is repeated M times to construct a group of balanced sets. To prevent the loss of information from the removal of data and to fully utilize all the data, the method to determine M is shown in Eq. (2.38). In the non-repeated random selector, sampling without replacement is used for sampling purposes to prevent "tied" data. The ceil function is used to round the value M toward positive infinity.

$$M = \text{ceil}\left(\frac{n_a}{n_b}\right) \tag{2.38}$$

- **Step 2: Correlation integration**: Corr$_i$, which is nonzero, is the correlation of a balance set i calculated with Eq. (2.26). Assume there are M balanced sets, the final correlation can be computed by Eq. (2.39) as below:

$$\frac{1}{\text{Corr}_{\text{final}}^2} = \frac{1}{M}\sum_{i=1}^{M}\frac{1}{\text{Corr}_i^2} \tag{2.39}$$

Table 2.5 presents a detailed algorithm for RCAF. The implementation and pseudocode were developed with MATLAB.

As depicted in Table 2.5, the computational complexity (CC) for RCAF is relatively low. According to Eq. (2.26), the CC for PPMC is linear [119] at $O(n)$ with data size n. Since RCAF consists of converting the majority class data into M data sets, with each data set having the size of the minority class, the CC for RCAF is approximately $O\left(M\left(\dfrac{n}{M}\right)\right)$ or $O(n)$. Although RCAF has a higher CC due to additional computations, e.g., Eqs. (2.38) and (2.39) and the requirement of more data storage, the improved correlation analysis under imbalanced data can justify the use of RCAF.

The Momentum Ratio R should be maximized as explained above. In Step 2 of RCAF, R is calculated with correlations from all balanced sets, as shown in Eq. (2.40). μ_{mse_i} denotes the μ_{mse} of each balanced set. μ_{me_i} denotes the μ_{me} of each balanced set. α_i is α of each balanced set.

$$\frac{1}{R_{final}^2} = \frac{1}{M}\sum_{i=1}^{M}\left[1 + \frac{\mu_{mse\_i} - \mu_{me\_i}^{2}}{\left[f(x_a) - f(x_b)\right]^b}\cdot\left(\frac{1}{\alpha_i} + \alpha_i + 2\right)\right] \tag{2.40}$$

For each balanced data set, since the number of data points in x_a and x_b are equal, $a_i = 1$. Equation (2.40) can be rewritten as Eq. (2.41).

$$\frac{1}{R_{final}^2} = 1 + \frac{4}{M\cdot\left[f(x_a) - f(x_b)\right]^2}\left(\sum_{i=1}^{M}\mu_{mse\_i} - \sum_{i=1}^{M}\mu_{me\_i}^2\right) \tag{2.41}$$

Assuming the sample size, i.e., n_a is large, the noise terms in Eq. (2.41) can be expressed as Eq. (2.42).

$$\begin{cases}\displaystyle\sum_{i=1}^{M}\mu_{mse\_i} = M\cdot\mu_{mse} \\ \displaystyle\sum_{i=1}^{M}\mu_{me\_i}^{2} = M\cdot\mu_{me}^2\end{cases} \tag{2.42}$$

By considering Eqs. (2.32), (2.41), and (2.42); Eq. (2.43) gives the equations of R for the original correlation and the new correlation. Note that the term α disappears in the Momentum Ratio under RCAF.

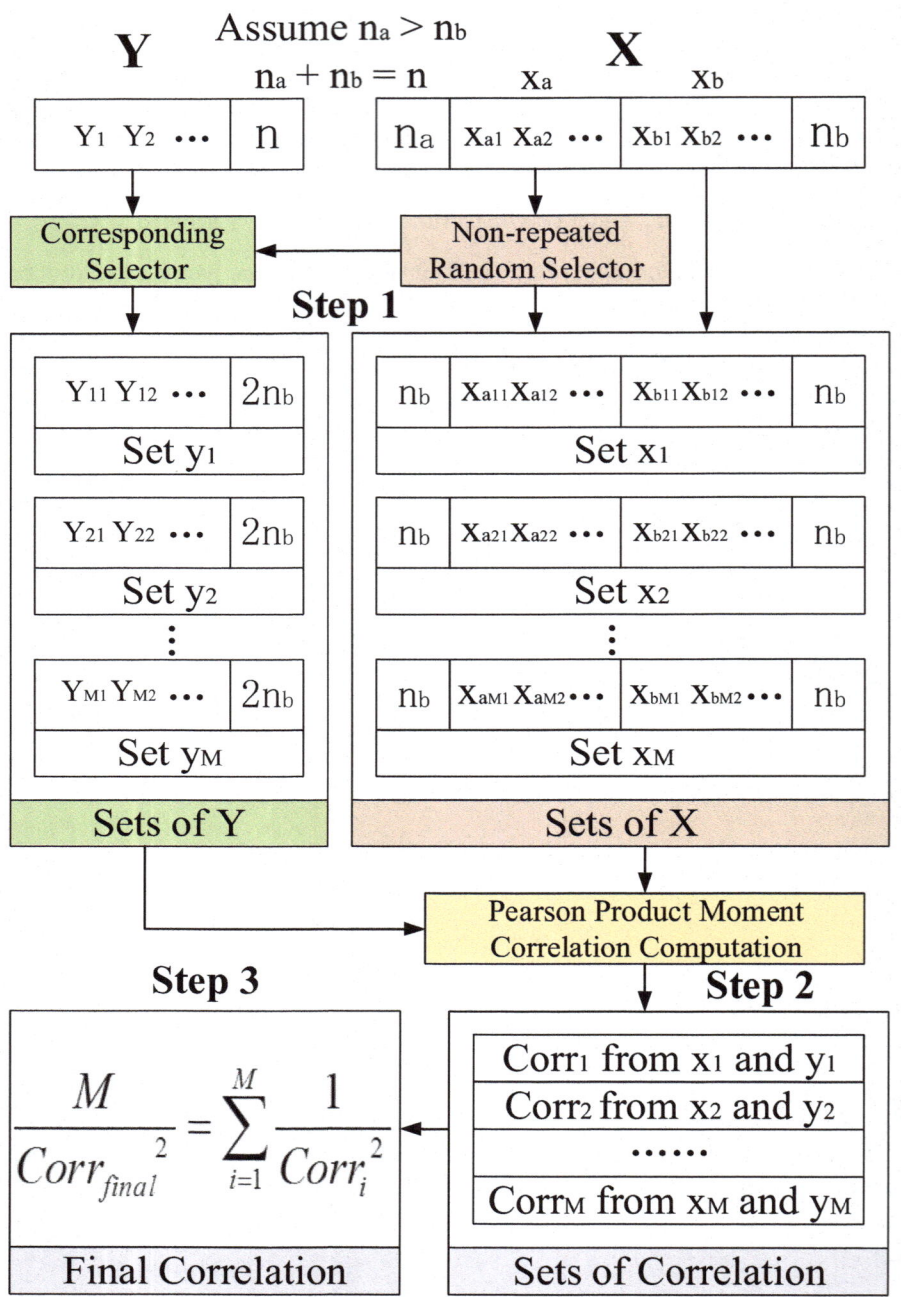

Fig. 2.22 Robust correlation analysis framework

Table 2.5 Algorithm for RCAF

Input:
$y_a = (y_{a1}, y_{a2}, y_{a3}, \ldots, y_{an})$;
$y_b = (y_{b1}, y_{b2}, y_{b3}, \ldots, y_{bn})$;
$n_a = \text{size}(y_a)$;
$n_b = \text{size}(y_b)$;
$x_a = \text{zeros}(n_a, 1) + 1$;
$x_b = \text{zeros}(n_b, 1) + 0$;
Output:
corr _ final: PPMC for x and y
Algorithm:
If ρ_{xy} is negative   % Use Eq. (2.26) to determine if the correlation is positive or negative.
sign $= -1$;
else
sign $= +1$;
end
If $n_a \geq n_b$ **then**
$M = \text{ceil}(n_a/n_b)$;
For counter $= 1 : M$
posi $= \text{randperm}(n_a, n_b)$;
$xk = x_a(\text{posi})$;
$yk = y_a(\text{posi})$;
$x = [xk; x_b]$;
$y = [yk; y_b]$;
cori$(1, \text{counter}) = \text{corr}(x, y)$; % Eq. (2.26)
cori$(1, \text{counter}) = 1. /(\text{cori}(1, \text{counter}).\,^\wedge 2)$;
end
else
$M = \text{ceil}(n_b/n_a)$;
For counter $= 1 : M$
posi $= \text{randperm}(n_b, n_a)$;
$xk = x_b(\text{posi})$;
$yk = y_b(\text{posi})$;
$x = [xk; x_a]$;
$y = [yk; y_a]$;
cori$(1, \text{counter}) = \text{corr}(x, y)$; % Eq. (2.26)
cori$(1, \text{counter}) = 1. /(\text{cori}(1, \text{counter}).\,^\wedge 2)$;
end
end
reg $= \text{mean}(\text{cori})$;
corr$_{\text{final}} = $ sign $* (1. /(\text{reg}.\,^\wedge 0.5))$;

$$\left\{ \begin{array}{l} \text{Original}: \dfrac{1}{R^2} = 1 + \dfrac{\mu_{\text{mse}} - \mu_{\text{me}}^2}{\left[f(x_a) - f(x_b) \right]^2} \left(\dfrac{1}{\alpha} + \alpha + 2 \right) \\[4mm] \quad\text{New}: \dfrac{1}{R_{\text{final}}^2} = 1 + \dfrac{\mu_{\text{mse}} - \mu_{\text{me}}^2}{\left[f(x_a) - f(x_b) \right]^2} \cdot 4 \end{array} \right. \tag{2.43}$$

$$\because 4 < \frac{1}{\alpha} + \alpha + 2$$

$$\therefore \frac{1}{R_{\text{final}}^2} < \frac{1}{R^2}$$

$$\therefore R_{\text{final}} > R$$

Base on Eq. (2.34), the correlations under RCAF are much more stable and slanting does not occur with respect to the increase of the imbalanced ratio. Figure 2.23 shows the simulation results. The imbalanced ratio increases as n_a increases. However, the correlations under RCAF do not have a large variation and the optimal value is maintained.

2.6   Real-Life Case Study: Correlation for Weather Conditions and Clearness Index

Weather condition is one of the major factors affecting the amount of solar irradiance reaching earth. As a consequence, one of the most important applications affected by solar irradiance due to weather perturbation is the Photovoltaic (PV) system. Weather condition changes affect the electrical power generated by a PV system with respect to time. Due to the nature of climate and the hemisphere of the earth, the number of samples for each weather condition, e.g., "Overcast" and "Heavy Rain," is expected to be disproportional for a given location.

The data structure for the correlation analysis is presented in Table 2.6. The data pairs in each row represent an observation. Column 1 represents the type of weather condition, i.e., 0 and 1 for weather conditions 1 and 2, respectively. Column 2 is the CI value.

The corresponding weather condition information (in string format) for the solar irradiance data in Johannesburg was obtained from Weather Underground [120]. There are 41 types of weather conditions in Johannesburg from 2009 to 2012. The sampling size of all weather conditions in Johannesburg can be found in Reference [A] in Acknowledgments. The same weather conditions can results in different CI values due to other perturbation effects that are factored out by the weather. The solar altitude angle range studied is between 0.8 and 1. The correlation results under the traditional approach and the novel correlation framework are provided in Figs. 2.24 and 2.25, respectively. The entire correlation matrix is a 41 × 41 square matrix.

The correlation between X and Y represents the variation of CI for the two weather transitions. A high correlation absolute value means the CI changes significantly with weather condition transitions. In contrast, if the absolute value of the correlation is low, CI changes slightly when the weather condition changes.

2.6.1   Clearness Index and Weather Conditions Statistical Analysis

The following section of this sub-section examines the correlation results in Figs. 2.24 and 2.25. To understand the uncertainty and stochastic properties of CI with respect to weather conditions, it is crucial to provide statistical measures and mathematical description of the random phenomenon for the variables.

The mean and standard deviation with error bars are presented in Fig. 2.26 for the weather conditions and CI for a solar altitude angle between 0.8 and 1.0. Bootstrapping is used to quantify the error in the statistics. The bootstrapped 95% confidence intervals for the population mean and standard deviation are calculated. Eight weather conditions selected from the correlation matrix are studied. The mean and standard deviation are calculated using Eqs. (2.44) and (2.45), respectively, for the weather conditions. s is the sample size of the weather condition. To compute

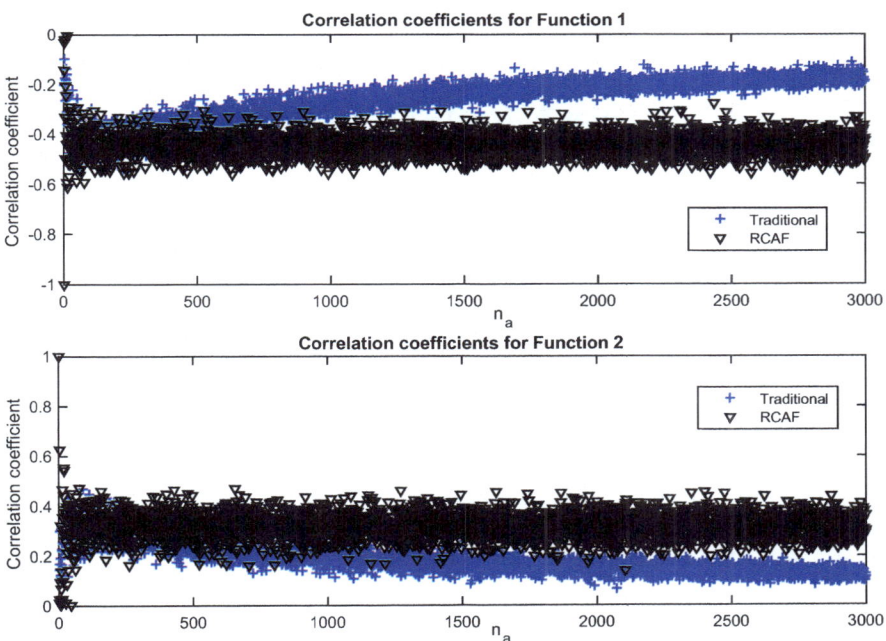

Fig. 2.23 Correlation comparison between traditional approach and RCAF

Table 2.6 Typical representation of a data set for the correlation analysis

Weather type (binary) $X = 0$ for weather type 1 $X = 1$ for weather type 2	$Y = CI$
1	0.71
1	0.69
0	0.43
1	0.61
0	0.32
1	0.54

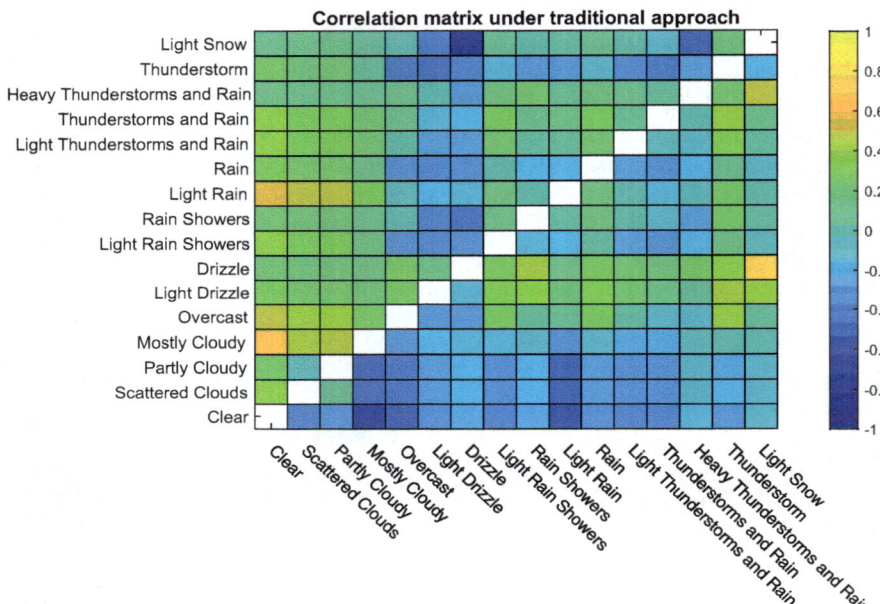

Fig. 2.24 Correlation matrix under traditional PPMC

the 95% bootstrap confidence interval of the mean and standard deviation, 2000 bootstrap samples are used.

$$w_{mean} = \frac{1}{s}\sum_{i=1}^{s} CI_i \qquad (2.44)$$

$$w_{sd} = \sqrt{\frac{1}{s}\sum_{i=1}^{s}\left(CI_i - w_{mean}\right)^2} \qquad (2.45)$$

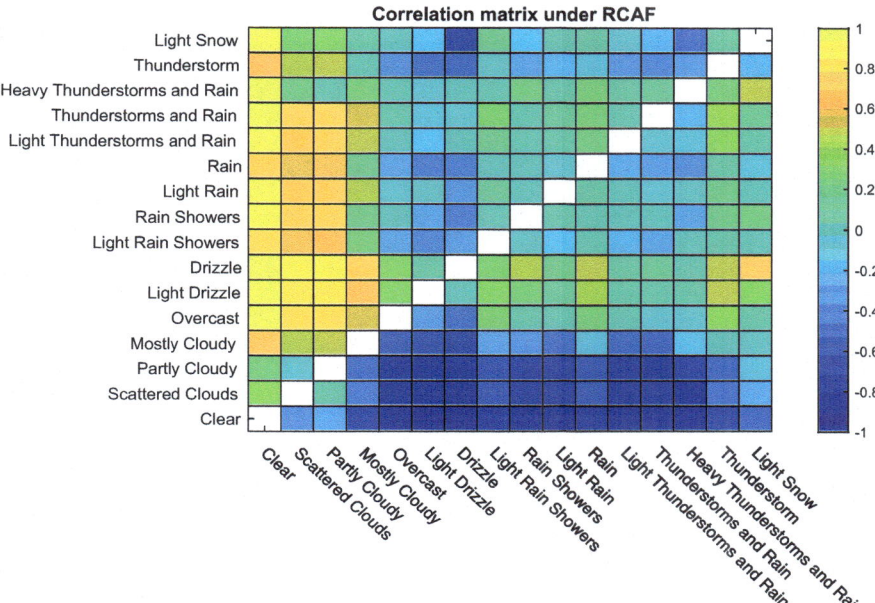

Fig. 2.25 Correlation matrix under RCAF

A graphical representation of the distribution of variables is presented in the histograms in Fig. 2.27. This effectively displays the probability distribution of CI for the weather conditions. The histogram shows that different weather conditions result in different distributions. The "Clear" case is a monomodal distribution with a peak at 0.8 CI, whereas "Mostly cloudy" has a peak at 0.3 CI. CIs are generally high for the "Clear" weather condition due to the frequency of high CI occurrences. In contrast, "Mostly Cloudy" has a high frequency of lower CI value occurrences.

Due to the highly stochastic nature of CI, as shown in the histogram, it is impossible to use a parametric method where an assumption of the data distribution is made. Kernel Density Estimation (KDE) is a non-parametric method to estimate the probability density function (pdf) of a random variable. KDE is a data smoothing problem where inferences about the population are made, based on a finite data sample. Let $(x_1, x_2, \ldots, x_n)$ be a sample drawn from distributions with an unknown density f. The kernel density estimator is:

$$\hat{f}_h(x) = \frac{1}{n}\sum_{i=1}^{n}G_h(x - x_i) = \frac{1}{nh}\sum_{i=1}^{n}G\left(\frac{x - x_i}{h}\right) \tag{2.46}$$

where n is the sample size. $G(\cdot)$ is the kernel function, a non-negative function that integrates to one and has a mean of zero. h is a smoothing parameter called the bandwidth and has the properties of $h > 0$.

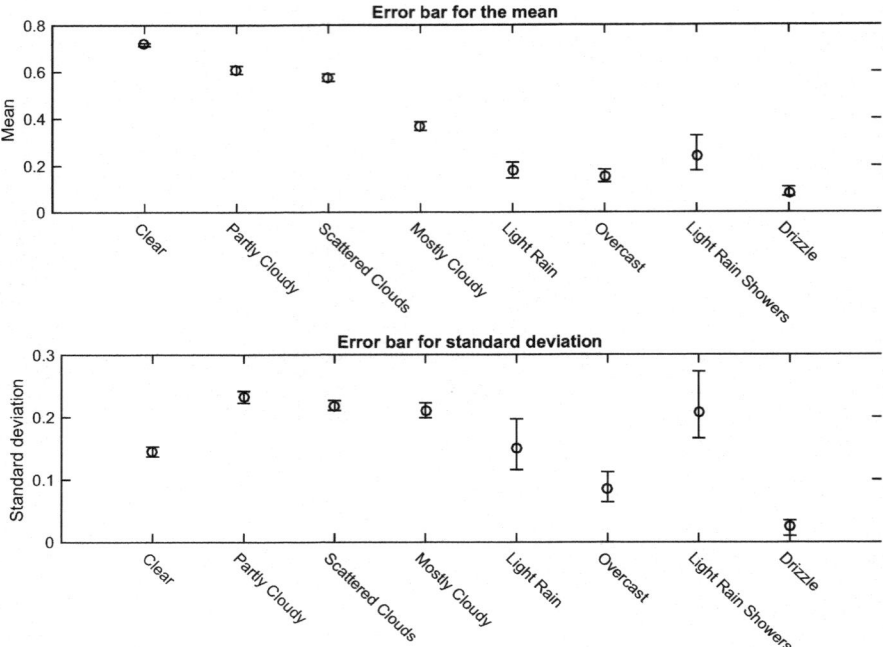

Fig. 2.26 Error bars for mean and standard deviation with eight types of weather conditions

The kernel smoothing function defines the shape of the curve used to generate the pdf. KDE constructs a continuous pdf with the actual sample data by calculating the summation of the component smoothing functions.

The Gaussian kernel is:

$$G(u) = \frac{1}{\sqrt{2\pi}} e^{-\frac{1}{2}u^2} \tag{2.47}$$

Therefore, the kernel density estimator with a Gaussian kernel is:

$$\hat{f}_h(x) = \frac{1}{nh} \sum_{\substack{j \neq i}}^{n} \frac{1}{\sqrt{2\pi}} e^{-\frac{1}{2}\left(\frac{x_j - x_i}{h}\right)^2} \tag{2.48}$$

The aim is to minimize the bandwidth, h. However, there is a trade-off between the bias of the estimator and its variance. The bandwidth is estimated by completing an analytical and cross-validation procedure. The bandwidth estimation consists of two steps:

1. Use an analytical approach to determine the near-optimal bandwidth;

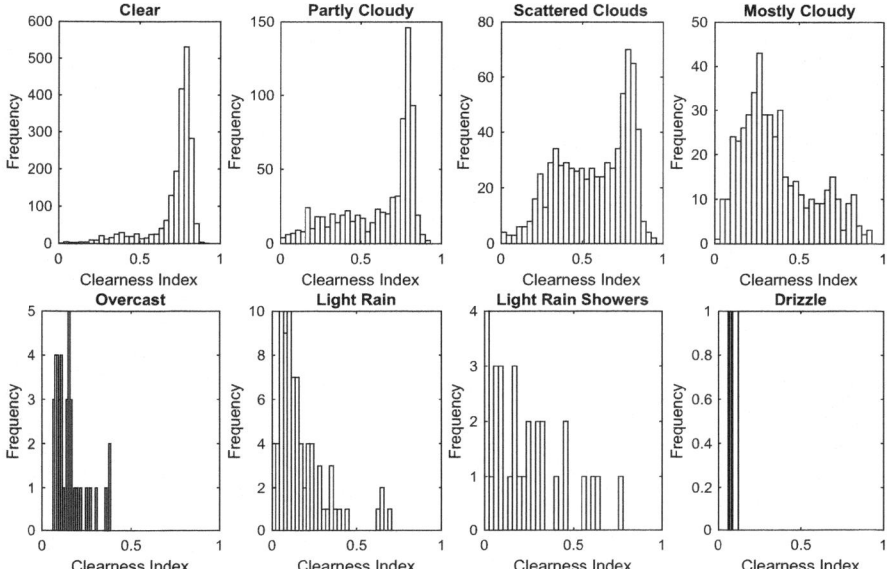

Fig. 2.27 Histograms of CI with respect to different weather conditions

2. Adopt a log-likelihood cross-validation method to determine the optimal bandwidth.

This adopted method has the advantage of avoiding the use of the expectation-maximization iterative approach to estimate the optimal bandwidth. The near-optimal bandwidth can be calculated with the analytical approach and could be further improved by using the maximum likelihood cross-validation method. This simplifies the estimation process and could potentially reduce the computational effort as this method is not an iterative approach.

Analytical method For a kernel density estimator with a Gaussian kernel, the bandwidth can be estimated with Eq. (2.49), the Silverman's rule of thumb [121].

$$h = \left(\frac{4\sigma^5}{3n} \right)^{\frac{1}{5}} \approx 1.06\sigma n^{-\frac{1}{5}} \tag{2.49}$$

where σ is the standard deviation of the data set. The rule of thumb should be used with care as the estimated bandwidth may produce an over-smooth pdf if the population is multimodal. An inaccurate pdf may be produced when the sample population is far from a normal distribution.

Maximum likelihood tenfold cross-validation method The maximum likelihood cross-validation method was proposed by Habbema [122] and Duin [123]. In essence, the method uses the likelihood to evaluate the usefulness of a statistical model. The aim is to choose h to maximize pseudo-likelihood $\prod_{i=1} \widehat{f_h}(x_i)$.

A number of observations $x_K = \{x_1, x_2, \ldots, x_k\}$ from the complete set of original observations x can be retained to evaluate the statistical model. This would provide the log-likelihood $\log(\hat{f}_{-k}(x_i))$. The density estimate constructed from the training data is defined in Eq. (2.50).

$$\hat{f}_{-k}(x_i) = \frac{1}{n_t h} \sum_{t \neq i}^{n_t} \frac{1}{\sqrt{2\pi}} e^{-\frac{1}{2}\left(\frac{x_i - x_t}{h}\right)^2} \tag{2.50}$$

where $n_t = n - n_k$. Let n_t and n_k be the number of sample data for training and testing, respectively. The number of training data will be the number of the entire sample data set minus the number of testing data. Since there is no preference for which observation is omitted, the log-likelihood is averaged over the choice of each omitted data sample, x_K, to give the score function. The maximum log-likelihood cross-validation (MLCV) function is given as follows:

$$\text{MLCV}(h) = \left(\frac{1}{n_k} \sum_{i=1}^{n_k} \log\left[\sum_{t \neq i}^{n_k} \frac{1}{\sqrt{2\pi}} e^{-\frac{1}{2}\left(\frac{x_i - x_t}{h}\right)^2} \right] - \log(n_k h) \right) \tag{2.51}$$

The bandwidth is chosen to maximize the function MLCV(h) for the given data as shown in Eq. (2.52).

$$h_{\text{mlcv}} = \underset{h>0}{\text{argmax}} \, \text{MLCV}(h) \tag{2.52}$$

KDE has been applied to compute the continuous pdf of CI for different weather conditions. Figure 2.28 shows the density estimation with the maximum log-likelihood cross-validation method for the "Clear" weather condition. The top figure shows the histogram and the density function fitted on the histogram. The bottom left figure shows the shape variation of kernel density with various bandwidths shaded in grey. The best bandwidth is highlighted in red. The bottom right figure shows the log-likelihood plot with respect to the bandwidth. The red circle identifies the bandwidth with the highest log-likelihood. The cross-validated pdf has a good fit with the histogram and has been confirmed with the log-likelihood. The optimal bandwidth estimation approach is shown to be effective and the density function gives a good representation of the histogram. The optimal bandwidth for the weather conditions can be found in Table 2.7.

The pdfs produced using KDE for the eight weather conditions are given in Fig. 2.29. Note that the pdf (such as for "Light rain") could be in the range of negative CI due to the nature of a fitted function. In practice, CI cannot be negative as this means the irradiance will have a negative value. This will give a negative value for solar power estimation. Hence, negative CI values should not be considered.

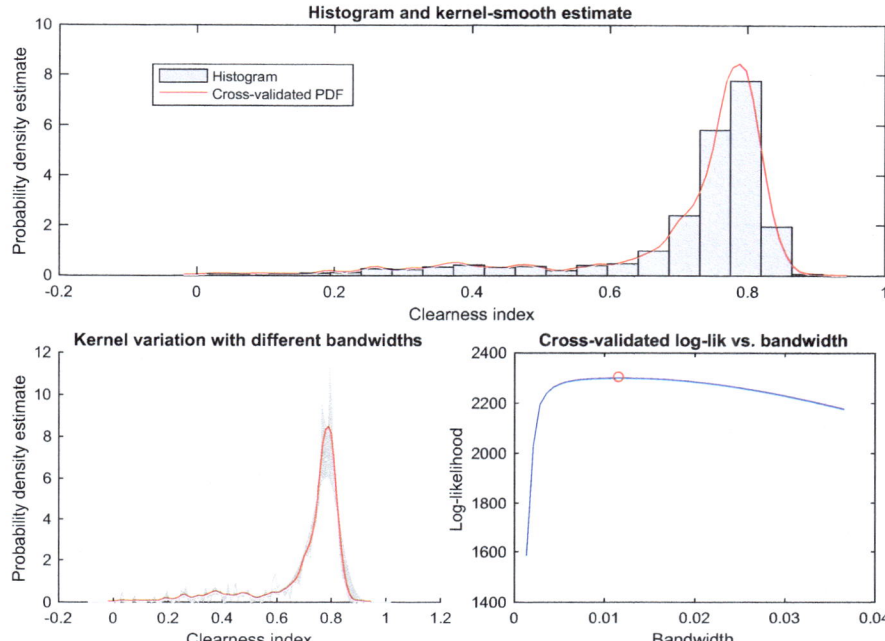

Fig. 2.28   Kernel density estimation for "Clear"

Table 2.7   Optimal bandwidth for PDFs

Weather condition	Optimal bandwidth h
"Clear"	0.0124
"Partly Cloudy"	0.0132
"Scattered Clouds"	0.0224
"Mostly Cloudy"	0.0313
"Light Rain"	0.0316
"Overcast"	0.0291
"Light Rain Showers"	0.1023
"Drizzle"	0.0260

2.6.2   Comparison of Sampling Techniques in Correlation Analysis

To compare the proposed framework with previous sampling methods for correlation analysis, the prominent sampling techniques: Synthetic Minority Over-Sampling Technique (SMOTE) and Adaptive Synthetic (ADASYN) sampling are employed in this study. SMOTE [124] was introduced in 2002 and is an over-sampling technique with K-Nearest Neighbors (KNN). First, the KNN is considered

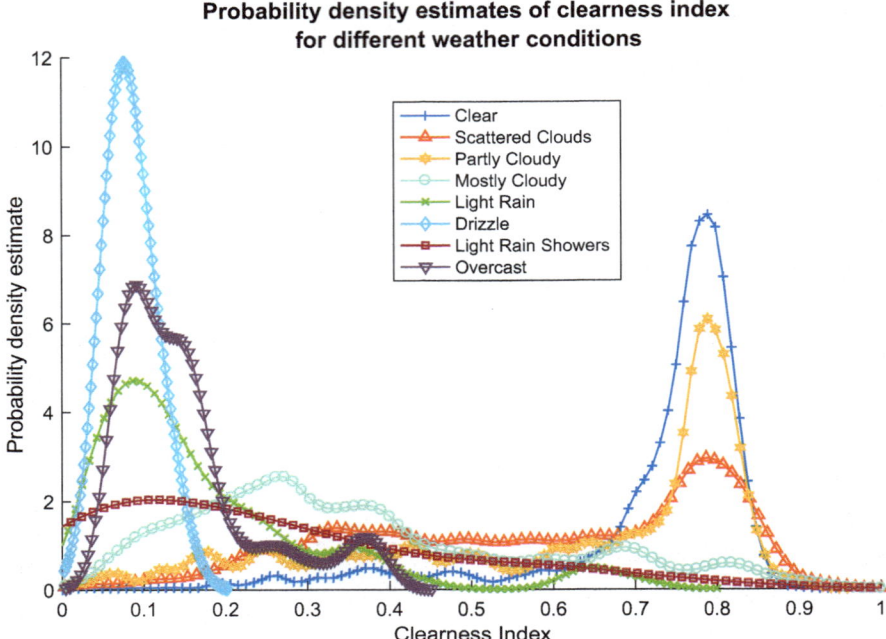

Fig. 2.29 PDF for various weather conditions

for a sample of the minority class. To create an additional synthetic data point, the difference between the sample and the nearest neighbor is calculated and multiplied with a random number between zero and one. The randomly generated synthetic data point will be within the two specific samples. In 2008, He et al. [125] introduced ADASYN for the over-sampling of the minority class. ADASYN is an improved technique that uses a weighted distribution for individual minority class samples depending on their level of learning difficulty. As such, additional synthetic samples are generated for minority class samples that are more difficult to learn. SMOTE generates an equal number of synthetic data points for each minority sample.

In this study, the number of nearest neighbors for SMOTE is produced according to the imbalanced ratio, as this suggests the number of data points needs to be generated. If the number of nearest neighbors for over-sampling is greater than five, under-sampling by randomly removing samples in the majority class will be similar; as the number of nearest neighbors would be too large for effective sampling [124]. In this work, the K-Nearest Neighbors for both ADASYN and SMOTE are considered to be five, which is the value used in the original work.

The constructed pdfs in Fig. 2.19 are useful for studying PPMC with different sampling methods. A sensitivity analysis is conducted to provide comparisons of the traditional approach and the RCAF approach. Data are generated from the pdf with random sampling. This analysis aims to understand the influence of the variation of data set size on correlation results. The size of the data set for each weather condition, at a solar altitude angle between 0.8 and 1.0, can be found in Reference [A] in Acknowledgments. The data set size for "Clear" is determined to be 1993 data points. A range of samples from 1 to 1993 is generated from the "Clear" pdf to study the impact of imbalanced data on correlation. Seven weather conditions are studied for this purpose. The data set size for the seven weather conditions is fixed throughout the analysis. As shown in Fig. 2.30, the correlation calculated with one data point for RCAF, SMOTE-under-sampling, and under-sampling is at perfect correlation, i.e., 1. This can be explained by the fact that the correlation between two data points at two different classes (except for the case where the two data points are equal) will be a perfect positive or perfect negative correlation.

As expected, the traditional PPMC and RCAF correlation at the end of the sensitivity analysis given in Fig. 2.30 can refer to the correlation of the correlation matrices in Figs. 2.24 and 2.25. The deviation between the correlation for all methods increases as the imbalanced ratio increases. This is also shown in Table 2.8. Additionally, the high standard deviation and mean error in Fig. 2.26 can result in a larger sampling range, and consequently will result in increased correlation inaccuracy.

The correlation reaches a steady state as the imbalanced ratio decreases, where the imbalanced ratio will have an insignificant effect on the correlation in the traditional approach. The SMOTE-under-sampling and ADASYN sampling methods are competitive with the proposed RCAF. However, SMOTE may generate data between the inliers and outliers. ADASYN focuses on generating more synthetic data points for difficult trained samples and may focus on generating from the outlier samples and deteriorate the correlation. Reference [126] suggests the previous sampling techniques should investigate outliers for optimal performance.

To quantify the variation in correlation with imbalanced data, Table 2.8 presents the standard deviation of the correlations with respect to different methods, as presented in Fig. 2.30. The correlation with one sample data is excluded in the standard deviation calculation since it can be considered an outlier as explained above.

2.6.3 Cluster Analysis of Weather Conditions

Classes with high correlation should be separated and in contrast, classes with weak correlation should be clustered together. According to the rule of thumb, a correlation of less than 0.3 [127] is considered a weak correlation. As shown in Fig. 2.24 and considering the case for "Clear", i.e., the column for "Clear," most of the correlations under the traditional approach are in the range 0–0.3. This signifies that they can be clustered as one weather group. However, the correlations computed with RCAF, as shown in Fig. 2.25, signify that only two other weather conditions, i.e., "Partly Cloudy" and "Scattered Clouds," are weakly correlated with "Clear." The following part employs two clustering approaches, K-Means and Ward's Agglomerative hierarchical clustering, to cluster weather conditions and understand the implications of the correlation results. However, since the number of data points

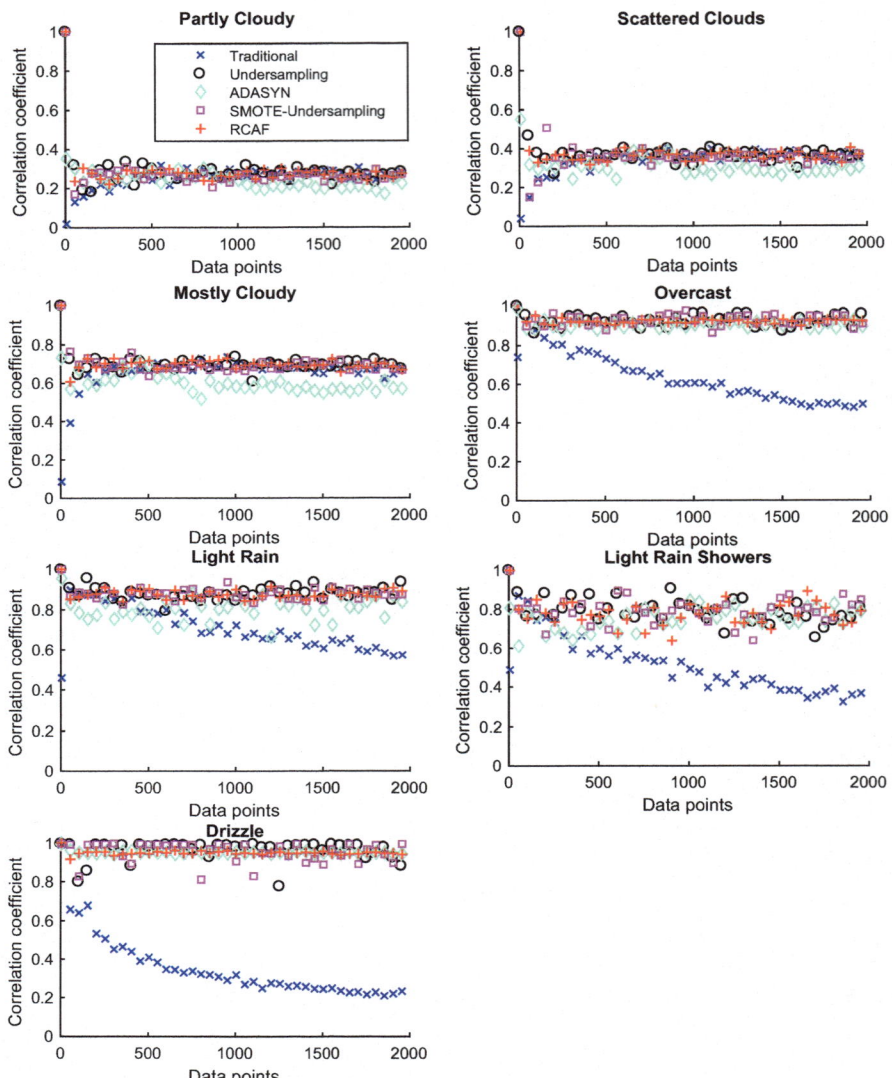

Fig. 2.30 Sensitivity analysis of correlation with no sampling (traditional) and different sampling methods

is different for the weather conditions, the mean calculated with Eq. (2.44) is used to duplicate an equal amount of data points to match the majority class, i.e., "Clear," for cluster analysis.

K-Means is an iterative unsupervised learning algorithm for clustering problems. The basis of the algorithm is to allocate the data point to the nearest centroid. The centroid is calculated as the mean value; based on the data in the cluster at the

Table 2.8 Standard deviation of correlation coefficients with imbalanced data

	Traditional	Under-sampling	ADASYN	SMOTE-Under-sampling	RCAF	Percentage difference between Traditional and RCAF (%)
"Partly Cloudy"	0.040	**0.026**	0.049	0.036	0.027	32.50
"Scattered Clouds"	0.047	0.030	0.035	0.035	**0.023**	51.06
"Mostly Cloudy"	0.057	0.025	0.041	0.030	**0.018**	68.42
"Overcast"	0.129	0.029	0.016	0.024	**0.012**	90.70
"Light Rain"	0.095	0.029	0.051	0.026	**0.020**	78.95
"Light Rain Showers"	0.122	0.066	0.069	0.050	**0.048**	60.66
"Drizzle"	0.129	0.069	**0.008**	0.044	0.009	93.02

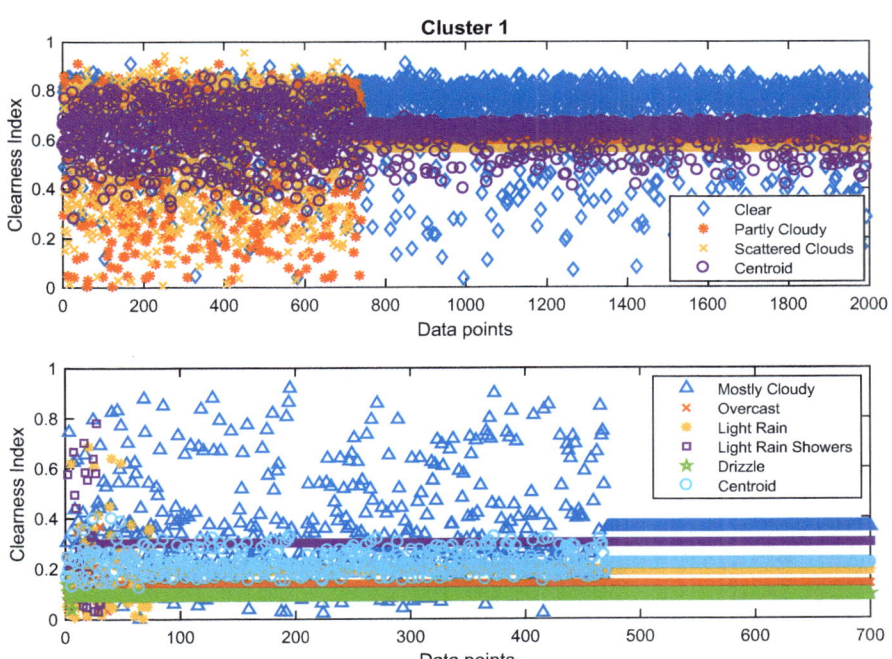

Fig. 2.31 K-Means clustering results for weather conditions

current iteration. The K-Means algorithm with Euclidean distance for time-series clustering can be referred to [128]. The K-Means clustering results for weather conditions with $K = 2$ is shown in Fig. 2.31. As shown, the CIs are generally higher for "Clear," "Partly Cloudy," and "Scattered Clouds" conditions. Due to the insufficient amount of data in minority classes, e.g., "Partly Cloudy," the values after the 740th data point will be denoted with the mean value of its data set. The mean value will not deteriorate the clustering results since the K-Means algorithm calculates the centroid as the mean value.

In Ward's Agglomerative hierarchical clustering [129], the clustering objective is to minimize the error sum of squares, where the total within-cluster variance is minimized. At each iteration, pairs of clusters are merged which leads to a minimum increase in total within-cluster variance. The results for the hierarchical clustering of weather conditions are depicted in Fig. 2.32. The weather conditions can be separated into two major branches with "Scattered Clouds," "Partly Cloudy," and "Clear" as one cluster. The results are consistent with the correlation results from RCAF.

The absolute value of the correlation may be very high if the sample size is extremely low, such as the case for "Heavy drizzle" in which only one data point is available. The correlation of "Heavy drizzle" under RCAF becomes 1 while the coefficient is less than 0.1 using the traditional approach. Numerous small sample balanced data sets are created in RCAF. A challenging research question that remains is that a severe lack of data points can be an issue for the correlation analysis. The limitations of RCAF and methods to overcome such issues need to be investigated.

The theoretical study of the imbalanced data effect on PPMC for continuous variables should be a focus in future work. This may provide a broader application in PPMC analysis and the method may be generalized.

The study of imbalanced data and noise in rank-order correlations will greatly benefit exploring relationships involving ordinal variables. PPMC measures the linear relationship between two continuous variables (it is also possible for one variable to be dichotomous as studied in this research) and Spearman-Rank measures the monotonic relationship between continuous or ordinal variables. Additionally, rank correlations such as Kendall's τ, Spearman's ρ, and Goodman's γ will be explored. Since a dichotomous variable is a special form of a continuous variable, i.e., by treating the continuous data as binary values, providing a mathematical deduction for the correlation measures with the continuous variable is challenging and will be future work.

2.7 Energy Storage for High Penetration of Solar

2.7.1 Electrical Energy Storage for PV Power System Applications

IHS Technology has reported that there is a total of 2 GW grid-connected energy storage projects worldwide by 2016, a 20% increase since the end of 2015 [130, 131]. The surge is due to government funding programs, EES costs reductions, and utility tenders [132]. Several countries aim to maximize their EES portfolio in order to counteract the adverse effect of solar PV systems. Hence, this leads to a sharp increase in the deployment of EES across the globe. In the United States, 111.8 MW of EES has been built by the end of 2015. The EES market in the US will exceed

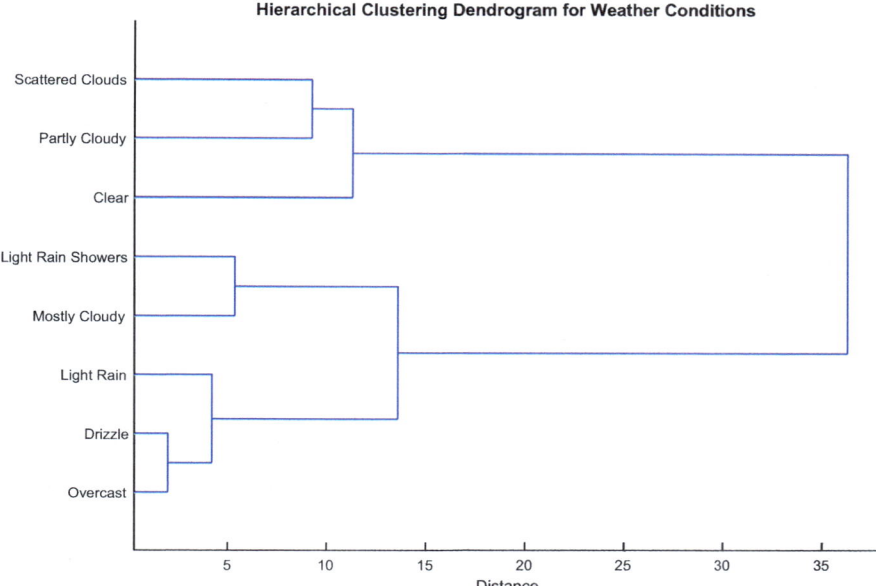

Fig. 2.32 Ward's agglomerative hierarchical clustering results for weather conditions

2 GW per year by 2021. This value is higher than the cumulative total of EES deployment in 2013 and 2014 [133]. India has ambitious targets for adopting renewable PV energy and one of the cornerstones to meet these challenges is to use energy storage technologies [134]. It is expected that 250 GW of EES will be built by 2030 [135]. In China, EES is already used for smoothing wind turbine output in wind farms. A demonstration project of 14 MW lithium iron phosphate battery system is fully constructed in Zhangbei, China [136]. The complete project is expected to have 110 MW of energy storage. China is aiming to deploy additional mature energy storage technologies into their grid in the near future. The country expects that EES performance could achieve significant breakthroughs by 2020, resulting in reduced investment costs. Figure 2.33 presents the installation energy storage capacity for worldwide, India and United Kingdom and United States. The scales of x axes (year) are different due to separate forecasting horizons from different sources.

The average cost of installing residential energy storage systems will fall from 1600 $/kWh in 2015 to 250 $/kWh by 2040 [139] and it is expected to see the price with a 70% reduction by 2030 [140]. Figure 2.34 presents the projected costs for EES. It shows that the costs of ESS could be constant at around 2030, at 350 $/kWh. The storage technology will be matured by then and the costs associated will be manufacturing and maintenance costs.

Unlike conventional generators which have the only use of creating electrical power and situates at generation level, electrical energy storage systems have a

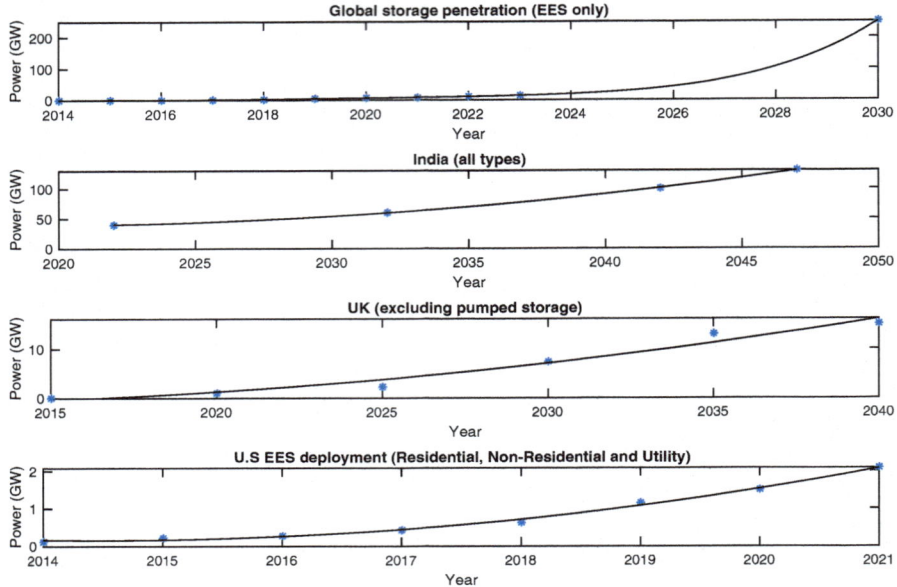

Fig. 2.33 UK, India, United States and global energy storage penetration [137, 138]

variety of use cases in a modern electric system. They could be found in the generation, transmission, and distribution levels of a power system [143, 144].

A study on the impact of short-term frequency stability of distributed autonomous micro-grid with EES is given in [145]. Improvements in the micro-grid short-term frequency stability could be achieved with a novel EES control scheme, by considering both an adaptive droop characteristic during battery state of charge (SoC)/depth of discharge (DoD) limitations and inertial response. The relationship is expressed in Eq. (2.53) [145].

$$\Delta f\left(s\right) = -\frac{G_{\mathrm{MG}}\left(s\right)}{1 + G_{\mathrm{MG}}\left(s\right) G_{\mathrm{EES}}\left(s\right)} \Delta P_{\mathrm{i}} \tag{2.53}$$

$\Delta f(s)$ is the micro-grid system's frequency deviation, ΔP_{i} is the input power disturbance from a renewable source, $G_{\mathrm{MG}}(s)$ and $G_{\mathrm{EES}}(s)$ are the frequency response for the micro-grid and the EES respectively. The frequency deviation could be minimized by increasing the frequency response from the EES.

Reference [146] investigated the trade-off between the storage capacity and outage probability due to the power imbalance between generation and demand. Under mild assumptions, the outage probability decreases exponentially with respect to

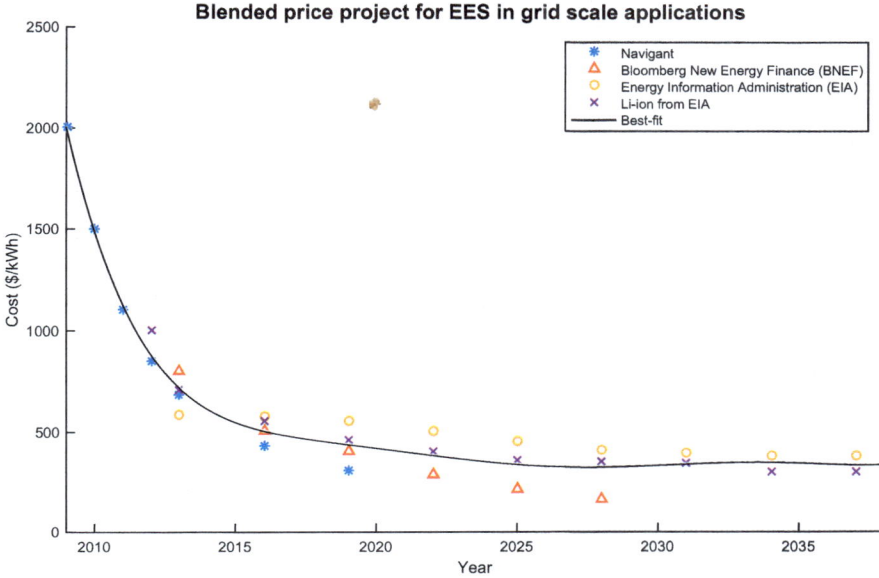

Fig. 2.34 Blended price project for EES in grid-scale applications [141, 142]

the square of the storage capacity. This finding implies that energy storage is an effective and economically viable solution to maintain the stability of a smart grid network, even in the presence of many volatile and intermittent renewable energy sources. The relationship between minimum storage capacity m and the target outage probability P_t is given in Eq. (2.54) [146]:

$$m \geq k_s \sqrt{\ln\left(\frac{1}{P_t}\right)}|V| \tag{2.54}$$

V is the cardinality of the set of uncertainty source and k_s is the ratio between the protection function and the degree of protection. The protection function is used to characterize the confidence region such that the resulting solution remains feasible when the random solution of the uncertain parameters belongs to the confidence region.

This sub-section focuses on EES for PV integration purposes. For PV system integration applications, the EES should have the following characteristics [147]:

- Having the energy and power capacity to meet the demands of unstable grid energy
- Robust to heavy cycling (charging and discharging)
- Very quick response time (milliseconds to seconds) and
- Susceptible to irregular full recharging

Table 2.9 Technical specifications for different EES technologies [143, 149, 151–156]

Type	Maximum size (MW)	Cycles at 80% DoD (×1000)	Expected useful lifetime (years)	Maximum DoD (%)	Round trip efficiency (%)
Vanadium RFB	10	10–13	15–20	100	65–85
Zn–Br RFB	2	5–10	5–15	60–70	72–80
Lead-acid	20–70	2–4.5	5–15	60–80	65–90
Li-ion	10	1.5–4.5	5–15	80–100	85–95
Ni–Cd	40	2–2.5	10–20	80	60–75
NaS	8	2.5–4.5	10–15	80–85	75–90

Table 2.10 Typical capital and operation and maintenance costs for different EES technologies [143, 149, 152, 155, 156]

Type	Capital cost ($/kWh)	Fixed O&M cost ($/kWh-year)	Variable O&M cost ($/MWh)
Vanadium RFB	530–675	3.8–19.4	0.22–3.14
Zn–Br RFB	200–595	3.6–7.7	0.34–2.25
Lead-acid	206–950	3.6–14.5	0.17–0.58
Li ion	527–1435	2.2–15.3	0.45–6.29
Ni–Cd	632–1256	4.5–26.9	Unknown
NaS	200–632	2.2–19.4	0.34–6.29

The EES that are suitable for PV integration applications is Redox flow battery (RFB), Lead-acid battery, Lithium-ion (Li-ion) battery, nickel–cadmium (Ni-Cd) battery and sodium–sulfur (NaS) battery. The following section gives an overview of these EES technologies. Tables 2.9 and 2.10 summarize the technical specifications and the capital and operating costs for different EES technologies.

Redox flow battery Flow batteries contain two electrolyte solutions in two separate tanks, circulated through two independent loops. When connected to a load, the migration of electrons from the negative to positive electrolyte solution creates a current. The sub-categories of flow batteries are defined by the chemical composition of the electrolyte solution; the most prevalent of such solutions are vanadium redox and zinc-bromine (Zn–Br). New zinc–iron (Zn–Fe) RFB, based on double-membrane triple-electrolyte design is estimated to have a system capital cost of under 100 $/kWh [148]. The low cost is achieved by a combination of high cell performance and the use of inexpensive materials (i.e., iron and zinc). The RFB technology shows a promising future and it is expected that it will be highly deployed for PV systems integration. The advantages of RFB are that the power and energy ratings are highly and independently scalable. Also, there is no degradation in energy storage capacity. The disadvantages are the relatively high balance of system costs and reduced efficiency due to rapid charge/discharge. The energy and power density is generally lower compared to other EES.

Lead-Acid battery Lead-acid batteries were invented in the nineteenth century and are the oldest and most commonly used EES; they are low cost and could be found in a range of applications, such as electric vehicles, off-grid power systems, and uninterruptible power supplies. Advanced lead-acid battery technology uses the standard lead-acid battery technology with the additional of ultra-capacitors. Lead-acid battery is a mature technology with established recycling infrastructure. However, it has an issue of poor ability to operate in a partially charged state, hence efficiency may be reduced with each charge. It has a relatively poor depth of discharge and a short lifespan.

Lithium-Ion (Li-ion) battery Li-ion batteries have historically been used in the electronics and advanced transportation industries. Li-ion batteries are increasingly replacing lead-acid batteries in many applications. They have the features of relatively high energy density, low self-discharge, and high charging efficiency. Li-ion storage systems remain relatively high costs compared to other available storage technologies.

Sodium-sulfur (NaS) battery NaS batteries are one of the most proven EES in MW scale, with projected total installations of 606 MW by 2012 [149]. NaS has a high energy density, high efficiency of charge/discharge and long cycle life, and is fabricated from inexpensive materials. The main issues with NaS are the need to operate at a high operating temperature of 300–350 °C, and the requirement for thermal management to maintain the ceramic separator and cell seal integrity, which otherwise crack at a lower temperature. The highly corrosive nature of the sodium polysulfides presents another challenge for ceramic insulator protection.

Nickel–cadmium (Ni–Cd) battery Ni–Cd batteries are among the oldest EES technologies that are further developed since the 1990s. Ni-Cd batteries have served in different applications uninterruptable power supply to telecommunications systems. The world's largest Ni-Cd battery at 27 MW rated power, and the US largest EES, has been running since 2003 in Fairbanks, Alaska, USA [150]. The technology is currently very expensive as compared to other EES technologies and there are severe issues with disposal handling associated with the toxicity of the heavy metals (Ni and Cd).

2.7.2 Storage Sizing

This sub-section presents a deterministic approach for sizing a solar photovoltaic (PV) and energy storage system (ESS) with Anaerobic Digestion (AD) biogas power plant (BPP) to meet a proportional scaled-down demand of the national load in Kenya, Africa. The aim is to achieve a minimal Levelized Cost of Energy (LCOE) for the system while minimizing the energy imbalance between generation and demand due to AD generator constraints and solar resources. This system also aims to maximize the sizing of PV to follow the future trend of high penetration of PV. LCOE for the system and Levelized Cost of Delivery (LCOD) are calculated for the hybrid energy system with the presence of energy storage. Four years of solar data collected from Johannesburg, Africa are used for system sizing purposes. An in-depth study of the optimization problem has been given and Particle Swarm Optimization with Interior Point Method is chosen for solar panel sizing. The optimal sizing ratio for the generation sources AD and PV is 2.4:5. The results show that the hybrid system will be cost-effective compared to the AD-only system when the discount rate drops below 8% with the current technology costs.

Determining the optimal solution for a stand-alone hybrid renewable energy systems optimization problem is a complicated task because of the high number of variables and the non-linearity in the performance of some of the system components [157]. The use of the AD Biogas power plant has not been included in the study. A model aimed to minimize the cost of the PV system according to the minimization of the PV array area and storage battery is presented in [158]. The method calculates the minimum number of storage days and the minimum PV array area. A comparison between stand-alone and hybrid system sizing is presented here. It is noted that the cost has not been evaluated for the system. The study uses average solar irradiance to determine the PV module characteristics which does not take the daily fluctuation of solar irradiance and daily energy storage requirement into account.

Reference [159] presents a PV-diesel hybrid power system with battery backup for a village with the computer package HOMER. HOMER [160] is an optimization software package which simulates different renewable energy sources system layouts and sized them on the basis of net present cost. It uses a sensitive analysis to consider different generation capacities and battery storage capacity to determine the optimal size of the system. The issue with this program is the high computational requirement, due to the large number of cases needed to be computed. The study required a total of 448,000 runs based on 28 sensitivities, where sensitivities are defined as the sizing control parameters such as the size of PV, diesel generator etc. Also, the software is of "Black Box" code utilization, where knowledge of its internal workings and optimization algorithm is unknown [161]. In the present study, it aims to use all renewables by replacing diesel with AD which is a controllable renewable.

An optimal sizing method for the wind-solar-battery hybrid system with stand-alone and grid-connected modes was proposed in [162]. A brute force technique is used to determine the optimal sizing by searching for the best combinations of the PV-wind-battery while satisfying the proposed constraints. Loss of power supply probability and the fluctuation rate of the total output of renewable sources relative to the average load power were calculated for every probable combination. The optimal combination is chosen with a minimum system cost. This method has a very high computation complexity and will scale up if detail sizing is required. A comparison of the sizing method has been made with HOMER, and the authors claimed the proposed approach has higher computational costs. Generators have not been considered for the sizing purposes

The techno-economic feasibility studies of utilizing PV-diesel–battery hybrid systems to meet the load of a residential building, with an annual electrical demand of 35.12 MWh and a commercial building with an annual electrical demand of 620 MWh are presented in [163, 164] respectively. HOMER software has been used to carry out the studies. It concluded that the hybrid system offers several benefits such as PV penetration is high; load can be fulfilled in the optimal way; diesel maintenance can be minimized and reliable power supply could be increased. The study uses monthly average daily solar global irradiance as input for the sizing purposes. The uncertainty aspect has not been included in the solar PV generation. The discount rate and financing costs have not been considered when performing economic analysis.

An optimal sizing methodology for a stand-alone and grid-connected PV-biomass hybrid energy system that serves the electricity demand of a typical village is presented in [165]. The results obtained show that the grid-connected hybrid system may be a cost-effective electrification solution for numerous villages in developing countries. However, in practice, it is impossible to be grid-connected in numerous locations, especially for remote areas. An energy storage system should be employed to overcome this issue. The technical constraints from the biomass gasifier have not been considered in the study and also the average global solar irradiance was used for the study.

A study on the optimal sizing of a hybrid wind-PV-diesel stand-alone power system is given in [166]. The consideration focused on the investment cost (installation and unit costs) and fuel cost minimization with constraints on the reliability requirement and CO_2 emission limit. The output power of the diesel generator ranges from 0 kW to the rated capacity, the technical and environmental constraints of diesel generator have been neglected in the study.

An algorithm for the economical design of a utility-scale photovoltaic power plant via compromising between the cost of energy and the availability of the plant was proposed in [167]. This sub-section introduces the effective Levelized Cost of Energy (ELCOE) index as the core of the proposed design algorithm. ELCOE is an improved index based on the conventional LCOE that includes the availability of a power plant in economical assessments. However, the ELCOE proposed did not consider the use of storage systems.

2.7.2.1 Context of the Sizing Problem

Due to the diurnal, stochastic effect of solar irradiance, and the constraints from the BPP, the optimal sizing has become a complicated issue. A schematic figure of the hybrid system to be sized is shown in Fig. 2.35. The hybrid system aims to dispatch the maximum available solar power at each instantaneous time interval to meet the load demand. The problem arises when the solar power starts increasing and decreasing during the morning and afternoon respectively, this will change the required output from the BPP.

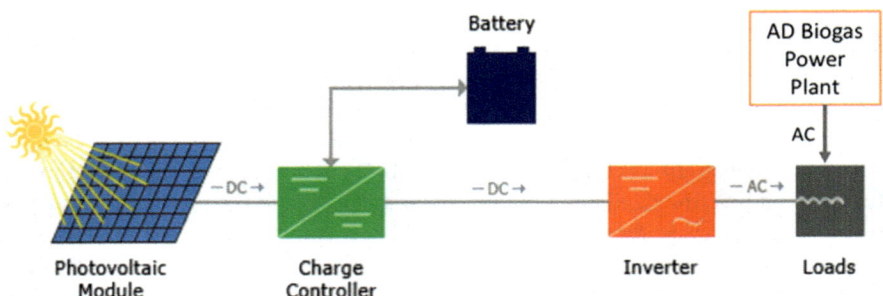

Fig. 2.35 Schematic diagram of the hybrid energy system

A technical constraint for partial load operation of gas turbine power plants is the minimum emissions-compliant load (EMCL). This is the lowest output at which the gas turbine power plant can operate and still meet environmental limits for nitrous oxides (N_2O) and carbon monoxide (CO) emissions. Operation at lower loads can result in reduced combustion temperature, less conversion of CO to CO_2 and potential emissions permit exceedances. The EMCL for most gas turbines is about 50% [168, 169] of full output. To enable a wider range of gas turbine output, manufacturers have introduced control systems designed to extend emissions-compliant turndown while minimizing efficiency impacts at part load. Part load is when the generator is at some specified load value below 100% of its rated capacity. The approach is to produce higher combustion temperatures at low loads. Higher combustion temperatures not only enhance the conversion of CO to CO_2 but also boost steam production and thus output from the steam turbine, improving overall part-load plant efficiency. As a result, some gas turbine models such as (Siemens SGT6-5000F) [170] can achieve emissions-compliant turndown to about 40% of baseload power [169, 171].

Additionally, if without enough cylinder pressure to maintain oil control at low loads, gas engines can develop ash deposits, a reduced detonation margin, and damaged engine components. Similar to diesel generator sets, deposit build-up on valves, spark plugs and behind piston rings can occur, which may cause cylinder liner polishing, power loss, poor performance, and accelerated component wear [172, 173].

In the afternoon, it is most likely there will be surplus energy. This energy should be stored and used to meet energy demands. In the late afternoon, the solar irradiance reduces and there is a need to increase the BPP output to meet the load. There are two occasions where there is not enough energy supply from both PV and BPP. BPP shuts down due to the operating constraint and not enough solar irradiance is available. Figure 2.36 presents a typical solar power curve from real-life irradiance data, the BPP power from AD, and a down-sized load curve. The BPP has been used to compensate the energy deficiency when solar energy is not available. The surplus and deficit energy are highlighted in green and purple, respectively.

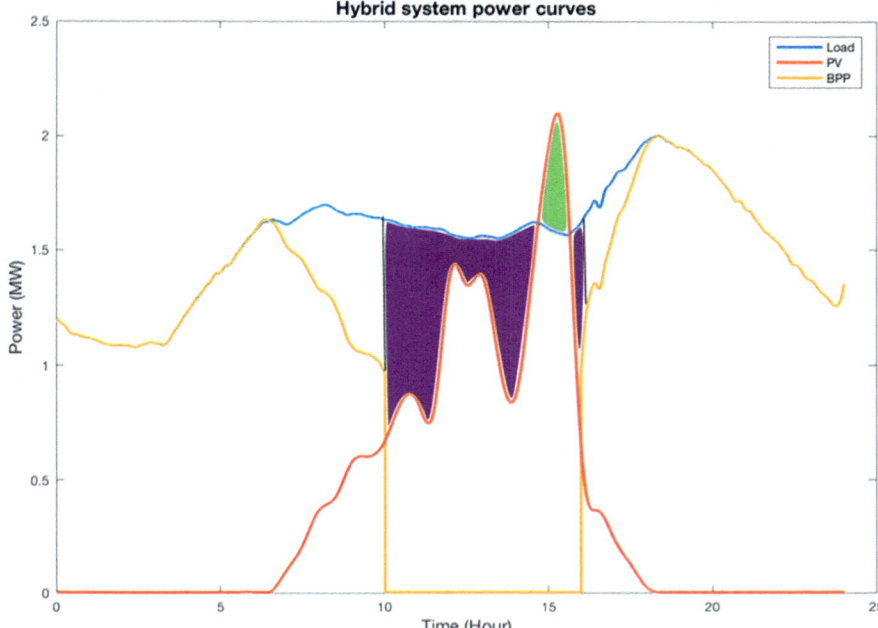

Fig. 2.36 Power curves of hybrid system

2.7.2.2 Optimization Problem Formulation and Methodology

Without loss of generality, optimal sizing is defined as the sizing to achieve minimal LCOE. For the standalone hybrid renewable power system, this is achieved by determining the balance of energy supply and demand. When the system is over-sized (surplus energy more than the deficit energy), energy wastage will occur and LCOE will increase. The undersized system will cause the energy imbalance issue and leads to a high risk in power supply security.

The objective function is:

$$\min_{\text{area}} \left| E_{\text{Battery}} - E_{\text{Deficit}} \right| \qquad (2.55)$$

where,

$$E_{\text{Battery}} = E_{\text{Surplus}} . \eta \qquad (2.56)$$

E_{Battery} is the energy produced from the PV system to be stored in the storage system to meet E_{Deficit} with round trip energy efficiency η considered.

$$E_{\text{Surplus}} = \int_{t=0}^{24} P_{\text{Surplus}} \, dt \tag{2.57}$$

E_{Surplus} is the additional energy produced from the PV system. P_{Surplus} is the instantaneous PV power subtracted by the instantaneous load demand when load demand is less than the PV power.

$$E_{\text{Deficit}} = \int_{t=0}^{24} P_{\text{Deficit}} \, dt \tag{2.58}$$

E_{Deficit} is the energy required from storage to meet the load that cannot be met by PV and AD during time t. P_{Deficit} is the deficit power when no solar or AD power is available to support load demand. The constraints for the AD system are given in Eqs. (2.59)–(2.61):

$$P_{\text{ADmin}} \leq P_{\text{AD}}(t) \leq P_{\text{ADmax}} \tag{2.59}$$

$$P_{\text{ADmin}} = P_{\text{ADmax}} * 0.4 \tag{2.60}$$

$$P_{\text{AD}}(t) = P_{\text{Load}}(t) - P_{\text{Solar}}(t) \tag{2.61}$$

It is assumed that BPP will not produce power when the power output drops below 40% of the rated capacity, as shown in Eq. (2.60). Equation (2.61) states that the output power from AD plant P_{AD} will be used to support the load demand after P_{Solar} has reached the output capacity during time t. Reasonable assumptions have been made for case studies that the rated capacity of BPP, P_{ADmax} is at 2.4 MW [174] with η at 70% [151, 155]. The output power of the solar panel, P_{Solar} is shown in Eq. (2.62).

$$P_{\text{Solar}} = \text{Irradiance}\left(\text{Wm}^{-2}\right) * \text{Area}\left(\text{m}^2\right) * \text{Efficiency}\left(\%\right) \tag{2.62}$$

2.7.2.3 Comparison of Optimization Methods

To have a better intuition of the optimization problem, Fig. 2.37 shows a plot of the objective function and the variable to be minimized for a case on 27th December 2012. Initially, as not shown in the figure, the objective function is at its minimum because the system's energy balance is achieved due to enough deployment of AD power to support the load demand. As the penetration of solar energy increases, AD will switch off due to the constraint given in Eqs. (2.59)–(2.61).

The solar power curves used for the case studies are from the practical irradiance data and not the clear sky model, hence it contains perturbations. This will influence the switching of AD in an unsystematic manner. The cost function will become

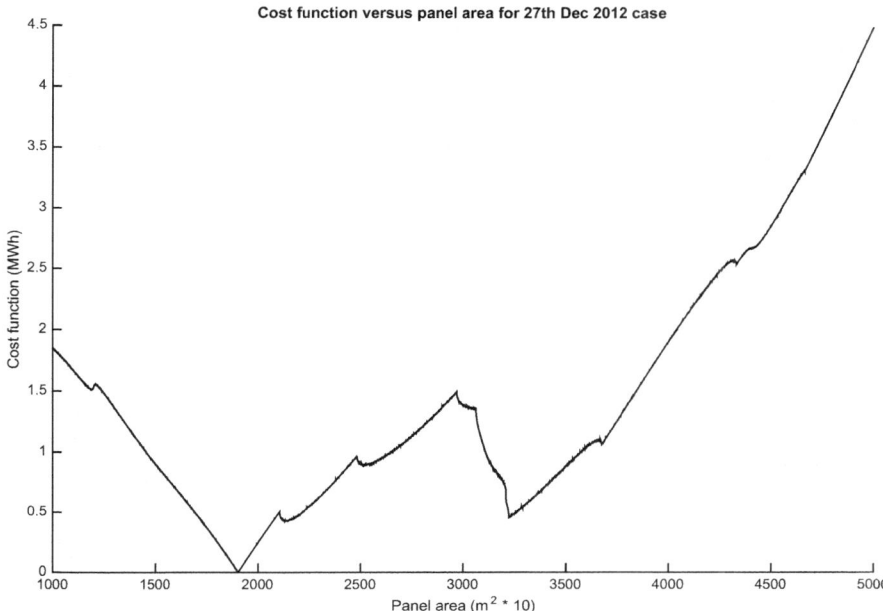

Fig. 2.37 Objective function with solar panel area variation for 27th December 2012 (LB case)

infinite as the panel areas approach infinity; this means there will be too much surplus energy. The optimization function is highly non-linear with multiple local minima.

Seven different types of optimization techniques have been studied for the optimal sizing problem. These are Interior Point Method (IP), Pattern Search (PS), Genetic Algorithm (GA), Genetic Algorithm with Interior Point Method (GAIP), Particle Swarm Optimization (PSO), Particle Swarm Optimization with Interior Point Method (PSOIP) and Simulated Annealing (SA). Two different search boundaries are used to study the optimization problem. The lower bound (LB) case is from 10,000–50,000 m² and the upper bound (UB) case is from 50,000 to 200,000 m². The reason for the LB to be 10,000 m² is to make sure the hybrid system has a reasonable penetration of solar PV energy. For IP, SA, and PS, the initial point needs to be predefined. For this study, initial points are 30,000 and 120,000 for LB and UB cases, respectively.

The results for 17th, 21st, and 22nd in Fig. 2.38 are errnoeuous as the cost function values are high and the resulting panel area is at the search boundary, i.e., 10,000 or 50,000 m². The x-axis represents the day for the daily case sizing and the y-axis is the cost function to be minimized. The optimal solution is not in the search boundary, hence the required panel area is significantly higher. This is due to the poor weather conditions of the day and the lack of solar irradiance. There is a requirement for more solar panels to provide enough solar energy. The majority of the results are similar with a few cases where there are high discrepancies. The

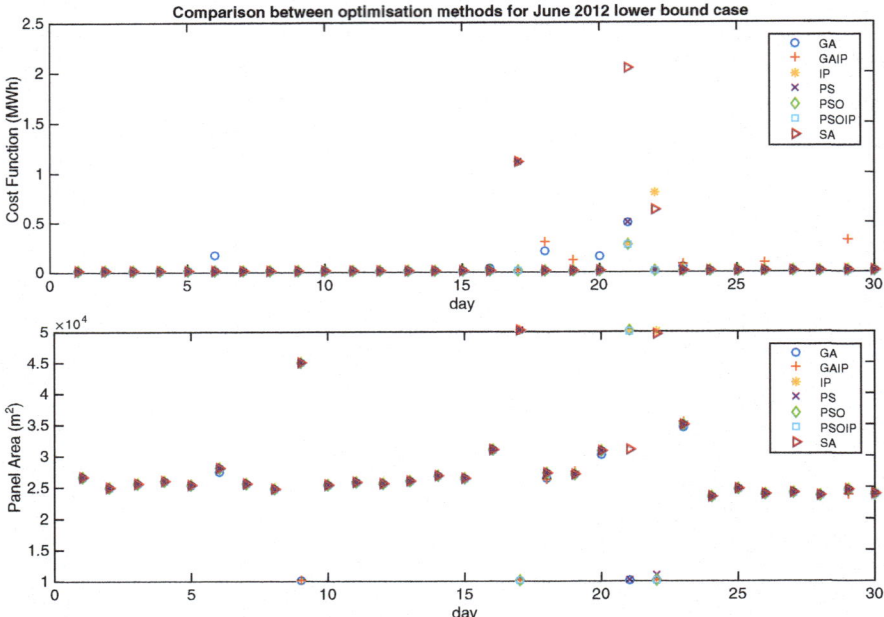

Fig. 2.38 PV system sizing for June 2012 LB case

optimization results for 4th June 2012 have been given in Fig. 2.39. It can be concluded that PSOIP has the lowest cost function value.

A UB case is used to size the solar panels for extreme weather conditions. Figure 2.40 shows the results for the June 2012 case. In the converse manner to Fig. 2.28, the cost function values are high for all days except for 17th, 21st, and 22nd. This signifies there are feasible solutions for these days and the cost function can be minimized. The optimization results for 22nd June 2012 are shown in Fig. 2.41.

By considering the factors of minimal cost function value and the ability to reach global optimal, it can be concluded that PSOIP is the best candidate for this optimization problem. However, it should be noted that the choice of the optimization algorithm is problem-dependent as explained in [161].

2.7.2.4   Optimization Framework and Result

The optimization process begins with data treatment and input of data to the optimization algorithm. Four years of solar irradiance data have been segmented into individual days to determine the required panel area for each day. LB and UB cases are computed for each day. If there is a feasible solution in the UB, this result will be replaced in the LB result as this signifies the required panel area is much higher. The 4 years of daily case optimization results are shown in Fig. 2.42. The red crosses in

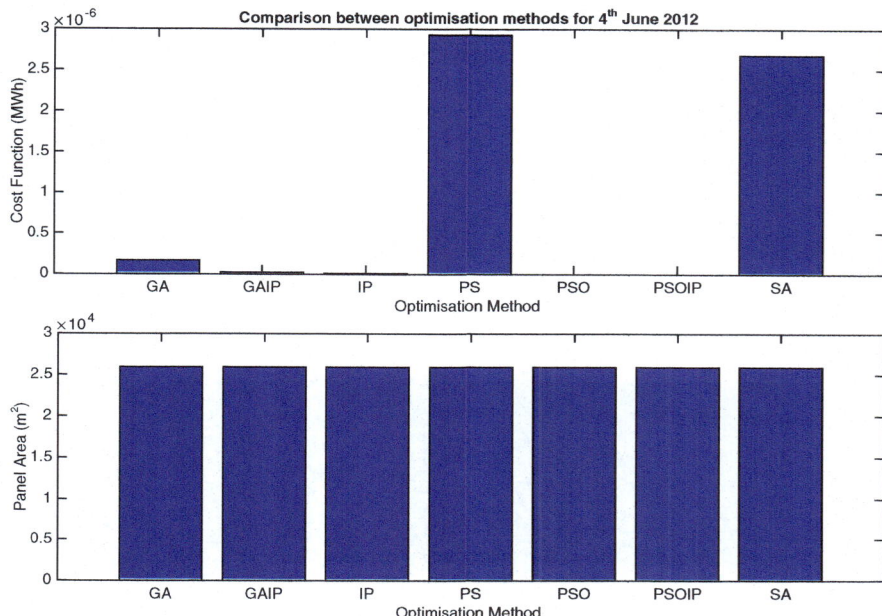

Fig. 2.39 Optimization method comparison for 4th June 2012

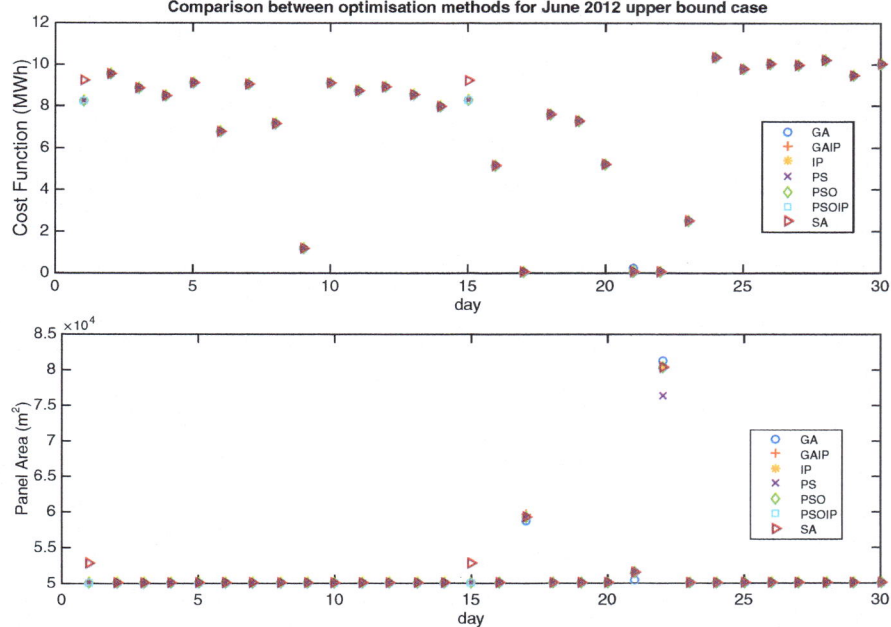

Fig. 2.40 PV system sizing for June 2012 UB case

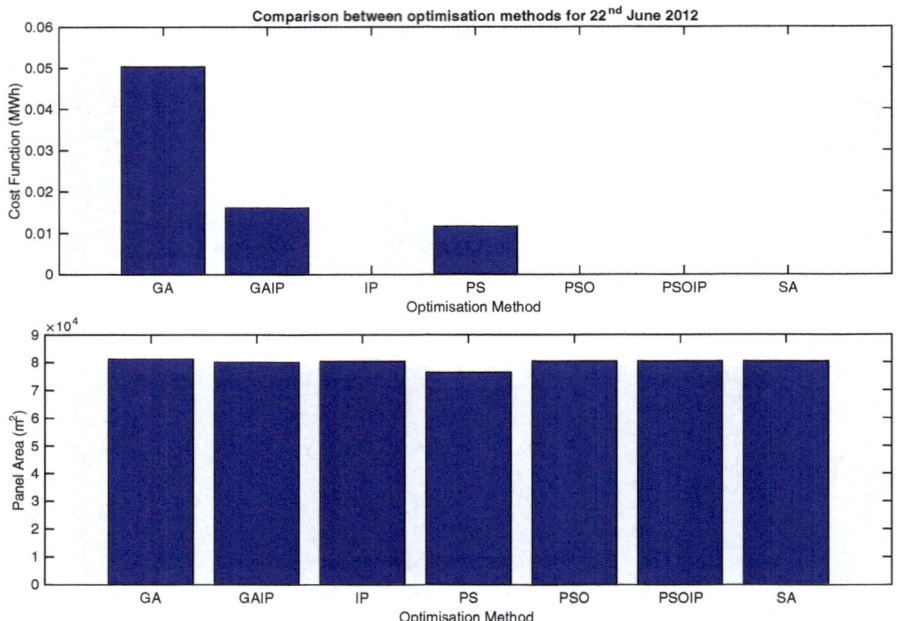

Fig. 2.41 Optimization method comparison for 22nd June 2012

the box plot mark the "outliers" of the sizing results, and have a large contrast to the mean. The explanation of the outliers is due to the low solar irradiance on the day and the requirement of a high panel area to produce enough surplus energy.

The search boundary needs to be divided into two known cases, namely, the LB case and the UB case. As shown previously, this is because the search boundary is too large for the optimization algorithms to converge and to determine the correct optimal point. PSO parameters such as the inertia range and minimum fraction neighbors have been tuned for the optimization algorithm to give accurate results. The inertia range determines the contribution rate of a particle's previous velocity to its velocity at the current time step. The proper selection of the neighborhood size affects PSO's trade-off between exploration and exploitation, and unfortunately, there is no formal procedure to determine the optimal size. The parameters and values for the PSO algorithm, for the daily case sizing are given in Table 2.18.

2.7.2.5　Component Sizing of Hybrid System

Sizing of Solar Panels

The solar panel to be used for the hybrid system is the Sharp ND-R250A5. It has an efficiency of 15% and has a rated power of 250 W/panel [175].

Annual Sizing Case Study

The daily solar irradiance profiles are connected as one annual profile for the annual sizing case study. The daily load curve is repeated according to the number of days in the annual case. The optimization is performed with PSOIP. Table 2.11 shows a consistent number of panels for the 4 years of data. It is safe to assume the number of panels required is 20,000 m^2 and the system is equivalent to 5 MW. When compared to the daily sizing case, it can be seen that the annual sizing case gives a "smoothing effect" since it averages the load and solar data first before the optimization performs. The implications of the energy balance are less well understood with this sizing approach.

Daily Sizing Case Study

Due to the stochastic effect and inconsistent solar irradiance level in different seasons, it is impossible to determine the exact rated capacity of the solar farm to provide enough supply to the grid. It is impractical to install a PV system that is capable of providing a solution to all events at all times; either the events would have to be very modest or the ESS is very large.

From Sect. 2.7.3.4, it is difficult to draw conclusions and insights from the box plot in Fig. 2.42. A method is proposed to determine the solar farm capacity by forming a histogram and considering the rated capacity of each day for the 4 years of data. The solar farm power rating for an individual day is calculated with Eq. (2.63).

$$P_{\text{SolarFarm day}} = P_{\text{Panel\_rated}} * N_{\text{PV day}} \tag{2.63}$$

$N_{\text{PV day}}$ is the panel area for the given day and $P_{\text{Panel\_rated}}$ is the rated capacity of the PV panel. Figure 2.43 shows the solar farm power rating for 4 years of the daily case in a histogram plot calculated with Eq. (2.63). This effectively displaying the probability distribution function for the required sizing of the solar panels. The calculated value is then rounded up to the nearest positive infinity. Most of the required capacity is in the range of 2–6 MW with few cases above 10 MW. This can be explained due to the poor weather and low irradiance. It will be uneconomical to size the PV system to provide solar energy to these extreme cases. The AD generator could be used to meet the energy requirement in this case.

Table 2.11 Annual case sizing results

Year	Panel area (m^2)	Rated capacity of PV Farm (MW)
2009	19,770	4.95
2010	19,851	4.97
2011	19,468	4.87
2012	19,051	4.77

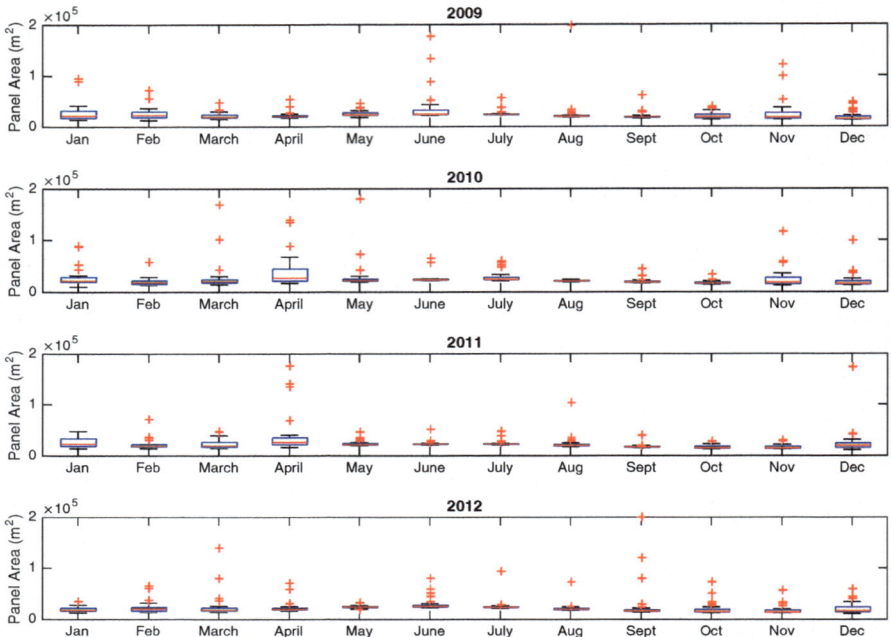

Fig. 2.42 Optimization results for the daily case PV system sizing

Table 2.12 provides the percentage at different solar power rating. At 2 MW, only 1.304% of days in 4 years have enough solar energy supply and up to 90% of days are covered when the sizing is at 5 MW. The capital cost increases significantly by increasing the solar farm by a megawatt. Since the total percentage cover increases slightly when above 5 MW and by taking the daily and annual case sizing results into consideration, it can be concluded that 5 MW is the best choice for sizing the solar farm. The deficit energy could be compensated by temporary running BPP at low efficiency.

Sizing of Storage

After the size of the PV farm has been determined, E_{Deficit} is calculated for each day to determine the energy required to be stored in the storage system. Table 2.13 shows the results for E_{Deficit} with the maximum value for the corresponding year. It shows that in 2011 the deficit energy is significantly lower and in 2009 has the highest deficit. Figure 2.44 shows the solar irradiance curves with the Julian day number for the maximum E_{Deficit} during the daily sizing case. The irradiance curve for 2011 has less fluctuation compared to the other 3 years. This could reduce the number of switching of the AD system and deficit energy could be reduced, resulting in a lower energy storage requirement.

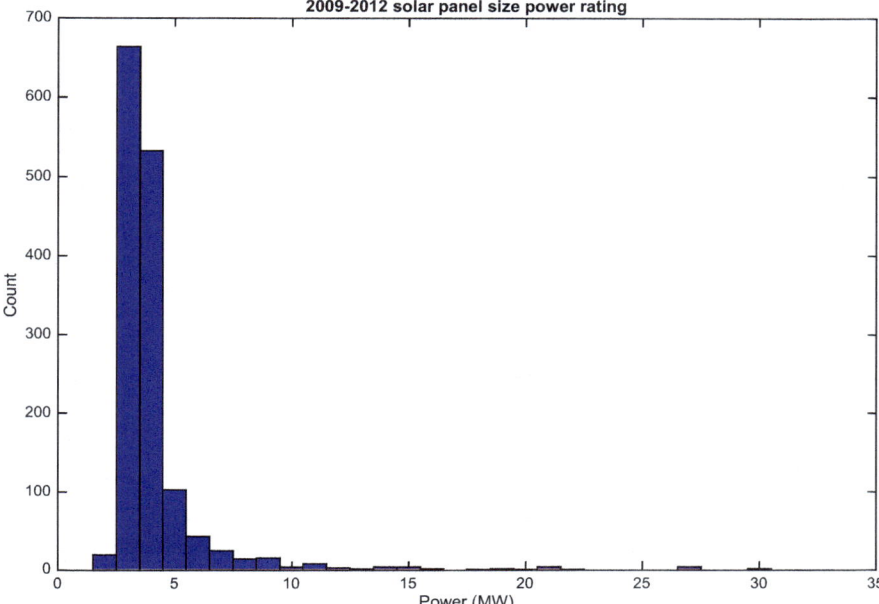

Fig. 2.43 Count for daily case solar panel sizing

Table 2.12 Percentage day covered at different power ratings

Power rating (MW)	Percentage for 4 years (%)	Accumulated percentage covered (%)
2	1.304	1.304
3	45.5731	46.8771
4	36.582	83.4591
5	7.0693	90.5284
6	2.8826	93.411
7	1.6472	95.0582
7+	4.9418	

Table 2.13 Maximum E_{Deficit} for 2009–2012

Year	E_{Deficit} (MWh)	E_{PV} (MWh)
2009	4.96	2.32
2010	4.82	2.15
2011	3.98	2.62
2012	4.71	2.01

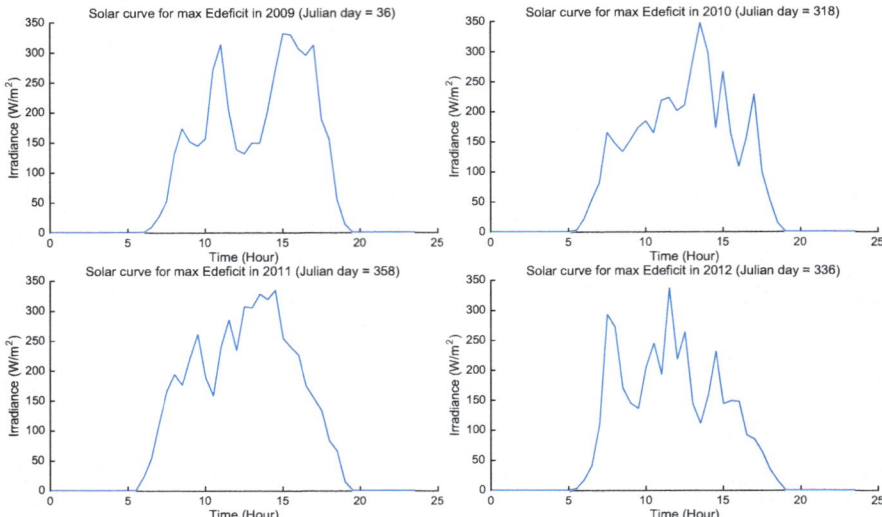

Fig. 2.44 Solar irradiance curves for 2009–2012 at maximum E_{Deficit}

The storage is to be sized at 5 MWh by considering that the E_{Deficit} is under 5 MWh for the 4 years of results

Sizing of Inverter and Controller

The following constraints in Eqs. (2.64) and (2.65) need to be fulfilled for the proposed hybrid system. The power output of total inverters and controllers need to be larger than the rated capacity of solar PV.

$$P_{\text{Inv}} N_{\text{Inv}} \geq P_{\text{solar}} \tag{2.64}$$

$$P_{\text{Con}} N_{\text{Con}} \geq P_{\text{solar}} \tag{2.65}$$

P_{Inv} and P_{Con} are the rated power of the inverter and controller respectively. N_{Inv} and N_{Con} are denoted for the number of inverters and controllers. For the hybrid system, the required number of inverters and controllers for the hybrid system is 1250 and 2500 respectively. The optimal sizing result for each component of the hybrid system is presented in Table 2.19.

2.7.2.6 Levelized Cost of Energy

LCOE is a measure of costs which attempts to compare different methods of electricity generation on a comparable basis. It is an economic assessment of the average total cost to build and operate a power-generating asset over its lifetime divided by the total energy output of the asset over that lifetime.

The economic projections on complex hybrid systems utilizing these three technologies are challenging and no comprehensive method is available for guiding decision-makers [176]. The authors claimed to have provided a new method of quantifying the economic viability of off-grid PV-battery-CHP systems by calculating the LCOE of the technology to be compared to centralized grid electricity. The proposed LCOE for the hybrid system is given in Eq. (2.66) below [176]:

$$
\text{LCOE} = \frac{I + \sum_{t=1}^{n} \dfrac{\left(I * i + O + F_{\text{chp}}\right)}{\left(1+r\right)^{t}}}{\sum_{t=1}^{n} E_{\text{tpv}} \left(1-d_{1}\right)^{t} + \dfrac{E_{\text{tchp}} \left(1-d_{2}\right)^{t}}{\left(1+r\right)^{t}}}
\tag{2.66}
$$

n is the lifetime of the hybrid system in years, r is the discount rate on the hybrid system per year, I is the total installation cost which includes the cost of solar PV, battery and the CHP module, i is the interest rate on the hybrid system for 100% debt financing. O is the total operation and maintenance cost. F_{chp} is the annual fuel cost of the CHP unit. E_{tpv} and E_{tchp} are the rated annual energy production from solar PV and CHP unit respectively. d_{1} and d_{2} are the degradation rates for solar PV and CHP unit respectively. The energy produced by the PV system is not discounted. It does not reflect the actual value of solar PV energy in the future. Cost implication due to storage has not been included in the analysis in detail. Although storage does not generate energy, the total energy production will be affected by storage due to round trip efficiency. This section presents the cost calculation of the system and comparisons for two different systems will be made. Table 2.14 gives the cost specification of the components for the hybrid system.

The general equation for LCOE [180] is given in Eq. (2.67). The cost and energy calculation of the system components is given in Eqs. (2.68)–(2.76).

$$
\text{LCOE} = \frac{\text{Lifecycle cost}\,(\$)}{\text{Lifetime energy production}\,(\text{kWh})}
\tag{2.67}
$$

$$
C_{\text{Redox}} = C_{\text{cap\_ESS}} + \sum_{t=0}^{n} \frac{C_{\text{O\&M\_ESS}t}}{\left(1+r\right)^{t}}
\tag{2.68}
$$

Table 2.14 Cost specification of the system

	PV (sharp ND-250QCS)	Vanadium redox flow battery (VRB)	AD biogas power plant	Inverter (Schneider electric XW4024)	Controller (outback FM 80)
Capital cost (Cap)	120 ($/unit) [175, 177]	760–1600 ($/ kWh) [151, 155]	$7.5M [174, 178]	812.05 ($/unit) [175]	335 ($/unit) [175]
Installation cost (Inst)	108 ($/unit) [175]	N/A	N/A	24.2386 ($/unit) [175]	6.7 ($/unit) [175]
O&M cost (O & M)	6 ($/unit/year) [175]	100–140 ($/ kWh) [151]	350 ($/ kW) [179]	2.43615 ($/unit/ year) [175]	1.005 ($/unit/ year) [175]

$$E_{\text{ESS}} = \eta \sum_{t=0}^{n} \frac{E_{\text{surplus}t}\left(1 - D_{\text{Redox}}\right)^t}{\left(1+r\right)^t} \tag{2.69}$$

$$C_{\text{pvsurplus}} = \left(C_{\text{cap\_pv}} + C_{\text{inst\_pv}} + \sum_{t=0}^{n} \frac{C_{\text{O\&M\_pv}t}}{\left(1+r\right)^t} \right) N_{\text{pvsurplus}} \tag{2.70}$$

$$C_{\text{pvdirect}} = \left(C_{\text{cap\_pv}} + C_{\text{inst\_pv}} + \sum_{t=0}^{n} \frac{C_{\text{O\&M\_pv}t}}{\left(1+r\right)^t} \right) N_{\text{pvdirect}} \tag{2.71}$$

$$E_{\text{pvdirect}} = \sum_{t=0}^{n} \frac{\left(E_{\text{direct}t}\right)\left(1 - D_{\text{pv}}\right)^t}{\left(1+r\right)^t} \tag{2.72}$$

$$C_{\text{AD}} = C_{\text{capAD}} + \sum_{t=0}^{n} \frac{C_{\text{O\&M\_AD}t}}{\left(1+r\right)^t} \tag{2.73}$$

$$E_{\text{AD\_total}} = \sum_{t=0}^{n} \frac{E_{\text{AD}t}}{\left(1+r\right)^t} \tag{2.74}$$

$$C_{\text{Inv}} = \left(C_{\text{cap}_{\text{inv}}} + C_{\text{inst}_{\text{inv}}} + \sum_{t=0}^{n} \frac{C_{\text{O\&M}_{\text{inv}}t}}{\left(1+r\right)^t} \right) N_{\text{inv}} \tag{2.75}$$

$$C_{\text{Con}} = \left(C_{\text{cap}_{\text{con}}} + C_{\text{inst}_{\text{con}}} + \sum_{t=0}^{n} \frac{C_{\text{O\&M}_{\text{con}}t}}{\left(1+r\right)^t} \right) N_{\text{con}} \tag{2.76}$$

In this study, the lifetime of the hybrid system is assumed to be 20 years, with the storage system degradation rate, D_{Redox} at 0.1% per year [181] and PV panel degradation rate, D_{PV} at 0.05% [167, 182]. There is an inverter lifetime extension to at least 20 years of full operation by 2013 and 30 years by 2020 [183]. The controllers and inverters are assumed to be replaced once during the lifetime of the system, and the replacement cost is the same as the capital cost [175]. $N_{pvdirect}$, $N_{pvsurplus}$, N_{Inv} and N_{Con} are the number of units of PV panels for generating energy for direct consumption, surplus energy for storage, inverters, and controllers respectively. E_{direct} is the energy generated from PV and directly supplied to the load without going through storage. E_{AD_total} is the total lifetime energy output from BPP powered by AD. $C_{pvsurplus}$ and $C_{pvdirect}$ are the total lifetime costs of PV generation that produce the surplus and direct consumption of energy for the system respectively.

2.7.2.7   Levelized Cost of Delivery

The term Levelized Cost of Storage (LCOS) was explored in [152], which is solely used for comparing storage technologies. The equation is of similar nature to LCOE. Levelized Cost of Delivery (LCOD) is proposed to compare the cost-effectiveness of storage for the system [184]. The LCOD is given in Eq. (2.77).

$$
\text{LCOD} = \frac{\sum_{t=0}^{n} \dfrac{C_{int}}{(1+r)^t}}{\eta \sum_{t=0}^{n} \dfrac{E_{int}}{(1+r)^t}} + \frac{\sum_{t=0}^{n} \dfrac{C_{ESSt}}{(1+r)^t}}{\eta \sum_{t=0}^{n} \dfrac{E_{int}}{(1+r)^t}}
\tag{2.77}
$$

By splitting Eq. (2.77) into two individual components, the final form of LCOE for the ESS is given in Eq. (2.78).

$$
\text{LCOD} = \frac{1}{\eta} \text{LCOE}\left(E_{surplus}\right) + \text{LCOS}
\tag{2.78}
$$

In practice, the energy flowing into ESS, E_{in} will be the surplus energy $E_{surplus}$. The cost of storing the surplus energy into ESS, C_{in} will be $C_{pvsurplus}$ plus the cost of the controller C_{Con}.

2.7.2.8   Levelized Cost of Energy for System

For a hybrid renewable and storage power system, the following LCOE relationship will hold:

$$\text{LCOE}_{\text{system}} = \frac{\sum_{t=0}^{n} \dfrac{C_{\text{system}\,t}}{(1+r)^t}}{\sum_{t=0}^{n} \dfrac{E_{\text{system}\,t}}{(1+r)^t}} \tag{2.79}$$

C_{system_t} and E_{system_t} are the total cost and total energy production from the system at year t respectively. The total cost of the renewable system is the sum of PV, BPP generation, power conversion, and storage costs. The total energy produced by the system is the energy output of ESS, the energy directly delivered to the load by PV, and the energy produced by BPP to support the energy deficit. Therefore, the LCOE for the system is given in Eq. (2.80).

$$\text{LCOE}_{\text{system}} = \frac{C_{\text{pvsurplus}} + C_{\text{ESS}} + C_{\text{pvdirect}} + C_{\text{AD}} + C_{\text{Inv}} + C_{\text{Con}}}{E_{\text{ESS}} + E_{\text{pvdirect}} + E_{\text{AD\_total}}} \tag{2.80}$$

2.7.2.9 Cost Analysis for PV Hybrid System

The LCOE at different discount rates 2, 8, 10, and 15% are studied for three case studies and the results are presented in Tables 2.15, 2.16, and 2.17. The AD-only system is a micro-grid system that generates energy solely by BPP, with no PV, ESS, inverter, and controller installed. The hybrid system is the micro-grid proposed in Fig. 2.36. Storage is an expensive component and also the energy stored in general is a small proportion as compared to generated energy, so the ratio between battery cost to its amount of stored energy will be bigger as compared to that LCOE$_{\text{system}}$. This is the reason that LCOD is significantly higher than the LCOE of the system. Using sensitivity analysis, the cross-over point for the system's LCOE can be determined. The results show that by considering the Redox storage at lower bound cost, the hybrid system could be cheaper than running an AD only system when the discount rate is below 8%. At a higher bound cost, the discount rate needs to be below 2%.

As reported in [151], the LCOS for the Redox storage system in renewable energy system integration is between 0.373 and 0.950 $/kWh, with a discount rate of 8%. Since there is a fixed cost, while both variable cost and energy are affected

Table 2.15 AD only case

r (%)	LCOE$_{\text{system}}$ ($/kWh)
2	0.383
8	0.403
10	0.409
15	0.427

Table 2.16 Hybrid system with VRB at lower bound cost

r (%)	LCOD ($/kWh)				LCOE$_{system}$ ($/kWh)			
	2009	2010	2011	2012	2009	2010	2011	2012
2	0.830	0.886	0.797	0.736	0.339	0.343	0.337	0.334
8	1.156	1.233	1.110	1.025	0.389	0.393	0.386	0.383
10	1.275	1.359	1.224	1.130	0.407	0.411	0.404	0.401
15	1.579	1.683	1.516	1.401	0.543	0.458	0.451	0.447

Table 2.17 Hybrid system with VRB at upper bound cost

r (%)	LCOD ($/kWh)				LCOE$_{system}$ ($/kWh)			
	2009	2010	2011	2012	2009	2010	2011	2012
2	1.210	1.292	1.161	1.070	0.358	0.362	0.356	0.353
8	1.753	1.871	1.681	1.550	0.418	0.423	0.416	0.412
10	1.949	2.080	1.870	1.724	0.440	0.445	0.438	0.434
15	2.456	2.621	2.356	2.172	0.497	0.502	0.494	0.490

equally by the discount rate, therefore LCOS increases as the discount rate increases. The LCOD of the system is much higher than LCOS when the cost of storing the energy is included. As reported in [185], the current discount rate for Solar PV and AD is 6–9% and 7–10% respectively. The discount rate for technologies that are supported by policy could be as much as 2–3% lower over the next decade and could fall by a further 1–2% by 2040. The LCOE for the hybrid system has been given the assumption that the capital cost of PV will be reduced by 50% as compared to that for PV system in [175] due to Swanson's law [177]. At high discount rates, capital intensive generation sources such as PV is at a disadvantage due to the value of energy and money is lower in the future.

In general AD-only system can have a smaller LCOE but may be different for a smaller discount rate. This is likely to be a future trend. If there are incentivizes for example, for equipment cost, there could be a reduction in capital cost and as such it could be better to have a hybrid system as this leads to a lower LCOE (Tables 2.18 and 2.19).

2.8 Standards, Recommended Practices and Guidelines

Some standards, recommended practices and guidelines related to Energy Storage and its applications are given below:

- IEEE 1679–2010—IEEE Recommended Practice for the Characterization and Evaluation of Emerging Energy Storage Technologies in Stationary Applications: provides a common basis for the expression of performance characteristics and the treatment of life testing data

Table 2.18 Parameters for PSOIP for daily case optimization

Max iteration	500
Inertia range	[0.1 1.0]
Self-adjustment weight	1.49
Social-adjustment weight	1.49
Function tolerance	1e − 6
Minimum fraction neighbors	0.5

Table 2.19 Summary of optimal size for the hybrid system

System component	Optimal size
PV	5 MW (20,000 units)
Vanadium Redox Flow battery (VRB)	5 MWh
Inverter (Schneider Electric XW4024)	1250 units (4 kW/unit)
Controller (Outback FM 80)	2500 units (2 kW/unit)

- IEEE 2030.2–2015—IEEE Guide for the Interoperability of Energy Storage Systems Integrated with the Electric Power Infrastructure: provides useful industry derived definitions for ESS characteristics, applications, and terminology that, in turn, simplify the task of defining system information and communications technology (ICT) requirements.
- IEEE P1547.9—Guide to Using IEEE Standard 1547 for Interconnection of Energy Storage Distributed Energy Resources with Electric Power Systems: provides information on and examples of how to apply the IEEE Std 1547, for the interconnection of Energy Storage Distributed Energy Resources.
- IEEE P2686—Recommended Practice for Battery Management Systems in Energy Storage Applications: includes information on the design, installation, and configuration of battery management systems in stationary applications, including both grid-interactive, standalone cycling and standby modes.
- IEEE P2814—Recommended Practices on Techno-economic Terminology for Hybrid Energy and Storage Systems.

2.9   Conclusion and Future Work

This section proposes a sizing methodology with a deterministic approach for a stand-alone high penetration PV system with support from ESS and AD biogas power plant. The costs have been calculated with the proposed levelized cost of energy methods and it shows that the hybrid energy system could be more economical than using a stand-alone AD biogas power plant.

The load curve is assumed to be the same for all days in the year due to the users have a consistent consumption in the small community. Future studies could consider when the load is irregular or less than the minimum AD output power. The conventional approach to sizing the power system is to use the cost of energy as the objective function. This approach could also be studied and comparisons could be made as to future work. State of charge, depth of discharge, and state of health of battery need to be considered in depth in the future work.

As technology advances, smartness is introduced to the cities and leads to transforming their infrastructure, systems, management, and operations to capitalize on new technologies and integrate connected solutions into how to operate and care for the citizens. Through advances in data collection and analytics, they can anticipate and respond to daily challenges like traffic flow and potential emergencies.

There is a need to incorporate sustainability and energy efficiency in developing smart city solutions. It is foreseen that the integration of solar power and energy storage is quickly is becoming a focal point of smart city planning. By looking at some of the innovative ways, smart city initiatives and decision-makers are adopting solar power and storage as a serious workable solution.

Acknowledgments The permission in using the contents of the following papers is very much appreciated.

A. C. S. Lai, F. Xu, Y. Tao, W. W. Y. Ng, Y. Jia, H. Yuan, C. Huang, L. L. Lai, Z. Xu and G. Locatelli, "A robust correlation analysis framework for imbalanced and dichotomous data with uncertainty," Information Sciences, vol. 470, pp. 58–77, 2019
B. C. S. Lai, Y. Jia, M.D. McCulloch and Z. Xu, "Daily clearness index profiles cluster analysis for photovoltaic system," IEEE Transactions on Industrial Informatics," 13(5), pp. 2322–2332, 2017
C. C. S. Lai, Y. Jia, L. L. Lai, Z. Xu, M.D. McCulloch and K. P. Wong, "A comprehensive review on large-scale photovoltaic system with applications of electrical energy storage," Renewable and Sustainable Energy Reviews, 78, pp.439–451, 2017
D. C. S. Lai, and M. D. McCulloch, "Sizing of stand-alone solar PV and storage system with anaerobic digestion biogas power plants," IEEE Transactions on Industrial Electronics, 64(3), pp.2112-2121, 2017

References

1. J. McLaren, Valuing the resilience provided by solar and battery energy storage systems (National Renewable Energy Laboratory, 2018). [Online]. https://www.nrel.gov/docs/fy18osti/70679.pdf
2. S. Chung, 100% renewable energy by 2050? Why wait?, Greenpeace.org, Sept. 2015. [Online]. http://www.greenpeace.org/international/en/news/Blogs/makingwaves/renewable-energy-revolution-2050-Shell/blog/54248/
3. C.S. Lai, Y. Jia, L.L. Lai, Z. Xu, M.D. McCulloch, K.P Wong, A comprehensive review on large-scale photovoltaic system with applications of electrical energy storage. Renew. Sustain. Energy Rev. **78**, 439–451 (2017)
4. "能源发展"十三五"规划," 国家发展改革委, 国家能源局 (2016). [Online]. http://www.ndrc.gov.cn/zcfb/zcfbtz/201701/t20170117_835278.html
5. V. Fthenakis, J.E. Mason, K. Zweibel, The technical, geographical, and economic feasibility for solar energy to supply the energy needs of the US. Energy Policy 37(2), 387–399 (2009)
6. X. Yang, Y. Song, G. Wang, W. Wang, A comprehensive review on the development of sustainable energy strategy and implementation in China. IEEE Trans. Sustain. Energy 1(2), 57–65 (2010)

7. Energy Research Institute National Development and Reform Commission, China 2050 high renewable energy penetration scenario and roadmap study (2015). [Online]. http://www.efchina.org/Attachments/Report/report-20150420/China-2050-High-Renewable-Energy-Penetration-Scenario-and-Roadmap-Study-Executive-Summary.pdf

8. Indian power sector.com, Jawaharlal Nehru national solar mission targets 20,000MW by 2022. [Online]. http://indianpowersector.com/electricity-regulation/national-solar-mission, visited on 16th May 2016

9. B. Mountain, P. Szuster, Solar, solar everywhere: opportunities and challenges for Australia's rooftop PV systems. IEEE Power Energy Mag. **13**(4), 53–60 (2015)

10. T. Stetz, J. von Appen, F. Niedermeyer, G. Scheibner, R. Sikora, M. Braun, Twilight of the grids: the impact of distributed solar on Germany's energy transition. IEEE Power Energy Mag. **13**(2), 50–61 (2015)

11. J. von Appen, M. Braun, T. Stetz, K. Diwold, D. Geibel, Time in the sun: the challenge of high PV penetration in the German electric grid. IEEE Power Energy Mag. **11**(2), 55–64 (2013)

12. K. Ogimoto, I. Kaizuka, Y. Ueda, T. Oozeki, A good fit: Japan's solar power program and prospects for the new power system. IEEE Power Energy Mag. **11**(2), 65–74 (2013)

13. A.Q. Huang, M.L. Crow, G.T. Heydt, J.P. Zheng, S.J. Dale, The future renewable electric energy delivery and management (FREEDM) system: the energy internet. Proc. IEEE **99**(1), 133–148 (2011)

14. L.L. Lai, Global Energy Internet and interconnection (IEEE Smart Grid Newsletter, 2015). [Online]. http://smartgrid.ieee.org/newsletters/october-2015/global-energy-internet-and-interconnection

15. SOLARGIS. [Online]. http://solargis.info/doc/free-solar-radiation-maps-GHI

16. Fact sheet: the solar star projects (SunPower Corporation, 2016). [Online]. https://us.sunpower.com/sites/sunpower/files/media-library/fact-sheets/fs-solar-star-projects-fact-sheet.pdf

17. Desert Sunlight Solar Farm (First Solar). [Online]. http://www.firstsolar.com/en/About-Us/Projects/Desert-Sunlight-Solar-Farm

18. Topaz Solar Farm (First Solar). [Online]. http://www.firstsolar.com/en/About-Us/Projects/Topaz-Solar-Farm

19. Case study: solar PV-hydro hybrid system at Longyangxia, China (International Hydro Association, 2015). [Online]. http://www.hydropower.org/blog/case-study-solar-pv-hydro-hybrid-system-at-longyangxia-china

20. Realtime generation of solar plants in Gujarat (Gujarat Energy Transmission Corporation Limited, 2016). [Online]. https://www.sldcguj.com/RealTimeData/GujSolar.asp

21. Neoen breaks ground on 300 MW French solar plant (PV Magazine, 2014). [Online]. http://www.pv-magazine.com/news/details/beitrag/neoen-breaks-ground-on-300-mw-french-solar-plant_100017099/#axzz4BLrsOucx

22. Agua Caliente Solar Project (First Solar). [Online]. http://www.firstsolar.com/en/About-Us/Projects/Agua-Caliente-Solar-Project

23. Copper Mountain Solar 3 (Cupertino Electric Inc.). [Online]. http://www.cei.com/our-work/copper-mountain-solar-3

24. Califonia Valley Solar Ranch (Energy.gov). [Online]. http://energy.gov/lpo/california-valley-solar-ranch

25. Antelope Valley Solar Ranch (Energy.gov). [Online]. http://www.energy.gov/lpo/antelope-valley-solar-ranch

26. V. Fthenakis, Considering the total cost of electricity from sunlight and the alternatives [point of view]. Proc. IEEE **103**(3), 283–286 (2015)

27. Paula Mints, Photovoltaic technology trends: a supply perspective. [Online]. http://www.idtechex.com/emails/5551.asp

28. R.W. Miles, G. Zoppi, I. Forbes, Inorganic photovoltaic cells. Mater. Today **10**(11), 20–27 (2007)

29. New world record for solar cell efficiency at 46% French-German cooperation confirms competitive advantage of European photovoltaic industry (Fraunhofer ISE, 2014). [Online].

https://www.ise.fraunhofer.de/en/press-and-media/press-releases/press-releases-2014/new-world-record-for-solar-cell-efficiency-at-46-percent

30. A. Willoughby, *Solar Cell Materials: Developing Technologies* (Wiley, 2014)
31. National Centre for Photovoltaics (NREL). [Online]. http://www.nrel.gov/ncpv/
32. B. O'regan, M. Grfitzeli, A low-cost, high-efficiency solar cell based on dye-sensitized. Nature **353**(6346), 737–740 (1991)
33. M.M. Lee, J. Teuscher, T. Miyasaka, T.N. Murakami, H.J. Snaith, Efficient hybrid solar cells based on meso-superstructured organometal halide perovskites. Science **338**(6107), 643–647 (2012)
34. M. Liu, M.B. Johnston, H.J. Snaith, Efficient planar heterojunction perovskite solar cells by vapour deposition. Nature **501**(7467), 395–398 (2013)
35. M. He, D. Zheng, M. Wang, C. Lin, Z. Lin, High efficiency perovskite solar cells: from complex nanostructure to planar heterojunction. J. Mater. Chem. A **2**(17), 5994–6003 (2014)
36. N.K. Noel et al., Lead-free organic–inorganic tin halide perovskites for photovoltaic applications. Energy Environ. Sci. **7**(9), 3061–3068 (2014)
37. S.N. Habisreutinger, T. Leijtens, G.E. Eperon, S.D. Stranks, R.J. Nicholas, H.J. Snaith, Carbon nanotube/polymer composites as a highly stable hole collection layer in perovskite solar cells. Nano Lett. **14**(10), 5561–5568 (2014)
38. J. Nelson, Polymer: fullerene bulk heterojunction solar cells. Mater. Today **14**(10), 462–470 (2011)
39. Z.M. Beiley, M.D. McGehee, Modeling low cost hybrid tandem photovoltaics with the potential for efficiencies exceeding 20%. Energy Environ. Sci. **5**(11), 9173–9179 (2012)
40. C.-H.M. Chuang, P.R. Brown, V. Bulović, M.G. Bawendi, Improved performance and stability in quantum dot solar cells through band alignment engineering. Nat. Mater. **13**(8), 796 (2014)
41. F. Katiraei, J.R. Aguero, Solar PV integration challenges. IEEE Power Energy Mag. **9**(3), 62–71 (2011)
42. A. Woyte, R. Belmans, J. Nijs, Fluctuations in instantaneous clearness index: analysis and statistics. Sol. Energy **81**(2), 195–206 (2007)
43. A. Maafi, S. Harrouni, Preliminary results of the fractal classification of daily solar irradiances. Sol. Energy **75**(1), 53–61 (2003)
44. R. Kumar, L. Umanand, Estimation of global radiation using clearness index model for sizing photovoltaic system. Renew. Energy **30**(15), 2221–2233 (2005)
45. A. Woyte, V. Van Thong, R. Belmans, J. Nijs, Voltage fluctuations on distribution level introduced by photovoltaic systems. IEEE Trans. Energy Convers. **21**(1), 202–209 (2006)
46. Y. Ghiassi-Farrokhfal, S. Keshav, C. Rosenberg, F. Ciucu, Solar power shaping: an analytical approach. IEEE Trans. Sustain. Energy **6**(1), 162–170 (2015)
47. S. Harrouni, A. Guessoum, A. Maafi, Classification of daily solar irradiation by fractional analysis of 10-min-means of solar irradiance. Theor. Appl. Climatol. **80**(1), 27–36 (2005)
48. T. Soubdhan, R. Emilion, R. Calif, Classification of daily solar radiation distributions using a mixture of Dirichlet distributions. Sol. Energy **83**(7), 1056–1063 (2009)
49. H. Khorasanizadeh, K. Mohammadi, N. Goudarzi, Prediction of horizontal diffuse solar radiation using clearness index based empirical models; A case study. Int. J. Hydrog. Energy **41**(47), 21888–21898 (2016)
50. C.M. Fernández-Peruchena, A. Bernardos, A comparison of one-minute probability density distributions of global horizontal solar irradiance conditioned to the optical air mass and hourly averages in different climate zones. Sol. Energy **112**, 425–436 (2015)
51. L. Wang, O. Kisi, M. Zounemat-Kermani, G.A. Salazar, Z. Zhu, W. Gong, Solar radiation prediction using different techniques: model evaluation and comparison. Renew. Sust. Energ. Rev. **61**, 384–397 (2016)
52. A. Sanfilippo, L. Martin-Pomares, N. Mohandes, D. Perez-Astudillo, D. Bachour, An adaptive multi-modeling approach to solar nowcasting. Sol. Energy **125**, 77–85 (2016)

53. L. Wang, W. Gong, M. Luo, W. Wang, B. Hu, M. Zhang, Comparison of different UV models for cloud effect study. Energy **80**, 695–705 (2015)
54. K. Bakirci, Models for the estimation of diffuse solar radiation for typical cities in Turkey. Energy **82**, 827–838 (2015)
55. T.E. Boukelia, M.-S. Mecibah, I.E. Meriche, General models for estimation of the monthly mean daily diffuse solar radiation (Case study: Algeria). Energy Convers. Manag. **81**, 211–219 (2014)
56. A. Peled, J. Appelbaum, Evaluation of solar radiation properties by statistical tools and wavelet analysis. Renew. Energy **59**, 30–38 (2013)
57. M. Muselli, P. Poggi, G. Notton, A. Louche, Classification of typical meteorological days from global irradiation records and comparison between two Mediterranean coastal sites in Corsica Island. Energy Convers. Manag. **41**(10), 1043–1063 (2000)
58. Z. Ren, W. Yan, X. Zhao, W. Li, J. Yu, Chronological probability model of photovoltaic generation. IEEE Trans. Power Syst. **29**(3), 1077–1088 (2014)
59. C. Tiba, A.N. Siqueira, N. Fraidenraich, Cumulative distribution curves of daily clearness index in a southern tropical climate. Renew. Energy **32**(13), 2161–2172 (2007)
60. T. Ayodele, A. Ogunjuyigbe, Prediction of monthly average global solar radiation based on statistical distribution of clearness index. Energy **90**, 1733–1742 (2015)
61. S. Buhan, Y. Özkazanç, Wind pattern recognition and reference wind mast data correlations with NWP for improved wind-electric power forecasts. IEEE Trans. Ind. Inf. **12**(3), 991–1004 (2016)
62. C.S. Ioakimidis, L.J. Oliveira, K.N. Genikomsakis, Wind power forecasting in a residential location as part of the energy box management decision tool. IEEE Trans. Ind. Inf. **10**(4), 2103–2111 (2014)
63. M.B. Ozkan, P. Karagoz, A novel wind power forecast model: statistical hybrid wind power forecast technique (SHWIP). IEEE Trans. Ind. Inf. **11**(2), 375–387 (2015)
64. Skye Instruments Ltd, SKS 1110 pyranometer. [Online]. http://www.skyeinstruments.info/index_htm_files/Pyranometer.pdf, visited on 25th August 2016
65. Skye Instruments Ltd, Solar radiation system for photo voltaics (2009). [Online]. http://www.skyeinstruments.info/index_htm_files/Solar%20Radiation%20System%20for%20Photovoltaics.pdf, visited on 25th August 2016
66. I. Rüedi, W. Finsterle, The World Radiometric Reference and its quality system, in *Proc. WMO Tech. Conf. on Meteorological and Environmental Instruments and Methods of Observation (TECO-2005)*, Bucharest, Romania, vol. 82 (2005), pp. 434–436
67. L. Wang, G.A. Salazar, W. Gong, S. Peng, L. Zou, A. Lin, An improved method for estimating the Ångström turbidity coefficient β in Central China during 1961-2010. Energy **81**(1), 67–73 (2015)
68. K. Scharmer, J. Greif, The European solar radiation atlas Vol. 1: Fundamentals and maps, École des Mines de Paris (2000)
69. T. Hove, E. Manyumbu, Estimates of the Linke turbidity factor over Zimbabwe using ground-measured clear-sky global solar radiation and sunshine records based on a modified ESRA clear-sky model approach. Renew. Energy **52**, 190–196 (2013)
70. F. Kasten, The Linke turbidity factor based on improved values of the integral Rayleigh optical thickness. Sol. Energy **56**(3), 239–244 (1996)
71. NASA surface meteorology and solar energy. [Online]. https://eosweb.larc.nasa.gov/cgi-bin/sse/grid.cgi, visited on 12th March 2016
72. P. Hedelin, J. Skoglund, Vector quantization based on Gaussian mixture models. IEEE Trans. Speech Audio Process. **8**(4), 385–401 (2000)
73. M.-S. Yang, C.-Y. Lai, C.-Y. Lin, A robust EM clustering algorithm for Gaussian mixture models. Pattern Recogn. **45**(11), 3950–3961 (2012)
74. T.W. Liao, Clustering of time series data—a survey. Pattern Recogn. **38**(11), 1857–1874 (2005)
75. U. Mori, A. Mendiburu, J.A. Lozano, Similarity measure selection for clustering time series databases. IEEE Trans. Knowl. Data Eng. **28**(1), 181–195 (2016)
76. R.C. de Amorim, C. Hennig, Recovering the number of clusters in data sets with noise features using feature rescaling factors. Inf. Sci. **324**, 126–145 (2015)

77. H. Izakian, W. Pedrycz, I. Jamal, Fuzzy clustering of time series data using dynamic time warping distance. Eng. Appl. Artif. Intell. **39**, 235–244 (2015)
78. F. Petitjean, A. Ketterlin, P. Gançarski, A global averaging method for dynamic time warping, with applications to clustering. Pattern Recogn. **44**(3), 678–693 (2011)
79. C. Zhu, D. Gao, Multiple matrix learning machine with five aspects of pattern information. Knowl.-Based Syst. **83**, 13–31 (2015)
80. J.F. Kolen, T. Hutcheson, Reducing the time complexity of the fuzzy c-means algorithm. IEEE Trans. Fuzzy Syst. **10**(2), 263–267 (2002)
81. C. Bouveyron, S. Girard, C. Schmid, High-dimensional data clustering. Comput. Stat. Data Anal. **52**(1), 502–519 (2007)
82. C. Bouveyron, C. Brunet-Saumard, Model-based clustering of high-dimensional data: a review. Comput. Stat. Data Anal. **71**, 52–78 (2014)
83. S. Agrawal, B. Panigrahi, M.K. Tiwari, Multiobjective particle swarm algorithm with fuzzy clustering for electrical power dispatch. IEEE Trans. Evol. Comput. **12**(5), 529–541 (2008)
84. X. Huang, Y. Ye, H. Zhang, Extensions of kmeans-type algorithms: a new clustering framework by integrating intracluster compactness and intercluster separation. IEEE Trans. Neural Netw. Learn. Syst. **25**(8), 1433–1446 (2014)
85. C.S. Lai, M.D. McCulloch, Sizing of stand-alone solar PV and storage system with anaerobic digestion biogas power plants. IEEE Trans. Ind. Electron. vol. 64, 2017, 2112–2121 (2017)
86. X. Wu, X. Zhu, G.-Q. Wu, W. Ding, Data mining with big data. IEEE Trans. Knowl. Data Eng. **26**(1), 97–107 (2014)
87. C.S. Lai, L.L. Lai, Application of big data in smart grid, in *2015 IEEE International Conference on Systems, Man, and Cybernetics (SMC)* (IEEE, 2015), pp. 665–670
88. G. Locatelli, M. Mikic, M. Kovacevic, N.J. Brookes, N. Ivanišević, *The Successful Delivery of Megaprojects: A Novel Research Method* (Project Management Institute, 2017)
89. H. He, E.A. Garcia, Learning from imbalanced data. IEEE Trans. Knowl. Data Eng. **21**(9), 1263–1284 (2009)
90. S. Wang, X. Yao, Multiclass imbalance problems: analysis and potential solutions. IEEE Trans. Syst. Man Cybern. B Cybern. **42**(4), 1119–1130 (2012)
91. Y. Xiao, B. Liu, Z. Hao, A sphere-description-based approach for multiple-instance learning. IEEE Trans. Pattern Anal. Mach. Intell. **39**(2), 242–257 (2017)
92. J.M. Malof, M.A. Mazurowski, G.D. Tourassi, The effect of class imbalance on case selection for case-based classifiers: an empirical study in the context of medical decision support. Neural Netw. **25**, 141–145 (2012)
93. M. Krstic, M. Bjelica, Impact of class imbalance on personalized program guide performance. IEEE Trans. Consum. Electron. **61**(1), 90–95 (2015)
94. D.-C. Li, C.-W. Liu, S.C. Hu, A learning method for the class imbalance problem with medical data sets. Comput. Biol. Med. **40**(5), 509–518 (2010)
95. X.-Y. Liu, J. Wu, Z.-H. Zhou, Exploratory undersampling for class-imbalance learning. IEEE Trans. Syst. Man Cybern. B Cybern. **39**(2), 539–550 (2009)
96. W.W. Ng, J. Hu, D.S. Yeung, S. Yin, F. Roli, Diversified sensitivity-based undersampling for imbalance classification problems. IEEE Trans. Cybern. **45**(11), 2402–2412 (2015)
97. D. Mease, A.J. Wyner, A. Buja, Boosted classification trees and class probability/quantile estimation. J. Mach. Learn. Res. **8**, 409–439 (2007)
98. M. Lin, K. Tang, X. Yao, Dynamic sampling approach to training neural networks for multi-class imbalance classification. IEEE Trans. Neural Netw. Learn. Syst. **24**(4), 647–660 (2013)
99. R. Batuwita, V. Palade, FSVM-CIL: fuzzy support vector machines for class imbalance learning. IEEE Trans. Fuzzy Syst. **18**(3), 558–571 (2010)
100. Y. Tang, Y.-Q. Zhang, N.V. Chawla, S. Krasser, SVMs modeling for highly imbalanced classification. IEEE Trans. Syst. Man Cybern. B Cybern. **39**(1), 281–288 (2009)
101. C. Diamantini, D. Potena, Bayes vector quantizer for class-imbalance problem. IEEE Trans. Knowl. Data Eng. **21**(5), 638–651 (2009)

102. G.M. Weiss, F. Provost, Learning when training data are costly: the effect of class distribution on tree induction. J. Artif. Intell. Res. **19**, 315–354 (2003)

103. Z.-H. Zhou, X.-Y. Liu, Training cost-sensitive neural networks with methods addressing the class imbalance problem. IEEE Trans. Knowl. Data Eng. **18**(1), 63–77 (2006)

104. X. Zhang, B.-G. Hu, A new strategy of cost-free learning in the class imbalance problem. IEEE Trans. Knowl. Data Eng. **26**(12), 2872–2885 (2014)

105. D.S. Yeung, J.-C. Li, W.W. Ng, P.P. Chan, MLPNN training via a multiobjective optimization of training error and stochastic sensitivity. IEEE Trans. Neural Netw. Learn. Syst. **27**(5), 978–992 (2016)

106. C. Seiffert, T.M. Khoshgoftaar, J. Van Hulse, A. Napolitano, RUSBoost: a hybrid approach to alleviating class imbalance. IEEE Trans. Syst. Man Cybern. A Syst. Hum. **40**(1), 185–197 (2010)

107. A. Rahman, D.V. Smith, G. Timms, A novel machine learning approach toward quality assessment of sensor data. IEEE Sensors J. **14**(4), 1035–1047 (2014)

108. S. Wang, X. Yao, Using class imbalance learning for software defect prediction. IEEE Trans. Reliab. **62**(2), 434–443 (2013)

109. F. Zhang, P.P. Chan, B. Biggio, D.S. Yeung, F. Roli, Adversarial feature selection against evasion attacks. IEEE Trans. Cybern. **46**(3), 766–777 (2016)

110. P. Mitra, C. Murthy, S.K. Pal, Unsupervised feature selection using feature similarity. IEEE Trans. Pattern Anal. Mach. Intell. **24**(3), 301–312 (2002)

111. R. Diao, Q. Shen, Feature selection with harmony search. IEEE Trans. Syst. Man Cybern. B Cybern. **42**(6), 1509–1523 (2012)

112. R. Diao, F. Chao, T. Peng, N. Snooke, Q. Shen, Feature selection inspired classifier ensemble reduction. IEEE Trans. Cybern. **44**(8), 1259–1268 (2014)

113. I.-S. Oh, J.-S. Lee, B.-R. Moon, Hybrid genetic algorithms for feature selection. IEEE Trans. Pattern Anal. Mach. Intell. **26**(11), 1424–1437 (2004)

114. Y. Liu, F. Tang, Z. Zeng, Feature selection based on dependency margin. IEEE Trans. Cybern. **45**(6), 1209–1221 (2015)

115. T.W. Chow, P. Wang, E.W. Ma, A new feature selection scheme using a data distribution factor for unsupervised nominal data. IEEE Trans. Syst. Man Cybern. B Cybern. **38**(2), 499–509 (2008)

116. Y. Yao, H. Tong, T. Xie, L. Akoglu, F. Xu, J. Lu, Detecting high-quality posts in community question answering sites. Inf. Sci. **302**, 70–82 (2015)

117. M.D. Ruiz, E. Hüllermeier, A formal and empirical analysis of the fuzzy gamma rank correlation coefficient. Inf. Sci. **206**, 1–17 (2012)

118. D.P. Francis, A.J. Coats, D.G. Gibson, How high can a correlation coefficient be? Effects of limited reproducibility of common cardiological measures. Int. J. Cardiol. **69**(2), 185–189 (1999)

119. Y. Liu, T. Pan, S. Aluru, Parallel pairwise correlation computation on intel xeon phi clusters, in *2016 28th International Symposium on Computer Architecture and High Performance Computing (SBAC-PAD)* (IEEE, 2016), pp. 141–149

120. Historical data, Weatherunderground.com. [Online]. https://www.wunderground.com/history/

121. B.W. Silverman, *Density Estimation for Statistics and Data Analysis* (CRC Press, Boca Raton, 1986)

122. J. Habbema, A stepwise discriminant analysis program using density estimtion, in *Compstat*, (Physica-Verlag, 1974), pp. 101–110

123. R.P.W. Duin, On the choice of smoothing parameters for Parzen estimators of probability density functions. IEEE Trans. Comput. **C-25**(11), 1175–1179 (1976)

124. N.V. Chawla, K.W. Bowyer, L.O. Hall, W.P. Kegelmeyer, SMOTE: synthetic minority oversampling technique. J. Artif. Intell. Res. **16**, 321–357 (2002)

125. H. He, Y. Bai, E.A. Garcia, S. Li, ADASYN: adaptive synthetic sampling approach for imbalanced learning, in *IEEE International Joint Conference on Neural Networks, 2008. IJCNN 2008.(IEEE World Congress on Computational Intelligence)* (IEEE, 2008), pp. 1322–1328

126. A. Amin et al., Comparing oversampling techniques to handle the class imbalance problem: a customer churn prediction case study. IEEE Access **4**, 7940–7957 (2016)

127. B. Ratner, The correlation coefficient: its values range between+ 1/– 1, or do they? J. Target. Meas. Anal. Mark. **17**(2), 139–142 (2009)

128. C.S. Lai, Y. Jia, M. McCulloch, Z. Xu, Daily clearness index profiles cluster analysis for photovoltaic system. IEEE Trans. Ind. Inf. **13**(5), 2322–2332 (2017)

129. F. Murtagh, P. Legendre, Ward's hierarchical agglomerative clustering method: which algorithms implement Ward's criterion? J. Classif. **31**(3), 274–295 (2014)

130. M. Boust, Grid-connected energy-storage projects in Pipeline to hit 2GW, led by US, China and South Korea (IHS Markit, July 2016). [Online]. https://technology.ihs.com/581101/grid-connected-energy-storage-projects-in-pipeline-to-hit-2gw-led-by-us-china-and-south-korea

131. A. Nowicki, US energy storage market to grow 9x by 2021 (SmartGridNews. com, June 2016). [Online]. http://www.smartgridnews.com/story/us-energy-storage-market-grow-9x-2021/2016-06-29

132. J. Runyon, Energy storage industry off and running in January 2016 (Renewable Energy World, Jan 2016). [Online]. http://www.renewableenergyworld.com/articles/2016/01/energy-storage-set-for-record-year-in-2016.html

133. M. Munsell, US energy storage market grew 243% in 2015, largest year on record (Greentech Media, Mar 2016). [Online]. http://www.greentechmedia.com/articles/read/us-energy-storage-market-grew-243-in-2015-largest-year-on-record

134. India Energy Security Scenarios 2047, User guide for India's 2047 energy calculator, electrical energy storage (EES). [Online]. http://indiaenergy.gov.in/docs/Storage%20 Documentation.pdf

135. Renewables and electricity storage: a technology roadmap for Remap 2030 (IRENA, 2015). [Online]. https://www.irena.org/DocumentDownloads/Publications/IRENA_REmap_ Electricity_Storage_2015.pdf

136. Technology roadmap: energy storage (International Energy Agency, 2014). [Online]. https://www.iea.org/publications/freepublications/publication/ TechnologyRoadmapEnergystorage.pdf

137. Future energy scenarios: GB gas and electricity transmission (nationalgrid, 2016). [Online]. http://fes.nationalgrid.com/fes-document/

138. R. Manghani, B. Simon, U.S Energy Storage Monitor: Q2 2016 (Energy Storage Association and gtmresearch, Jun 2016). [Online]. http://energystorage.org/system/files/resources/gtm_ research_-_esa_q2_2016_presentation_2016_06_14_final.pdf

139. T. Kenning, Residential storage costs will fall 84% globally by 2040—BNEF (Energy Storage News, Jun 2015). [Online]. http://www.energy-storage.news/news/ residential-storage-system-costs-to-fall-by-84-globally-by-2040-bnef

140. K. Zipp, Energy storage prices expected to drop 70% by 2030 (Solar Power World, Jan 2016). [Online]. http://www.solarpowerworldonline.com/2016/01/ideal-power-partners-with-austin-energy-on-us-doe-funded-projects-to-integrate-solar-pv-and-storage-for-commercial-sites/

141. Industry solar: crossing the charm (Deutsche Bank Markets Research, 2015). [Online]. https://www.db.com/cr/en/docs/solar_report_full_length.pdf

142. IRENA, Battery storage for renewables: market status and technology outlook (2015). [Online]. http://www.irena.org/documentdownloads/publications/irena_battery_storage_ report_2015.pdf

143. F. Díaz-González, A. Sumper, F. DÃaz-GonzÃ, O. Gomis-Bellmunt, *Energy Storage in Power Systems* (Wiley, 2016)

144. B. Robyns, B. Francois, G. Delille, C. Saudemont, *Energy Storage in Electric Power Grids* (Wiley, 2015)

145. I. Serban, R. Teodorescu, C. Marinescu, Energy storage systems impact on the short-term frequency stability of distributed autonomous microgrids, an analysis using aggregate models. IET Renew. Power Gener. **7**(5), 531–539 (2013)
146. K. Yang, A. Walid, Outage-storage tradeoff in frequency regulation for smart grid with renewables. IEEE Trans. Smart Grid **4**(1), 245–252 (2013)
147. K. Zipp, What is the best type of battery for solar storage? (Solar Power World, 2015). [Online]. http://www.solarpowerworldonline.com/2015/08/what-is-the-best-type-of-battery-for-solar-storage/
148. K. Gong et al., A zinc–iron redox-flow battery under $100 per kW h of system capital cost. Energy Environ. Sci. **8**(10), 2941–2945 (2015)
149. B. Zakeri, S. Syri, Electrical energy storage systems: a comparative life cycle cost analysis. Renew. Sust. Energ. Rev. **42**, 569–596 (2015)
150. A. Poullikkas, A comparative overview of large-scale battery systems for electricity storage. Renew. Sust. Energ. Rev. **27**, 778–788 (2013)
151. Lazard, Lazard's levelized cost of storage analysis V1.0. [Online]. https://www.lazard.com/media/2391/lazards-levelized-cost-of-storage-analysis-10.pdf (Visited on 15th April 2016), 2015
152. World Energy Resources, E-storage: shifting from cost to value Wind and solar applications (World Energy Council, 2016)
153. G. Locatelli, E. Palerma, M. Mancini, Assessing the economics of large Energy Storage Plants with an optimisation methodology. Energy **83**, 15–28 (2015)
154. J. Leadbetter, L.G. Swan, Selection of battery technology to support grid-integrated renewable electricity. J. Power Sources **216**, 376–386 (2012)
155. X. Luo, J. Wang, M. Dooner, J. Clarke, Overview of current development in electrical energy storage technologies and the application potential in power system operation. Appl. Energy **137**, 511–536 (2015)
156. Energy storage: tracking the technologies that will transform the power sector (Deloitte, 2015). [Online]. http://www2.deloitte.com/content/dam/Deloitte/us/Documents/energy-resources/us-er-energy-storage-tracking-technologies-transform-power-sector.pdf
157. J.L. Bernal-Agustín, R. Dufo-López, Simulation and optimization of stand-alone hybrid renewable energy systems. Renew. Sust. Energ. Rev. **13**(8), 2111–2118 (2009)
158. S.H. El-Hefnawi, Photovoltaic diesel-generator hybrid power system sizing. Renew. Energy **13**(1), 33–40 (1998)
159. S. Rehman, L.M. Al-Hadhrami, Study of a solar PV–diesel–battery hybrid power system for a remotely located population near Rafha, Saudi Arabia. Energy **35**(12), 4986–4995 (2010)
160. Homer Energy. [Online]. http://www.homerenergy.com/software.html
161. O. Erdinc, M. Uzunoglu, Optimum design of hybrid renewable energy systems: overview of different approaches. Renew. Sust. Energ. Rev. **16**(3), 1412–1425 (2012)
162. L. Xu, X. Ruan, C. Mao, B. Zhang, Y. Luo, An improved optimal sizing method for wind-solar-battery hybrid power system. IEEE Trans. Sustain. Energy **4**(3), 774–785 (2013)
163. S. Shaahid, M. Elhadidy, Economic analysis of hybrid photovoltaic–diesel–battery power systems for residential loads in hot regions—a step to clean future. Renew. Sust. Energ. Rev. **12**(2), 488–503 (2008)
164. S. Shaahid, M. Elhadidy, Technical and economic assessment of grid-independent hybrid photovoltaic–diesel–battery power systems for commercial loads in desert environments. Renew. Sust. Energ. Rev. **11**(8), 1794–1810 (2007)
165. S. Singh, S.C. Kaushik, Optimal sizing of grid integrated hybrid PV-biomass energy system using artificial bee colony algorithm. IET Renew. Power Gener. **10**(5), 642–650 (2016)
166. Y.-Y. Hong, R.-C. Lian, Optimal sizing of hybrid wind/PV/diesel generation in a stand-alone power system using Markov-based genetic algorithm. IEEE Trans. Power Delivery **27**(2), 640–647 (2012)
167. Z. Moradi-Shahrbabak, A. Tabesh, G.R. Yousefi, Economical design of utility-scale photovoltaic power plants with optimum availability. IEEE Trans. Ind. Electron. **61**(7), 3399–3406 (2014)

168. B. Igoe, Dry low emissions experience across the range of Siemens small industrial gas turbines, *An unpublished report of Siemens Industrial Turbomachinery Limited, UK* (2011)
169. Wartsila, Combustion engine vs. gas turbine: part load efficiency and flexibility. [Online]. http://www.worldenergyoutlook.org/media/weowebsite/energydevelopment/2012updates/measuringprogresstowardsenergyforall_weo2012.pdf, visited on 25th August 2016
170. Siemens, Reliable and powerful—economical, safe-investment packages SGT6-PAC 5000F/SCC6-PAC 5000F (2015). [Online]. http://www.energy.siemens.com/hq/pool/hq/power-generation/gas-turbines/SGT6-5000F/SGT6-5000F%20PAC_LowRes.pdf, visited on 25th August 2016
171. TMI Staff & Contributors, Design and operating considerations for combined cycle plants (2011). [Online]. http://www.energy.siemens.com/hq/pool/hq/power-generation/gas-turbines/SGT6-5000F/SGT6-5000F%20PAC_LowRes.pdf, Turbomachinery Magazine, visited on 25th August 2016
172. B. Jabeck, The impact of generator set underloading (2015). [Online]. https://forums.cat.com/t5/BLOG-Power-Perspectives/The-Impact-of-Generator-Set-Underloading/ba-p/69719, Caterpillar Inc, visited on 25th August 2016
173. Noria Corporation, Lubricating natural gas engines. [Online]. http://www.machinerylubrication.com/Read/29018/natural-gas-engines, Machinery Lubrication, visited on 25th August 2016
174. R&D Construction Ltd, R&D wins Renewable Project for a 2.4 Megawatt Anaerobic Digestion Plant. [Online]. http://www.randdconstruction.co.uk/rd-wins-renewable-project-for-a-2-4-megawatt-anaerobic-digestion-plant, visited on 1st March 2016
175. A. Hassan, M. Saadawi, M. Kandil, M. Saeed, Modified particle swarm optimisation technique for optimal design of small renewable energy system supplying a specific load at Mansoura University. IET Renew. Power Gener. 9(5), 474–483 (2015)
176. A.S. Mundada, K.K. Shah, J. Pearce, Levelized cost of electricity for solar photovoltaic, battery and cogen hybrid systems. Renew. Sust. Energ. Rev. 57, 692–703 (2016)
177. R.M. Swanson, A vision for crystalline silicon photovoltaics. Prog. Photovolt. Res. Appl. 14(5), 443–453 (2006)
178. The Association for Decentralised Energy, First AD biogas plant opens in Kenya (2015). [Online]. http://www.theade.co.uk/first-adbiogas-plant-opens-in-kenya_3412.html, visited on 18th March 2016
179. National Renewable Energy Laboratory (NREL), CREST cost of energy model: anaerobic digestion V1.4. [Online]. https://financere.nrel.gov/finance/content/crest-cost-energy-models, visited on 18th March 2016
180. S.B. Darling, F. You, T. Veselka, A. Velosa, Assumptions and the levelized cost of energy for photovoltaics. Energy Environ. Sci. 4(9), 3133–3139 (2011)
181. I. Pawel, The cost of storage—how to calculate the Levelized Cost of stored Energy (LCOE) and applications to renewable energy generation. Energy Procedia 46, 68–77 (2014)
182. National Renewable Energy Laboratory (NREL), CREST cost of energy model: photovoltaic V1.4. [Online]. https://financere.nrel.gov/finance/content/crest-cost-energy-models, visited on 18th March 2016
183. G. Petrone, G. Spagnuolo, R. Teodorescu, M. Veerachary, M. Vitelli, Reliability issues in photovoltaic power processing systems. IEEE Trans. Ind. Electron. 55(7), 2569–2580 (2008)
184. C.S. Lai, M.D. McCulloch. Levelized cost of electricity for solar photovoltaic and electrical energy storage. Applied Energy, 190, pp.191–203 (2017)
185. Prepared for the committee on climate change, Discount rates for low-carbon and renewable generation technologies (Oxera, 2011). [Online]. http://www.oxera.com/Latest-Thinking/Publications/Reports/2011/Discount-rates-for-low-carbon-and-renewable-genera.aspx, visited on 15th March 2016

Chapter 3
Blockchain Applications in Microgrid Clusters

3.1 Introduction

Renewable energy is a potential solution to environmental pollution and resource exhaustion problems caused by fossil fuel generation [1, 2]. The increasing amount of renewable energy calls for a clean energy trading mechanism. Due to the intermittent and non-dispatchable characteristics of the renewable energy sources (RESs), it is difficult for the wholesale market to trade in real-time [3] and thus brings great challenges to the safe and stable operation of power systems.

To investigate the smart city development, this chapter presents a smart grid architecture to enhance the energy distribution ability for various stakeholders including grid operators, prosumers, and consumers. In the architecture, the nodes denote the energy exchanging points and the monetary transactions are regulated by Blockchain. Mobile application is implemented to provide the stakeholders access to the Blockchain network. This chapter focuses on the microgrid. Microgrids integrate renewable energy locally and promote renewable energy utilization and electrical power supply reliability in a distributed manner [4]. Microgrids are developed with the inclusion of but not limited to advanced control methods, power electronics, microelectronics, and big data analytics.

Integrated fault-tolerant Information and Communications Technology (ICT) systems can improve consumers' quality of services and guarantee their needs whilst minimizing cost and resource consumption.

Microgrids are highly dependent on information technologies and high-performance telecommunications networks for the exchange of data between nodes. The stringent requirements require network automation and Artificial Intelligence (AI).

Distributed generation could be achieved as primary energy resources including wind and solar are widely available. Microgrid as a portion of an intelligent smart power network can maximize renewable energy resources usage and simultaneously

C. S. Lai et al., *Smart Grids and Big Data Analytics for Smart Cities*, https://doi.org/10.1007/978-3-030-52155-4_3

meet the consumers' energy demand. The microgrid does not need to be geographically large and can be local (e.g., home or small community) with photovoltaic panels, energy storage, and household appliances.

Microgrids have bi-directional power and data flow and are capable of monitoring power plants and individual appliances. Distributed computing and communications allow the delivery of real-time information and enable the instant supply and demand balance at the local level. New communication protocols, electronic devices, and ICT technologies promote microgrid transformation. The following features can be achieved:

- The integration of heterogeneous power generation and energy storage systems with universal interoperability standards to support "plug-and-play" convenience.
- The advent of new electricity markets and business models allow prosumers to export their energy resources to secure revenue and reduce energy cost.
- Real-time monitoring and system diagnostics enhance power quality, reduce energy utilities financial loss, and enables automated maintenance.
- Technical performance enhancements include the increased load factors, reduced system power losses, and system outage time.

Microgrids play a critical role to address global warming and to meet the ever-increasing energy demand including high penetration of electric vehicles. Microgrids consist of the following main functions:

- Active Network Management (ANM): Consist of software, automation, and control systems that monitor the grid in real-time to guarantee the system operates within limits.
- Automatic Voltage Control (AVC): Voltage and reactive power of the system buses are within the defined values and minimize the power loss.
- Advanced Metering Infrastructure (AMI): Includes communication networks in various levels of the infrastructure hierarchy, smart meters, Meter Data Management Systems (MDMS), software application platforms, and interfaces to transmit and process data.
- Dynamic Line Rating (DLR): Transmission owners can determine thermal capacity and estimate line rating in real-time.
- Phasor Measurement Unit (PMU): Electronic devices that measure AC phasors and synchronize these measurements with Global Positioning System (GPS).
- Reactive Power Compensation (RPC): Power electronic devices for the control and compensation of reactive power.

In distributed generations, new business models are needed to manage RES and load consumption. Microgrid clusters or multi-microgrids [5, 6] are made up of many adjacent interconnected microgrids within a certain region, which could supply energy to each other for the optimal RESs usage. This timely review focuses on blockchain applications in microgrids and microgrid clusters.

The common approach of utilizing renewable energy locally is peer-to-peer (P2P) energy trading among agents within microgrids and microgrid clusters [7–9].

Traditional consumers could only buy electricity that is transmitted over a long distance from energy suppliers. P2P energy trading makes renewable energy balancing possible, in which prosumers (generate and consume electricity agents) not only meet their own electricity demand with RESs but also sell surplus energy to other consumers who are in short of supply within microgrids and microgrid clusters, instead of feeding the surplus energy into the power grid [10]. Energy trading between prosumers and consumers is called P2P energy trading [11]. Noted that P2P energy trading is not only in closed form and consumers could purchase electricity from the main grid and other microgrids within a microgrid cluster. The transition to a renewable energy era requires a clean energy market trading mechanism, to incorporate technologies upgrade and secure energy and monetary transactions [12]. Apart from these, systematic verification also needs to be implemented for the security and efficiency of electricity trading.

Blockchain technology uses Internet of Things (IoT) to facilitate negotiation between agents for distributed energy transactions. With wireline or wireless data links distributed across the mesh network, meaningful real-time services will be accessible to consumers, such as information about over energy usage. Consumers can automatically respond to their needs. Blockchain-based ledger has the advantage of letting consumers and vendors energy transactions, whilst actors have no access to each other's identity.

Blockchain technology is very suitable for P2P energy trading as the decentralized structure of the blockchain naturally matches with the implementation of control and business processes in microgrid [13]. Nakamoto [14] created blockchain technology and the Bitcoin system. The P2P electronic money system, i.e., Bitcoin is the original application of blockchain technology. The blockchain is a type of database technology. It is revolutionary to find a simple and resourceful method that blockchain guarantees the underlying data remains true as time progresses [15]. A blockchain is combined with cryptographically linked blocks. The newly created block is linked to preceding blocks to avoid information being tampered. Although blockchain initially focuses on recording transaction logs in the blockchain, contents in each block can record other data or even logistic information. Moreover, a Smart contract [16] is a certain algorithm defined by users and contains some operational procedures. A smart contract could run a specific program automatically, which could complete negotiation, settlement, and payment between prosumers and consumers automatically [17]. These characteristics make blockchain ideal ICT for P2P energy trading.

Blockchain technology composes a list of principal functionalities, such as:

- *Monitoring*—When signed in, the user can view a list of parameters in real-time, such as live energy usage of the domestic appliance; the energy consumed by the network (microgrid); the energy output by the photovoltaic panels; ratio of energy used from local production and microgrid.
- *Trading*—The user can examine the amount of energy stored with the energy storage, and choose an available energy supplier to receive the corresponding daily sales rate. Once the supplier has been identified, the user can set the amount

of energy to be sold. The user can permit the sale transaction after identifying the revenue for the transaction. The transaction will be recorded in the Blockchain ledger.

- *Recording*—The user can record the data regarding system operation (e.g., energy export and import from the microgrid) and transactions with energy sale.

Cities can be smarter by adopting innovative ICT solutions for accumulating and processing big data created by IoT devices, wearable devices, and sensor networks. A dedicated simple mobile application will promote the consumer to participate in the grid, share information, and buy/sell energy between the involved nodes (energy providers and consumers) with Blockchain ledger.

In 2018, the Chinese blockchain industry white paper [18] stated that blockchain is leading a new round of technological and industrial revolutions around the world. A survey published by the German Energy Agency [19] also stated that 21% of the respondents think blockchain is a revolutionary technology for the energy industry while 60% believe that it will be further disseminated. A report published by the UK Government Chief Scientific Adviser [20] stated that blockchain would help the government to reduce fraud in transactions, enhance security against cyber-attack, and reduce the cost of paper-intensive processes. Blockchain technology has many advantages as follows:

1. Decentralization: With each participant holding a decentralized ledger that records all the transaction information, each participant could get complete logs of transactions, which will guarantee transparent transactions.
2. Automation: Smart contract, a programmable code stored in the blockchain could be executed automatically when the agreements between different parties are met and record changes in the decentralized ledgers.
3. Security: With distributed consensus algorithms, cryptographic hash functions, and public-private key cryptography, it is possible for secure transactions between different parties and the ledgers would not be tampered by malicious nodes or cyber-attack, which will maintain database security.
4. Compatibility: Decentralized ledger will not only store the logs of transactions but also the information such as usage of resources and Internet of Things (IoT) devices.

It should be noted that this chapter is different from the preceding review wrote by Andoni et al. [12], which focuses on blockchain applications of the whole energy sector and gives a list of recent blockchain research projects and startups. In comparison to Ref. [11] which focuses on comparisons between different microgrid trials and projects, this chapter provides a comprehensive review of blockchain applications in microgrids. In addition, this review focuses on state-of-the-art researches of blockchain applications, while Ref. [21] focuses on analyzing different business models of P2P electricity trading. This review focuses on blockchain applications in microgrids and microgrid clusters and introduces the schematic operational mechanism about the P2P energy trading model with blockchain. The main contribution of this chapter can be summarized as follows:

1. The drivers and technical background of blockchain for microgrid P2P energy trading are analyzed.
2. The state-of-the-art of blockchain applications in the energy sector are reviewed.
3. The schematic operational mechanism of the P2P energy trading model is introduced. A novel smart contract-based hybrid P2P energy trading model with cryptocurrency is described.
4. The future applications of blockchain in microgrids are envisioned.
5. The advantages of using blockchain with microgrids are summarized and the challenges with blockchain-based P2P energy trading are discussed.

The rest of this chapter is organized as follows. Section 3.2 presents the motivations for blockchain and P2P energy trading. Section 3.3 provides the fundamental of blockchain, such as the technical framework and key elements. Section 3.4 gives an overview and classification of state-of-the-art blockchain applications in the energy sector. Section 3.5 introduces the schematic operational mechanism of the P2P energy trading model and presents a new smart contract-based hybrid P2P energy trading model with cryptocurrency. Section 3.6 envisions the future applications of blockchain in microgrids. Section 3.7 discusses the advantages of using blockchain with microgrids and the challenges for blockchain-based P2P energy trading. Section 3.8 concludes the chapter.

3.2 Motivations for Blockchain and P2P Energy Trading

The increasing number of distributed energy resources, electrical energy storage (EES) systems, and smart meters serve as the underlying motivation for blockchain and P2P energy trading. This section presents the historical and future development of these resources to demonstrate that there is a need for blockchain in P2P energy trading.

3.2.1 Distributed Renewable Energy Resources and Electrical Energy Storages

The technological advancement (e.g., improved efficiencies) in distributed RESs would help to establish a distributed energy market. P2P energy trading is the most popular form of the distributed energy market with blockchain. As fossil fuel electricity generation declined for the fifth consecutive year in 2017, there was also a boom for renewable electricity generation such as solar (+21.9%) and wind (+15.1%) [22]. As reported by the International Energy Agency (IEA) as shown in Fig. 3.1, solar power showed a 40% development in power generation in 2017 with respect to 2016. To meet the IEA's sustainable development scenario (SDS) target, it needs average yearly growth of 17% from 2017 to 2030 [23]. Figure 3.2 shows the

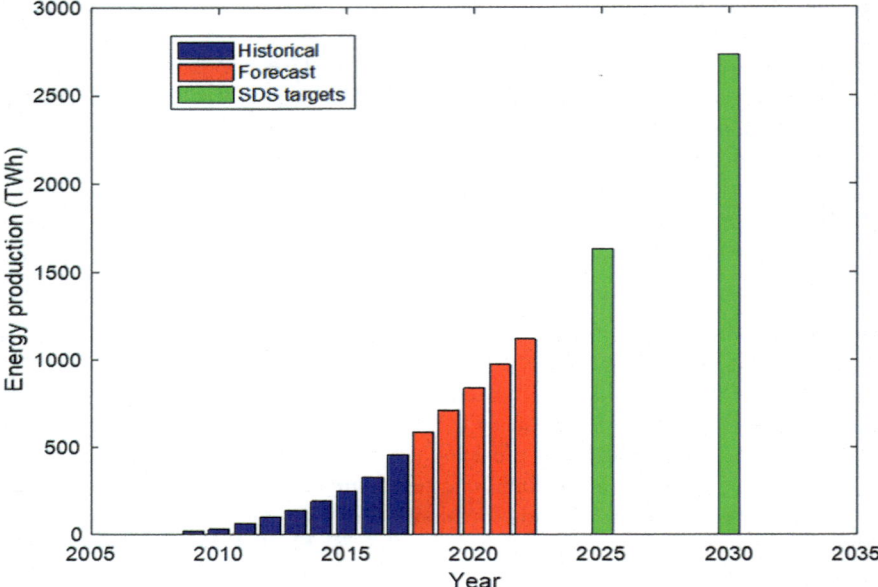

Fig. 3.1 Solar power generation on historical development and targets [23]

global cumulative residential PV installations [24] and there will be a boom for residential solar PV in the coming-future energy sector [25].

Moreover, battery prices have decreased by 22% from 2016 to 2017 and are continuing to reduce [26]. The average residential energy storage system installation cost is estimated to reduce from \$1600/kWh in 2015 to \$250/kWh by 2040 [27, 28]. The global installed capacity of EES is increasing rapidly to alleviate the adverse effects of PV systems [28].

The development of distributed RES and EES provides the basis for blockchain development in the distributed energy market.

3.2.2 Smart Meters and Wireless Communication

A smart grid is integrated by intelligent control, monitoring, and communication of energy consumption data. Smart meters are vital components of a smart grid by helping consumers to minimize electricity cost and consumption in real-time, and accurate billing. Compared to the traditional automatic meter reading (AMR), Advanced Meter Infrastructure (AMI) with smart meters enables an efficient way to control and communicate different participants [29]. The smart meter deployment is growing rapidly in the world as shown in Fig. 3.3 [30].

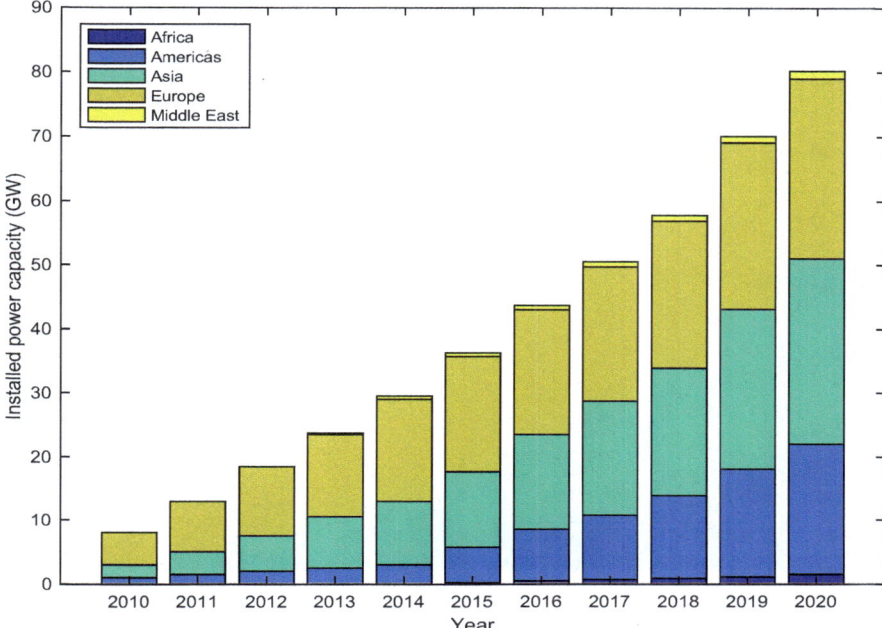

Fig. 3.2 Global cumulative residential PV installations [24]

The wireless sensor networks (WSNs), such as the Zigbee devices network is used in smart grid communication [31]. With great security and reliability, the Zigbee devices are widely used for monitoring energy systems and energy management of buildings and homes [32].

Blockchain is a great tool to deal with the problems of security and privacy concerns during the share of information and data authentication [33]. Minoli et al. [34] claimed that blockchain is important in IoT environments. Reference [35] proposed a secure energy trading system within an industrial IoT environment through consortium blockchain.

With the increasing deployment of IoT devices, we could foresee that IoT devices will facilitate the blockchain technology in the energy sector. Reference [36] provides a communicating power supply (CPS), by which electricity metering, computation, and communication between IoT devices are achieved with a very low cost. CPS provides a promising way to enable energy management of buildings or other entities for great energy savings. As discussed earlier, using blockchain with smart meters or other IoT devices would bring significant benefits to all participants.

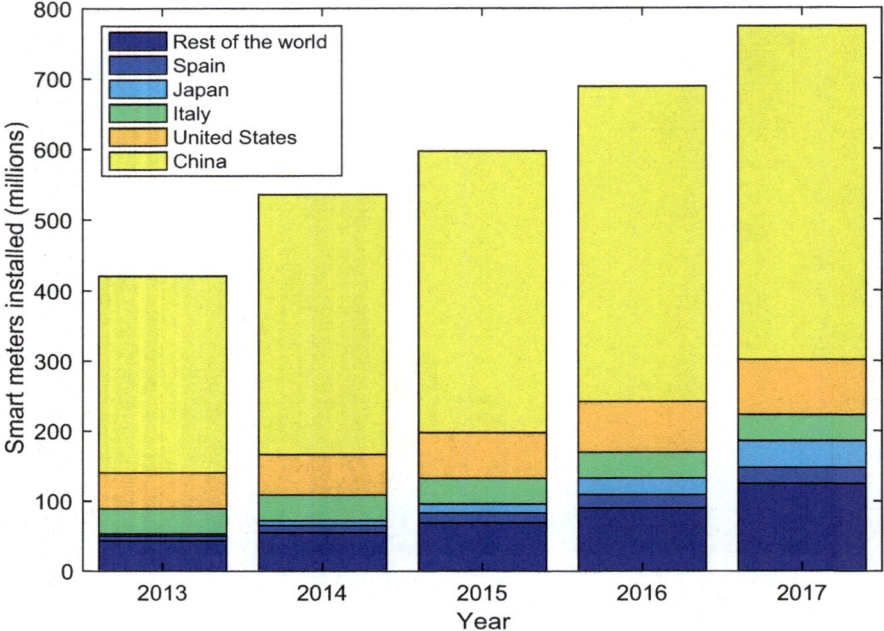

Fig. 3.3 Global cumulative installations of smart meters [30]

3.3   Fundamentals of Blockchain in Microgrids

Blockchain is a novel technology and the fifth disruptive innovation following mainframes, personal computers, internet, and social network [37, 38]. In this section, we examine the technical background of blockchain technology, focusing on the technical framework, the operational mechanism of blockchain, and the contents of a block. Comparisons of different consensus mechanisms and the classification of blockchain in admission mechanisms are given.

3.3.1   The Blockchain Framework

Blockchain uses distributed consensus algorithms, hash function, and asymmetric cryptography which is suitable for the Internet where there is a trust issue. Without third-party interventions, blockchain could ensure the data in the blocks stay true over time by solving the challenges of double-spending [39] and the Byzantine Generals Problem [40]. Byzantine Generals Problem will lead to different nodes with different ledgers, which breaks the rule of consistency. There is a group of scattered client nodes contained in the blockchain network, and each node holds a distributed database that is the record of all transactions with the characteristics of security, tamper-proof, and decentralization [41]. The transaction is firstly verified when transaction data is converted into a "data block" or "block". Based on the confirmation mechanism for the blockchain, the transaction was irreversibly confirmed after a continuous verification of 6 blocks [42, 43]. With the linked block in a chronological structure, blockchain would record all the transaction information that have accomplished and use cryptography technologies to protect data integrity and tamper-proof.

As Fig. 3.4 shows, the framework of the blockchain contains six layers consisting of data, network, consensus, incentive, contract, and application layers [44]. The details of the individual layer are discussed as follows:

- Data layer: contains all transaction data which is stored in the blocks. The layer secures the data with asymmetric cryptography [45], to encrypt data and timestamp the block to ensure the chronological sequencing order of the transaction. Merkel tree [46] verifies the data integrity and to ensure the data is non-tampered.
- Network layer: contains the whole P2P network of all nodes. The transmission and verification mechanism of the data are defined in this layer [47].
- Consensus layer: includes all consensus mechanism algorithms, which are core technologies of blockchain for solving the problem of how to achieve consistency

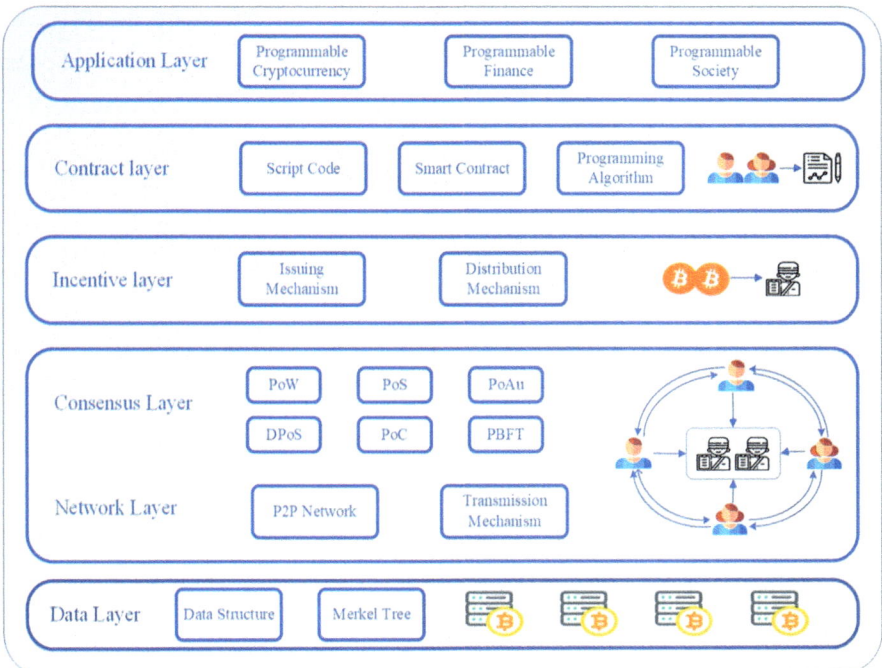

Fig. 3.4 The blockchain framework

of data in distributed scenarios [48]. Consensus mechanism [49] ensures the blockchain nodes reach the consensus in the whole network on the block information. The recent block is included in the blockchain. The blockchain information saved by the nodes is consistent and immune from malicious attacks [50]. The most famous consensus mechanisms include Proof-of-Work (PoW), Proof-of-Stake (PoS), and Practical Byzantine Fault Tolerance (PBFT) [51]. The implementation of these consensus mechanisms and in general the operational protocols of blockchain systems are agreed from the platform community and executed by different platforms automatically. More advanced projects led by companies, often develop their own platform, with their own consensus (and sometimes their own cryptocurrency, etc.)

- Incentive layer: incents nodes in the network to be miners or validator nodes, who record or validate the transactions and information in the new block according to record regulations. When the miner or validator node completes its task and the new block is verified by the network, the validator or miner will get the monetary rewards of newly generated cryptocurrencies and the transaction fees paid by the traders. The transaction fees depend on how rapidly the transactions are confirmed by the miners [52].
- Contract layer: contains all kinds of script code, programming algorithms, and smart contracts. The smart contract is a critical aspect of the blockchain, which could be implemented with the Ethereum platform [53, 54]. Ethereum is a public and open-source distributed computing platform that allows smart contract functionality.
- Application layer: represents the applications that are derived from blockchain such as Bitcoin. Blockchain has been used in several industries including finance, logistic, IoT, and energy [55].

3.3.2 Blockchain Operational Mechanism: A Case Study with Bitcoin

This section presents the blockchain operational mechanism with a Bitcoin case study. The Bitcoin system is a peer-to-peer network that stores all transaction information in the blockchain. The first block of Bitcoin is known as the Genesis block. The Bitcoin system completes the first P2P electronic cash system where participators could trade with each other without any intermediaries, as blockchain will secure transactions [56]. Bitcoin is one of the cryptocurrencies that can be traded in the exchange for fiat currencies. It is decentralized without third-party control in the bitcoin network [57]. Bitcoin has reached its maximum total market capitalization of around 273.62 billion USD in 2017 [58], while the market capitalization is only about 0.04 billion USD in 2012.

Anyone can join the Bitcoin network with the same technical standards and include their own transaction information to extend the blockchain. The blockchain operational mechanism is shown in Fig. 3.5. A transaction will be submitted to the Bitcoin network while participant A transmits bitcoins to participant B. If the transaction is viable, then the node will broadcast the transaction to other nodes through verification. Secondly, validated nodes, also called miners, would collect all the transaction information in the past 10 min and add them to a block with a timestamp of each transaction. Validator nodes would compete with each other based on the computational power to solve a cryptographical puzzle for adding a new block to the blockchain. The challenge of solving the problem is a variable parameter, which is determined by the network setting [59]. The average speed of generating a new block is 10 min/block. Bitcoin network will re-calculate and set a new difficulty value at every 2016 blocks (Eq. 3.1) [41, 60]. The miner will announce the solution to the entire network when it solves the puzzle and other nodes will confirm the solution. The new block is directly included in the blockchain if there is only one legitimate block. Multiple branches will occur when there is more than one block generated simultaneously. In this case, the chain with the most workforce or to be the longest will eventually win. Other branches of the blockchain will be isolated and eventually eliminated.

Step 1: Peer A would like to transmit bitcoins to peer B;
Step 2: Validator nodes compete with each other to earn the right to add a new block to the blockchain;
Step 3: The newly generated block is transmitted to all the nodes for verification;
Step 4: The block successfully passes through the verification;
Step 5: The block is included to the blockchain;
Step 6: Peer B gets bitcoins transmitted from peer A.

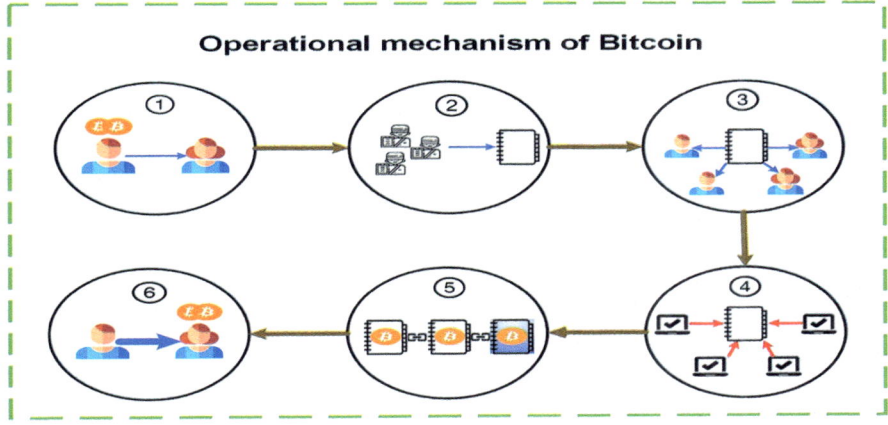

Fig. 3.5 The operational mechanism of the Bitcoin system

3.3.3   The Block Content

The contents in a block (especially the first block) represents the blockchain mechanism. A block is classified into two components including block header and block body (Fig. 3.6) [61]. A block header data takes up 80 bytes of storage space, which contains information of Bitcoin protocol version, pre-block hash, Merkel tree root, time-stamp, difficulty, and nonce. Pre-block hash is received from the prior block with the cryptographic hash function, i.e., Secure Hash Algorithm SHA256 [62]. Merkel tree root is the digest of transaction information stored in the block body through hash function [63]. Time-stamp presents the time when the block was created. Blockchain defines the difficulty to limit the time spent for generating a new block [64]. Nonce is used for adjusting hash outputs of the block header to satisfy the difficulty limit [65].

According to the blockchain operational mechanism, pre-block hash and time-stamp provide chronological linking of blocks. The block header can represent the whole block with the Merkle tree root and can be transmitted and processed independently. Although the transaction information stored in the blockchain is in a chronological sequencing order, each node could only check the last block to validate that all blocks have not been tampered and simplify the verification process.

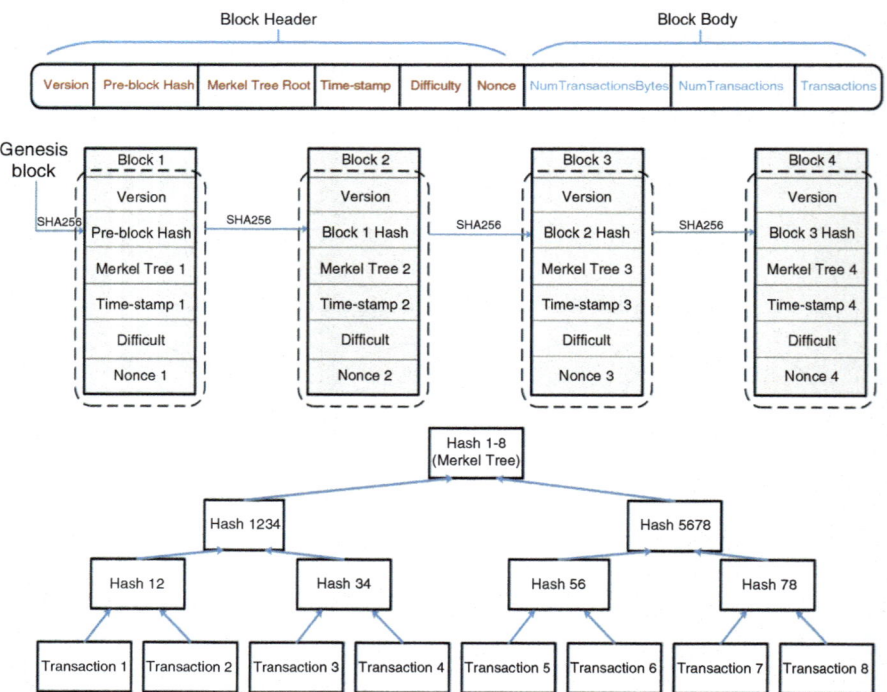

Fig. 3.6 The contents of a block in the blockchain

As the difficulty is adjusted at every 2016 blocks, X_n is the time (in seconds) of the nth block being mined. Then the difficulty is adjusted at X_{2016i} times for $i = 1, 2,$ .... The new difficulty value is based on the previous difficulty value calculated with Eq. (3.1) [66].

$$D_i = D_{i-1} * \left(\frac{2016\left[\text{blocks}\right] * 10\left[\frac{\text{min}}{\text{block}}\right] * 60\left[\frac{\text{s}}{\text{min}}\right]}{X_{2016i} - X_{2016(i-1)}\left[\text{s}\right]} \right) \tag{3.1}$$

where D_i and D_{i-1} (unitless) are the difficulty in the ith segment $(X_{2016(i-1)}, X_{2016i})$ and $(i-1)$th segment $(X_{2016(i-2)}, X_{2016(i-1)})$, respectively. The difficulty stipulates that a block is valid only if the hash output of the block is less than the difficulty. The hash output will eventually become less than the difficulty by adjusting the time-stamp, Merkel tree root, and nonce. It should be noted that the average block generation rate of 10 min/block refers specifically to the Bitcoin system that uses the PoW consensus mechanism. In other systems such as Ethereum, the average block time is much faster and at the rate of 15–20 s. There are energy blockchain projects aiming for close to real-time clearing and execution, such as the Tobalaba test network build by the Energy Web Foundation where block time varies between 3 and 10 s [67].

3.3.4 Consensus Mechanisms

It is crucial to maintain credible information in a distributed ledger. In a distributed network, when two parties communicate with each other, it would be difficult to confirm whether the message is correct or has not been tampered, which is a famous problem called Byzantine Generals Problem. There is a need for different consensus mechanisms to obtain consistency in the distributed ledgers of different nodes.

There are many different consensus mechanisms that would determine the performance of blockchain, not only in scalability but also speed to reach consensus. Therefore, different applications require different consensus mechanisms. Famous consensus mechanisms consist of Proof of Stake (PoS), Proof of Work (PoW), Practical Byzantine Fault Tolerance (PBFT), and Delegated Proof of Stake (DPoS). In PoW, miners compete to solve complex mathematical problems to receive rewards such as for Bitcoin. The node has a higher chance to become the block validator in PoS mechanism when there is more stake; this statement applies in PPcoin [67], Nxt [68], and BlackCoin [69]; As for DPoS mechanism, witnesses would validate all signatures and generate blocks of information, which is applied in BitShares [70]. PBFT is used for solving the Byzantine Generals Problem, which is applied in Hyperledger [71] and Tendermint [72]. Details of different consensus algorithms on architectures, performances, and applications could be found in [73, 74].

3.3.5   Admission Mechanism in Blockchain

There are three categories of blockchain considering the admission mechanism namely public blockchain, consortium blockchain, and private blockchain [75]. Accordingly, blockchain can also be categorized into two types based on the permission privilege: permissioned blockchain that includes private blockchain, consortium blockchain, and non-permissioned blockchain, i.e., public blockchain [76]. A public blockchain is transparent and open. Any individuals or groups can transact in the public blockchain and compete for billing rights. Ethereum and Bitcoin are typical representations of the public blockchain. A consortium blockchain is semi-public that applies to a certain group or organization. It is necessary to pre-specify several nodes as validated nodes. The mission of generating blocks is accomplished by all defined validated nodes. Other nodes in the consortium blockchain could trade with each other but without billing rights. One of the famous applications in consortium blockchain is the Hyperledger project initiated by Linux Foundation [77]. The private blockchain is in closed form. A private blockchain is a distributed ledger and reversible, which is exclusive to companies or individuals [78]. Only internal transactions or information within the companies or groups are recorded in a private blockchain. The categories of blockchain in the admission mechanism are presented in Fig. 3.7.

In summary, this section reviews blockchain considering blockchain framework, operational mechanism, consensus mechanism, and admission mechanism. These technologies would change the energy sector significantly.

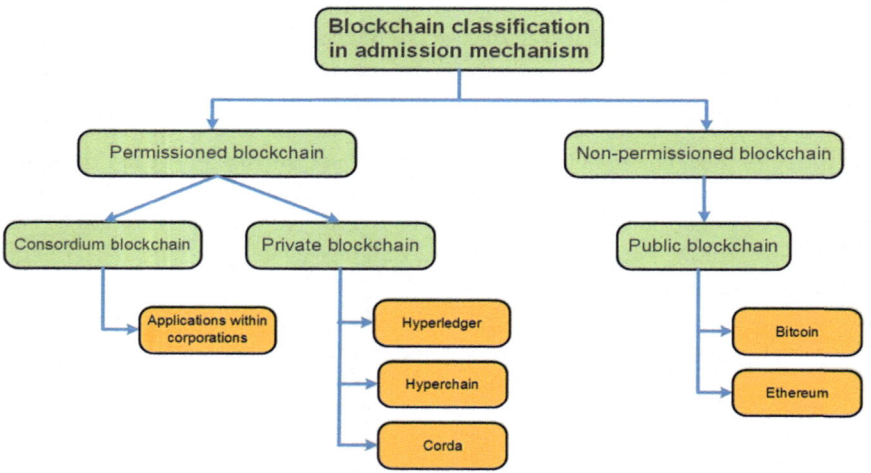

Fig. 3.7 Categories of blockchain in admission mechanism

3.4 Recent Research of Blockchain in the Microgrids

This section provides a state-of-the-art review of blockchain applications in microgrids. This review spans from operational mechanisms to applicational innovations of blockchain. These focuses are classified into four categories: P2P energy trading between consumers and prosumers, vehicle-to-vehicle energy trading, carbon emission trading, and energy demand-side management.

Figure 3.8 presents the percentage share of worldwide blockchain initiatives in the energy sector between March 2017 and March 2018 [79]. As shown in Fig. 3.8, most blockchain initiatives focused on five applications, i.e., P2P transactions, grid transactions, energy financing, sustainability attribution, and electric vehicles. P2P transactions are the dominant application.

3.4.1 P2P Energy Trading Between Prosumers and Consumers

Currently, most low-carbon electricity is transmitted across the grid [6]. Recently, there were projects demonstrated that microgrids promote renewable energy local utilization and enhancing power supply reliability and efficiency. Successful projects include Brooklyn Microgrid operated by Transactive Grid in Brooklyn, New York [80], and Power Ledger project operated in Australia [81] Power Ledger

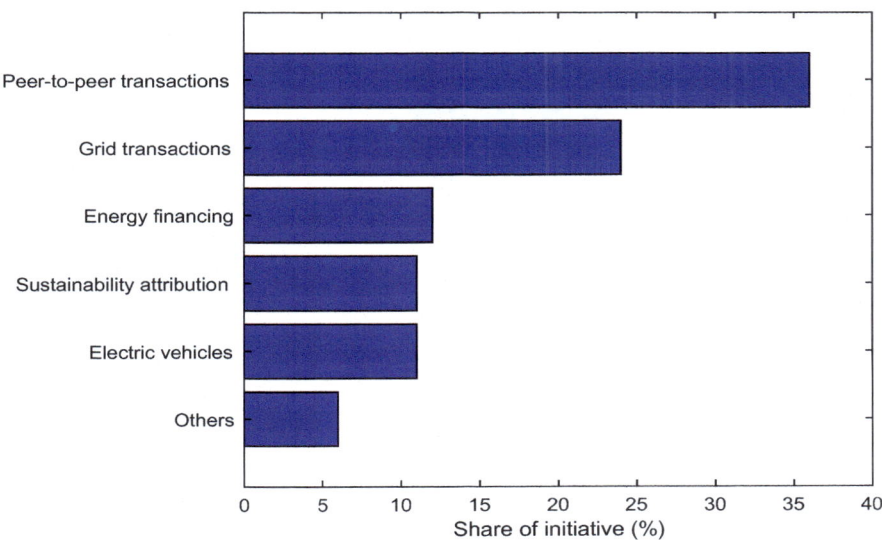

Fig. 3.8 Percentage share of blockchain initiatives in the energy sector worldwide between March 2017 and March 2018 [79]

(POWR). These projects demonstrate that microgrids will bring economic interests to both consumers and prosumers [82].

POWR [81] is an Australian start-up that focuses on the use of distributed energy sources (DERs) like solar and wind energy inexpensively. Different from traditional methods, Power Ledger appeals for P2P energy transactions by recording the consumption and generation in real-time and trading in a transparent platform that allows participants to select the power source. POWR is a transparent and secure energy trading platform that uses two different tokens, i.e., POWR and Sparkz to trade energy between consumers and prosumers. POWR is the utility token that allows participants access to and use the platform, while Sparkz represents the tokenized value of electricity, which is used for energy transactions [83]. The dual token ecosystem model is created to synchronize the ecosystem globally. The transaction platform is powered by a public Ethereum blockchain with a smart contract.

The focuses in P2P energy trading could be categorized into two categories (based on the trading models and the optimization methods) as follows:

- Trading models: Literature [84] has listed and discussed the motivations, challenges, and market structures of three different P2P trading markets, including the full P2P market, community-based market, and hybrid market. In brief, it means that participants or communities could merely trade with each other, or participants and communities could trade in a hybrid way. Literatures [7, 85] proposed a four-layer system architecture of the P2P energy trading model with three dimensions, which are categorized by key functions involved in P2P energy trading, the size of the peers participating in P2P energy trading, and time sequence of the P2P energy trading process. The simulation results demonstrate that P2P energy trading could minimize the energy exchange with the utility grid. Local generation and demand would be more balanced by increasing a variety of peers in a microgrid, which would enhance the advantages of P2P energy trading.

- Optimization methods: The authors in [86] analyzed two separate competitions include (1) price competition among the sellers and (2) choice of seller among the buyers. A new game-theoretic model is presented to deal with these two problems. To maximize the utility of participants as well as social welfare, an efficient double auction mechanism was proposed in [87]. A smart contract is also an efficient technique to implement P2P energy trading, literature [88, 89] proposed an auction model incorporating smart contract for energy trading between prosumers and consumers. Literature [17] deems smart contract as an energy supplier which would complete negotiation, settlement, and payment automatically and balance the system through settlement procedure. Different from earlier, there are many papers [90–92] that are focused on physical impacts in the microgrid. Di Silvestre et al. [92] proposed a new method for calculating the energy losses and attribution during energy transactions between consumers and prosumers. Similarly, Liu et al. [90] aimed at minimizing the overall energy cost and P2P energy sharing losses in the distribution network.

In summary, there are many publications focus on P2P energy trading with blockchain whatever in electricity trading models and market optimization methods. More different trails with different market mechanisms need to have experimented. More different factors that affect P2P energy trading should be considered, such as power lines congestion and participants' privacy protection.

3.4.2   Vehicle-to-Vehicle Energy Trading

Electric vehicles (EVs) can charge their batteries through localized vehicle-to-vehicle energy trading. In general, we could classify the different EVs with different usage patterns into two sets: EES could satisfy the users or not. EVs that could not satisfy their power demand could trade with other EVs that have unused electricity, by which it would enhance the energy utilization efficiency and avoid the energy losses caused by long-distance electricity transmission. In addition, it is also an effective method to minimize the impact of charging from the main grid and reduce the payments for users in the day time [93]. Blockchain technology will enable vehicle-to-vehicle energy trading in preserving privacy and securing payments.

Literature [94] proposed a localized vehicle-to-vehicle electricity trading mechanism among plug-in hybrid EVs. The work illustrated that it could use consortium blockchain to obtain trustful and secure electricity trading. Within consortium blockchain, there are multiple pre-specified nodes called local aggregators (LAGs) to publicly audit the transaction data and perform the consensus process. LAGs act as energy brokers and auctioneers to do statistics of the energy demand and supply and carry out iterative double auction among EVs to increase social welfare.

Literature [95] presented a new vehicle-to-vehicle energy trading system between EVs, which aimed to minimize the impact of the charge of EVs on the power grid during the peak period in the day time. The work focused on two sets of drivers: those who finished their daily trips with energy surplus in their batteries and those required to charge their vehicles during some daily stops. Two optimization algorithms are proposed to determine the best charging schedules and optimal P2P delivery prices. The case study shows that it will reduce the total daily energy cost by up to 71% through vehicle-to-vehicle energy trading with the optimization algorithms.

Considering mobile charging vehicle-to-vehicle (MCV2V), vehicle-to-vehicle (V2V) and grid-to-vehicle (G2V) scenarios, Huang et al. [96] proposed an optimal charging planning framework to maximize users' satisfaction and reduce users' cost. The framework is based on consortium blockchain to guarantee the privacy and security of electricity trading. A novel improved Non-dominated Sorting Genetic Algorithm is developed to solve the optimization problem.

In summary, most of the V2V energy trading has considered the EV electricity usage optimization and minimize the electricity cost for the EV owners. But more experiments should test the performance effects introduced by V2V energy trading on the EV batteries. What's more, how to match different trading EVs should also be considered.

3.4.3 Carbon Emission Trading

Carbon credits are the tokens as permission for the entities to emit the greenhouse gas. Entities with excess credits could sell to other entities that have emitted excess greenhouse gas [97]. As such, different entities would have a different cost for carbon emission. Carbon emission trading is an efficient way to reduce emissions and motivate entities to upgrade technology [98]. But how to process the carbon emission information is a difficult problem. Blockchain would provide a new method to manage the information of carbon emission trading and avoid fraud issues [99]. All information on transactions would be recorded in the distributed ledgers and could not be tampered [100].

There are many companies and startups developing blockchains for the carbon market, such as Power Ledger, CarbonX [101], IBM [102], and Veridum [103]. Power Ledger creates a trading platform for carbon trading. Enabled by blockchain and smart contracts, entities could trade carbon credits or certificates between different organizations in a secure way. The immutable distributed ledger would help to promise the credibility of an asset [81]. Energy Blockchain Labs cooperated with IBM created a carbon asset development platform [104]. The trading platform aims to ensure that the data is traceable, transparent, and visible to all stakeholders in real-time. It will help not only participants to track their carbon footprints in the distributed ledger but also regulators to easily monitor that participants meet their carbon reduction goals. CarbonX is the first P2P personal carbon trading company in the world that aims to motivate people against climate change by rewarding low-carbon behaviors of individuals. People who propose carbon-friendly decisions including riding a bicycle as an alternative to driving and using electrical appliances instead of a fossil fuel one would get the rewards of CarbonX Tokens (CxTs) [105]. Blockchain can build a secure and transparent marketplace for personal carbon trading.

Literature [106] demonstrated that blockchain technology has the potentials to easily track the circulation of carbon credits from generation via ownership trading to ultimate redemption. In this way, three main signs of progress would be achieved including (1) regulators could easily audit the trials; (2) traders may significantly reduce the related time and cost; and (3) renewable energy producers convert their credits into money instantly after electricity generation.

Literature [107] proposed a blockchain enhanced emission trading scheme for the manufacturing industry with the aim of minimizing the carbon emission. The carbon emission could be measured and recorded in the distributed ledgers with the characteristics of transparent, secure, and immutable. A novel evaluation technique is proposed to examine the advantages and disadvantages of the proposed system considering four aspects, including supply, labor, wastes, and energy.

Different from the above-mentioned papers, a blockchain-enabled reputation-based emission trading system is proposed in [108]. Participants with a high reputation have the chance to pick a desirable trade offer and to complete the trade quicker. Blockchain technology would provide a rigid and transparent record of permits and reputation. The case study concludes that the proposed model is a feasible scheme to implement emission trading.

In summary, blockchain would help to manage the carbon emission trading information, which would provide a transparent and secure market for fair transactions. But how to match the carbon emission trading entities and maximize the environmental benefits should be further researched.

3.4.4 Energy Demand-Side Management

Demand-side management (DSM) assists the power balance of the power system with a series of measures at the consumption side [109]. DSM avoids expensive spending on building new power plants and delay to upgrade the transmission lines. Literature [110] gave a review of DSM techniques, including frequency regulation, direct-load control, demand bidding, and Time-of-use pricing. The study gives a conclusion on the advantages of DSM, including to reduce the generation margin and transmission grid investment and increase operational efficiency and improve distribution network investment efficiency.

Blockchain technology would influence the DSM in two ways:

1. introduce a new virtual cryptocurrency for trading;
2. blockchain enables P2P energy trading.

With the first approach, literature [111, 112] proposed that prosumers in the smart grid could trade with each other using a new digital currency, NRGcoin. Prosumers who generate renewable energy are rewarded with a certain amount of NRGcoins that could trade for energy or trade for fiat money on the exchange market. The rewarded ratio of the NRGcoin is variable according to local supply and demand, which strives prosumers and consumers to balance supply and demand, i.e., achieve demand-side management. Because the more energy supply meets demand, additional NRGcoins prosumers would get and the less NRGcoins consumers would pay.

With the second approach, literature [113] presented a new way to keep supply and demand balancing over P2P energy trading. Blockchain enables P2P energy trading but also causes a problem of energy balance in the network. This paper proposed a game-theoretic model for DSM that considered storage and supply constraints in the form of power outages.

In addition, Liu et al. [114] presented an energy sharing model among neighboring PV prosumers and proposed an internal price and cost model of prosumers. Although the model has not integrated the blockchain into the P2P energy trading model, this paper presented another method to achieve DSM through an energy-sharing model with a price-based demand response.

Most of the research work on Blockchain is concentrated on how to increase the efficiency of energy utilization and maximize the utility of participants. As is known to all, blockchain is a new-born technology and has many aspects to improve, including technology fields and people acceptance. It will be of great benefit for us to improve blockchain technology and carry out more experiments in energy fields (Table 3.1).

3.5 The Schematic Operational Mechanism of P2P Energy Trading Model in Microgrid Clusters

This section presents the schematic operational mechanism of P2P energy trading in microgrids and microgrid clusters. The comparisons of different P2P energy trading frameworks are given in Table 3.2. In this section, the authors present a framework that includes four different layers for the operational mechanism of the P2P energy trading model. The framework consists of a smart contract-based hybrid P2P energy trading model with cryptocurrency.

3.5.1 The Framework and Components of the P2P Energy Trading Model

The framework of the P2P energy trading model presents how to implement P2P energy trading in the real world. The framework consists of four parts as shown in Fig. 3.9. Different from the frameworks showed in Table 3.2, this framework gives a comprehensive presentation of how to apply blockchain into the P2P energy trading model in microgrids and microgrid clusters. The four-layer framework shows the schematic operational mechanism of the blockchain-based P2P energy trading model.

3.5.1.1 Physical Layer

The physical layer is liable for physical connections, sensing, and gathering the information between prosumers and consumers in the model. The main parts of the physical layer include transmission lines, converters, and smart meters. Transmission lines and converters are deployed on the premises of participants to achieve power transmission. Smart meters are the main sensors for tracking the energy generation and consumption. Noted that literature [91] had mentioned that several transactions occurring at the same time can create heterogeneous operation conditions, it must

Table 3.1 The classification of blockchain applications in the energy sector

Blockchain applications	References	Traditional methods	Benefits	Challenges
P2P energy trading	[7, 9, 35, 86, 115, 116]	1. Buy energy from the energy retailers 2. Consume the self-generated renewable energy	1. Avoid the long-distance transmission of electricity 2. Promote electrical power supply reliability 3. Decrease carbon emission by using renewable energy	1. It is difficult to predict the renewable generation and consumption 2. High investment to upgrade the infrastructure
Vehicle-to-vehicle energy trading	[94, 95, 117]	1. Charge the power in the charge stations 2. Charge the power at homes or corporations that have installed charging piles	1. Reduce the impact of charging from the main grid 2. Promote electrical power supply reliability	1. Potential to reduce the total life of batteries 2. Guarantee to safety energy transmission
Carbon emission trading	[107, 108, 118]	1. Trade the carbon credits in the carbon exchange	1. Provide a new method to manage the information of carbon emission and avoid fraud issues 2. Facilitate the management of the carbon trading market 3. Facilitate the P2P personal carbon trading	1. Attract more persons to join the carbon trading market 2. It is difficult to meter the amount of carbon emission of personal behaviors
Virtual cryptocurrency	[111, 119]	1. Use fiat currency	1. Provide a new method to facilitate supply and demand balance	1. Protect the security of cryptocurrency
Blockchain enables energy demand-side management	[113, 120]	1. Direct-load control 2. Time-of-use pricing 3. Demand bidding	1. Guarantee the seamless and secure implementation of demand-side management	1. Coordinate the preference of energy use of all the consumers 2. Coordinate the energy distribution from power suppliers

propose an efficient method to attribute the energy losses to an individual transaction. When the power is transmitted in the transmission lines, the information detected by smart meters will help to calculate the power losses during the transmission. In addition, microgrids should link with each other. One or several common

Table 3.2 The advantages and limitations of different P2P energy trading frameworks

Reference	Focus	Descriptions
[80]	Classification of layers	Microgrid setup; Grid connection; Information system; Market mechanism; Pricing mechanism; Energy management trading system; Regulation
	Major components	Microgrid; Superordinate grid; Blockchain; Energy supply system; Microgrid energy market; Energy management trading system; Legislative rules
	Advantages	The layers are able to establish a secure blockchain-based decentralized microgrid energy market with the whole operational processes and implementation procedures.
	Limitations	The market design needs additional examination. The socio-economic incentives of community members to participate in localized energy markets need to be studied further to adapt the market design to achieve an efficient allocation of local energy generation.
[7]	Classification of layers	Power Grid layer; ICT layer; Control layer; Business layer
	Major components	Physical components of the power system; Communication devices, protocols, applications, and information flow; Control functions; Peers, suppliers, distribution system operators (DSOs) and energy market regulators
	Advantages	The layers can reduce the energy transmission between the utility grid and microgrid and balance local demand and generation.
	Limitations	Have not considered how to protect the security of P2P energy trading with blockchain. Since P2P energy trading is a typical type of decentralized energy transaction, it will lead to insecurity in transactions and a lack of transparency. Blockchain would be a promising tool to deal with this problem.
[121]	Classification of layers	Physical layer; Cyber layer
	Major components	Blockchain; IoT; Cloud; Energy entities in all energy generation, transmission, and delivery side
	Advantages	The layers are able to facilitate the data acquisition and data exchange by popular sensing technology and wireless sensor network. Decentralized and distributed processing environment achieves data processing in a decentralized manner.
	Limitations	Current energy laws, policy, and energy trading systems need reform to support P2P energy trading. The proposed framework also has the prospective to change the prosumers' and consumers' energy consumption behaviors. The consumption changes will provoke conflicts between social dissatisfaction and economic performance.

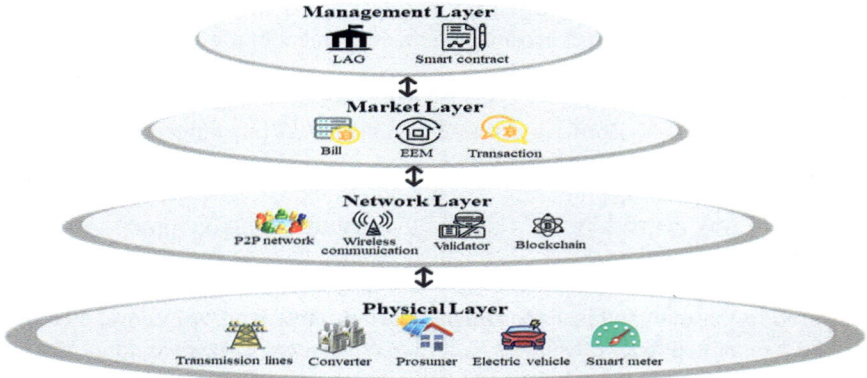

Fig. 3.9 The framework of the P2P energy trading model

couple points should be set for microgrids toward the main grid. It will help to improve the efficient usage of energy and balance the generation and demand within microgrids and microgrid clusters.

3.5.1.2   Network Layer

The network layer is responsible for information communication and value transmission. Prosumers and consumers within microgrids would get access to the P2P energy trading market. A secure and transparent communication method is necessary for the P2P energy trading model. Blockchain is an ideal instrument for P2P communication and transaction. Blockchain enables many secure distributed ledgers that are unalterable and holds by each participant. When a transaction is finished, the earned value would transmit to a certain account of prosumer through blockchain. More precisely, we could choose different types of blockchain and different consensus mechanisms mentioned in Sect. 3.2 of this chapter for a certain blockchain network. The measured and monitored data of smart meters would transmit to blockchain and the data would write into the blocks in the blockchain.

3.5.1.3   Market Layer

The market layer is liable for the market mechanism and business model. The market mechanism means how to implement transactions between prosumers and consumers. In general, prosumers and consumers could trade in the electricity exchange market (EEM). Participants would submit their orders with the necessary information, including participant ID, the amount of surplus or demand energy, the available or demand time for the energy, etc. Noted that the prices for the energy would be set at an upper limit market price and a lower limit market price sometimes, which are corresponding to utility price and feed-in tariff respectively. It would be an efficient method to attract prosumers and consumers to join the EEM within the microgrid.

As for the business model, it means the procedures for implementing P2P energy trading. In general, the procedures could be classified into three parts, including bidding, implement, and settlement. Participants would submit their bid orders and ask orders in the bidding procedure. The bid orders and ask orders would be matched in a certain way defined by EEM. Noted that not all of the bid orders would find their matched ask orders. So, LAGs will act as managers to deal with this problem by trading with other microgrids or the main grid. The corresponding bills will be delivered to these unmatched consumers. During implement procedure, electricity will be transmitted from prosumers to certain consumers. LAGs would monitor the transmission network in real-time and keep the energy balanced. In the settlement period, energy bills will give away to each participant. Blockchain will implement payments according to the energy bills with the cost of energy balance.

3.5.1.4   Management Layer

The management layer is liable for the security of energy supply and balance of generation and consumption under the network constraints. Renewable energy would facilitate self-sufficient energy supply and reduce the reliance on the main grid, which enhances the security of energy supply. Energy trading within the microgrid cluster is also an efficient way to secure energy supply. The price responding mechanism is suitable for participants to maximize their revenues and minimize energy bills. The more energy supply matches the demand, the less should be paid by the consumers for a certain amount of energy. What's more, while the network constraints are violated, LAGs would balance the network by rejecting the orders or introducing energy outside the network, etc.

The main components of a P2P energy trading model could be classified into six parts as follows:

- *Blockchain* is the main ICT of the P2P energy trading model. With blockchain, two main aims would be achieved: reduce the entrance threshold for participants and secure the information and property, which makes P2P energy trading a reality. P2P energy trading is a decentralized market. Blockchain technology offers a chance to establish a decentralized market and makes decentralized decisions.
- *Smart meters* are deployed to measure the production and consumption of renewable energy of each participant in real-time and transmit the data to the blockchain system. Moreover, smart meters will help to calculate the power losses during the transmission.
- *EEM* is necessary for participants to submit their amount of supply and demand and corresponding bid and ask prices, which provides a marketplace and market access to trade local renewable energy. What's more, EEM is also a marketplace for LAGs to supervise the balance of the supply and demand and offer a signal

for a certain microgrid to purchase the electricity from other microgrids or the main grid. When renewable energy is oversupplied in a microgrid, the microgrid could trade with other microgrids that have a short supply within the microgrid cluster through LAGs or feed into the main grid directly. All of the transactions are implemented in the EEM.

- *LAGs* can help to conduct energy trading and electricity transmission. As electricity is not a virtual commodity, there are specific physical constraints for transmission lines and generators. Each microgrid would need a LAG for energy management. What's more, LAGs would act as a microgrid operator on behalf of a certain microgrid to trade with other LAGs of microgrids within the microgrid cluster and conducts transactions between microgrids and the main grid.
- *Transmission lines and converters* must be deployed in different renewable sources. All of the renewable sources would link with each other to transmit electricity. Converters would be used for controlling the transmission direction of the current.
- As for *renewable energy sources*, the energy trading model could combine PV energy generation and wind power generation. Considering the size and structure of the system, the PV plant is the most common renewable energy source within a microgrid.

3.5.2 A Smart Contract-Based Hybrid P2P Energy Trading Model with Cryptocurrency

The recent publications on P2P electricity trading can be found in Table 3.3 of Appendix. Driven by the enabling renewable energy localized usage maximization and enhancing energy supply reliability by P2P energy trading, the authors developed a smart contract-based hybrid P2P energy trading model with cryptocurrency, named "localized renewable energy certificate (LO-REC)". This model could be conducted with the same framework as proposed in Sect. 3.5.1.

As energy trading and electricity transmission between different microgrids could go through LAGs, P2P energy trading could be conducted within a microgrid or microgrid cluster. Prosumers could not only trade electricity with consumers in a microgrid but also trade with another microgrid within a microgrid cluster through LAGs, which make up the hybrid P2P energy trading market. Meanwhile, LAGs would act as managers to deal with the problem of unmatched orders by trading with other microgrids or the main grid.

Traditional renewable energy certificate (REC) is a type of carbon credits with the ambition of facilitating the usage of renewable energy and reducing carbon emission. REC is an instrument that proves that renewable electricity used by electricity consumers. A REC is issued when one megawatt-hour (MWh) of electricity is generated and delivered to the electricity grid from a renewable energy resource [122]. Different from traditional REC, LO-REC is another type of currency which is applied in decentralized renewable energy trading with the same ambition of the REC mentioned earlier. LO-REC is proposed to appeal to consume generated renewable energy locally instead of feeding renewable energy into the main grid and act as the cryptocurrency of P2P energy trading.

Prosumers and consumers could trade electricity with each other in the EEM using the LO-REC. The orders are matched (based on Sect. 3.5.1) in the EEM. LAGs manage the energy supply and balance the electricity generation and consumption under the network constraints. Noted that EEM and LAGs are service providers that facilitate the hybrid P2P energy trading in this model.

A smart contract could perform the interface between prosumers and consumers, which is an efficient way to implement the P2P electricity trading mechanism without any third-party oversight and complete the transactions automatically. A set of rules for the match of the orders, settlement, and payment could be defined in the smart contract and implement automatically. There are five key procedures for the smart contract to achieve its functions as follows:

1. Receive the bid orders and ask orders for transactions in time slot t;
2. Wait for a certain time for acquiring all of the orders;
3. Match the orders within a microgrid;
4. LAG on behalf of a certain microgrid to trade with another LAG in microgrid clusters or main grid for unmatched orders;
5. Settlement for all of the orders.

The schematic operational mechanism of the smart contract is shown in Fig. 3.10.

The pricing of the bid and ask orders could be defined based on the supply and demand ratio of the electricity [114]. It is an efficient method to keep the balance of the power system through the demand response. Participants could adjust their usage time of the electric appliances to maximize their pay off. During the pricing time period, a non-cooperative game model could be used to optimize the hybrid

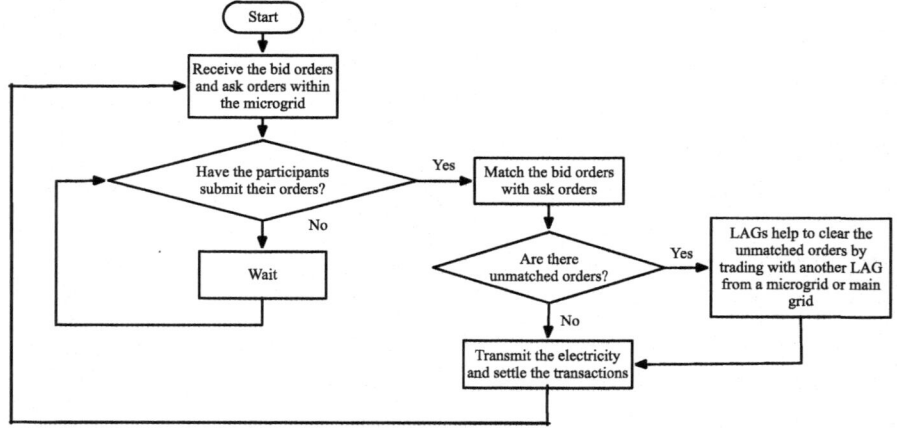

Fig. 3.10 The schematic operational mechanism of the smart contract

P2P energy trading model [7]. A Nash equilibrium is a solution to the model for each participant to maximize their pay off. Noted that other typical pricing mechanisms and optimization methods for the P2P energy trading model could refer to the references listed in Table 3.3 of Appendix.

From the future perspective, the hybrid P2P energy trading model could be integrated with the carbon credits to encourage the reduction of greenhouse gas emissions. It is an efficient way to reduce carbon emission by utilizing renewable energy instead of fossil fuel, which is an activity of carbon offsetting. The amount of renewable energy generated and consumed could be measured by smart meters, which paves the way for the future decentralized personal carbon trading. Producers who generate renewable energy and consumers who consume renewable energy locally would be rewarded with a certain amount of carbon credits. It provides a win-win situation to achieve benefits for the energy consumers and the environment. More systematic analysis and case studies are necessary for application in the future. A schematic diagram of the hybrid P2P energy trading model is given in Fig. 3.11.

3.6   Envision Future Applications of Blockchain in Microgrids

This section presents a prospect of potential applications of blockchain in microgrids and microgrid clusters based on state-of-the-art research. This review gives an analysis of three areas that blockchain may have significant impacts on.

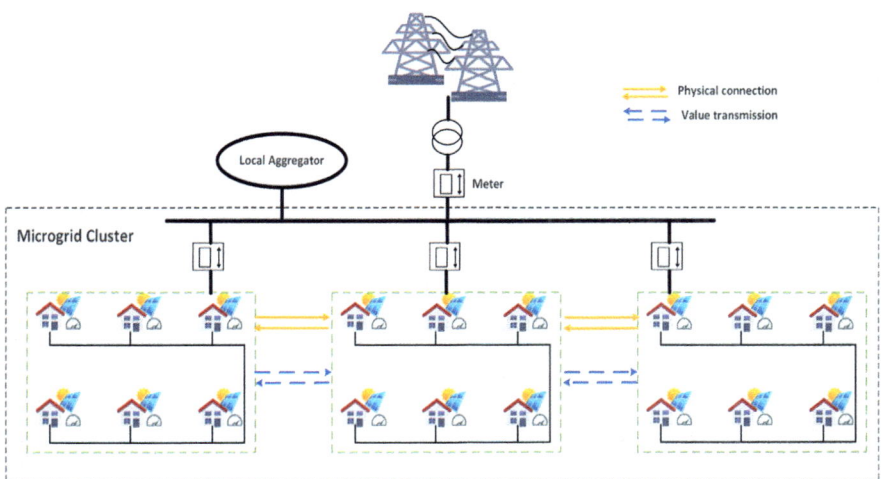

Fig. 3.11 A schematic diagram of the hybrid P2P energy trading model for microgrids and microgrid clusters

3.6.1 Machine-to-Machine (M2M) Micropayments

With smart IoT devices and autonomous electric vehicles, there is a challenge to achieve auto-payments between different autonomous devices. For example, when commercial electric autonomous vehicles run out of energy on the road, these vehicles can find other electric vehicles or unmanned charging piles to purchase the power for themselves. In this scenario, blockchain will automate billing and settlement for transactions with transparent and secure features. It will significantly improve the efficiency of automated IoT devices and autonomous electric vehicles without human intervention.

Reference [123] proposed a novel charging and billing mechanism for electric autonomous vehicles (EACs) to charge their batteries. This paper proposed a vision of M2M micropayments and implemented a proof-of-concept based on Tangle [124], which is an underlying technology of the IOTA [125] cryptocurrency. This paper illustrated a framework to achieve M2M micropayments, which consists of three major layers: (1) physical and user layer, which is responsible for sensing and gathering information concerning the charging process; (2) network and communication layer, which chooses Tangle as decentralized database and IOTA support flash transactions; and (3) services layer, which provides charging services for EACs and data insight for service providers.

Reference [126] also focused on using IoT devices to achieve M2M trading. The work presented a summary of ideas and applications to introduce transactional functionality to the IoT. Then the paper showed an M2M electricity market in the context of the chemical industry between two electricity producers and one electricity consumer in detail.

Machine to machine trading is a trend of share economy in microgrids, especially for commercial electrical autonomous vehicles. It is necessary to execute a trial that includes larger networks and implement a more sophisticated pricing model.

3.6.2 Combined Artificial Intelligence and Blockchain for Smart Buildings Energy Optimization

A smart building uses different sensors to collect data about heating, air conditioners, and lightings, etc. It uses automated processes to control the building's operation according to business functions and services. Sensors can collect information concern the status of various devices of the building, and make a better decision of using available resources with higher efficiency while incorporating the desired behaviors [127]. With more renewable energy sources deployed in smart buildings, such as rooftop PV panels, smart buildings would have the potential to operate in a more efficient way and decrease the energy cost [128]. Smart buildings can use AI to analyze meteorological data, holiday schedules, including public transport and traffic information to reduce energy losses and improve the efficiency and comfort of occupants. For example, Verdigris [129], an IoT startup, applied AI technology to building management. It can use AI algorithms to offer predictive analysis and anomaly detection of the buildings.

In the scenario of smart buildings like office buildings, hospitals, and stadiums, it could use blockchain technology to introduce a novel business model for P2P energy trading by selling excess energy. Combining AI and blockchain would create a new approach to manage renewable energy and lower the energy cost in smart buildings.

3.6.3 Full Life-Cycle Management for Energy Storage Systems

It is important to obtain exact knowledge of EV battery utilization. EV manufacturers like Tesla [130] or BYD Auto [131] use lithium-ion batteries to supply electric power. But these batteries must be carefully operated, or it will lead to loss of capacity, reduced efficiency, and lifetime. Operating conditions affect battery performance including cycling rate and temperature. Therefore, the performance of the battery can vary greatly. Blockchain would be suitable for accurately recording the conditions that the battery has faced, with the characteristics of being transparent and immutable. Combining these immutable records in the blockchain with the physics-based model of batteries can obtain the current health of the batteries through accurate state estimation and performance prediction. In addition, it will help the insurance company to price premiums and EV leasing dealers to price the vehicles fairly according to actual usage and operating conditions [15].

In September 2016, Wanxaing Group unveiled the "Innovative Energy City" plan [132]. The company has received many visions of the project, including a project proposed by Fujitsu Research and Development Center Co., Ltd. Suzhou Branch with main ideas about battery life cycle management of EVs based on blockchain technology. The proposal is about saving battery's history in blockchain for immutable and traceable and using a smart contract to evaluate its value for the fair transaction.

There are many battery application scenarios in microgrids, such as store sufficient energy produced by PV panels on the rooftop and EV power supply. It is important to manage the health of these batteries. Currently, most of the companies focus on how to use blockchain to optimize the usage of batteries such as Sonnen [133] and WePower [134] but do not explicitly consider the health of batteries. Companies should find an efficient and effective way to obtain information on the health of batteries.

3.7   Standards, Recommended Practices, and Guidelines

Blockchain establishes an open, transparent, and secure P2P network, which has significant positive effects on the usage of distributed energy sources. Blockchain applications in microgrids will not only achieve value exchange but more importantly, it achieves P2P information exchange in a more secure, efficient, and transparent fashion. This paper showed the synergy of microgrid and blockchain could achieve the following:

1. Blockchain could keep the transactions between prosumers and consumers secure and remove the intermediaries to prevent single-point-failure;
2. Transparent and secure decentralized ledgers could trace the origins of energy consumed or supplied, deliver prices and supply-demand information to participants and reduce operating cost by improving business processes;
3. A smart contract could enable automated and decentralized microgrid energy management to achieve the balance of demand and supply [113] and provides prosumers and consumers optimal bid or ask prices automatically [135];
4. By using self-generated power, consumers could receive a higher revenue by purchasing deficient electricity from prosumers at a reduced price than utility;
5. Reduce transmission losses by achieving self-sufficient and use energy in a more efficient and low-carbon manner.

Previous works have implied that blockchain-based P2P energy trading between prosumers and consumers could provide environmental benefits and cost-saving for the participants. Moreover, P2P energy trading can provide economic benefits for both prosumers and consumers who engage in energy trading. In a community energy system, not everyone can join this network as it has topology and geographical limitations. In this way, prosumers and consumers will have restricted access to the energy trading network, which would help to protect the security of the transactions and keep the privacy of participants secure. Moreover, a permissioned blockchain system would be required to restrict access to private data of consumers.

Blockchain-based P2P energy trading is suitable for high renewable penetration systems. Nevertheless, it also leads to some challenges with respect to security, efficiency, and regulation. Firstly, P2P energy trading is a good way to consume renewable energy locally, but there is a challenge on how to manage these unpredictable and irregular electricity flows between different prosumers and consumers. Secondly, with intermittent renewable energy generation, there is a potential for contract violation between bid orders and ask orders. Although we could buy electricity from the grid to meet the electricity demands, there are challenges for the power plants to schedule the generation sources and the power grid to optimize the power flow. Thirdly, there is a lack of adequate market regulations and grid management of these decentralized energy trading.

Moreover, from the utility companies' perspective, the usage of renewable energy and decentralized energy trading could be seen as a threat. However, it also provides a new opportunity for these utility companies to earn profits by providing ancillary services to decentralized energy trading. For example, the utility companies could construct the necessary grid infrastructures for decentralized energy trading and provide the grid management and earn the service fee from their consumers.

For future research, this comprehensive review brings the following points to be solved in the future:

- It must be considered that different market mechanisms and optimization methods suit different scenarios of P2P energy trading. For example, different domains would have different types of load profiles or even have different climates, which should propose diverse pricing strategies to suit different scenarios.
- To expand the P2P energy trading in a larger territory, the security problems caused by diverse bi-directional power flows should be considered seriously. The methods for local aggregators to keep the power flows and voltage in the safety range should be proposed in the future.
- A series of reforms and regulations should be designed for the future P2P energy trading market. For example, the third party that regulates the decentralized energy trading market should be set up and the related service fee required needs to be examined.
- The evaluation methods of different P2P energy trading models. The evaluation criteria should consider the cost-saving for the participants, the flexibility of the proposed model for different scenarios (e.g., renewable penetration), and the security and reliability of the power system. The evaluation results would serve as the basis for the selection of the most suitable P2P energy trading model for different domains.
- Big data technologies in blockchain are related to data accuracy and security enhancement. These features are core aspects of the blockchain model. Data sharing will become easier and more common as accountability and security are ensured. Researchers examined the complexities including the continuous expansion of a blockchain system with big data technologies.
- Machine learning techniques can be used to identify any abnormal and illicit activities that might be happening on the blockchain in real-time. Machine learning and blockchain have many synergies and interactive applications. The two technologies can work together for data mining and security enhancement.

3.8 Conclusions

This chapter presents a comprehensive review of blockchain applications in microgrids and microgrid clusters. The development of decentralized renewable energy sources, storages, and smart meters, etc. provide the basic motivations for decentralized electricity trading and blockchain applications in microgrids. The state-of-the-art review of blockchain applications in the energy sector is reviewed. Among these research works and trials, the most promising application is P2P energy trading for prosumers and consumers. We present a four-layer framework to demonstrate the operational mechanism of blockchain-based P2P energy trading. Under this framework, we discuss a smart contract-based hybrid P2P energy trading model with cryptocurrency named localized renewable energy certificate (LO-REC) to maximize renewable energy localized usage and enhance energy supply reliability. Furthermore, future applications of blockchain in microgrids are envisioned and the techniques for blockchain application in P2P energy trading are discussed. As low carbon energy will be the primary energy source, this chapter paves the way for a number of future researches about blockchain applications in microgrids.

Appendix

Table 3.3 A systematic review of P2P electricity trading with blockchain

	This chapter	Li et al. [35]	Zhang et al. [7]	Tushar et al. [116]
Research context	Proposed a smart contract-based hybrid P2P energy trading model. The LO-REC is proposed to appeal to consume renewable energy locally.	A secure energy trading system named "energy blockchain" and a credit-based payment scheme to reinforce frequent and fast energy trading is proposed.	Proposed a four-layer hierarchical system architecture model and determined the important technologies in P2P energy trading. Blockchain is not discussed.	Reviewed the adoption of game-theoretic approaches in P2P energy trading, as an effective and viable energy management solution.
Optimal pricing methods	Noncooperative game.	Proposed a Stackelberg game.	Noncooperative game.	Discussed noncooperative and cooperative games for smart energy management.
Price consensus mechanism	The pricing of the bid and ask orders are according to the electricity supply and demand.	Energy aggregators set transaction prices according to the present energy market.	Each peer submits its bid/ask order with the traded price to the energy trading market.	Negotiate the energy transactions and prices among peers within the energy network.
P2P energy trading domain	Microgrids and microgrid clusters.	Microgrids, energy harvesting network, vehicle to a grid network.	Microgrid.	Distributed energy resources and storage, service, and vehicle to a grid network.
P2P energy trading architecture	Prosumers within a microgrid or microgrid cluster.	N/A	Prosumers and consumers in a microgrid.	Prosumers in a P2P network.

(continued)

Table 3.3 (continued)

	This chapter	Li et al. [35]	Zhang et al. [7]	Tushar et al. [116]
Objective function	To minimize the electricity cost of prosumers and maximize the local usage of renewable energy.	To maximize the economic benefits of credit banks.	To maximize its own economic benefits for each peer.	To maximize the social welfare for each peer: to maximize the payoff to each user and the revenue of each utility company.
Findings	P2P energy trading within the microgrid and microgrid cluster has the benefits of cost-efficient, enhancing the security of energy supply, and reducing carbon emission.	"Energy blockchain" can be used in general P2P energy trading scenarios, and avoid to use a trusted intermediary. The proposed scheme is efficient and secure in the industrial internet of things.	P2P energy trading is able to enhance the local balance of energy consumption and generation for a low voltage grid-connected microgrid. The increased diversity of load and generation profiles can make energy balancing better.	Energy-trading distribution mechanism in P2P networks is needed that avoids privacy and security threats to the end-users and sellers.

	Kang et al. [94]	Thakur and Breslin [136]	Park et al. [137]	Noor et al. [113]
Research context	To enhance transaction privacy and security protection, a consortium blockchain method is presented to achieve localized P2P electricity trading for plug-in hybrid electric vehicles.	Microgrids form a coalition and a microgrid could trade renewable energy with another microgrid instead of trading with the utility grid.	A blockchain-based P2P energy transaction platform is developed to establish a secure P2P trading environment within a smart home environment.	A demand-side management model incorporating storage components is proposed to align supply and demand.
Optimal pricing methods	Proposed an iterative double auction mechanism.	N/A	N/A	Non-cooperative game.
Price consensus mechanism	Local aggregators work as energy brokers for electric vehicles to execute energy bidding.	A peer signals a transaction to transfer funds to other peers.	Each energy transaction is conducted between consumers and prosumers. The prices are affected by the level of supply and demand.	The price of electricity depends on the amount of energy consumed and/or time of day of energy utilization.

(continued)

Table 3.3 (continued)

	Kang et al. [94]	Thakur and Breslin [136]	Park et al. [137]	Noor et al. [113]
P2P energy trading domain	Vehicle to vehicle energy trading.	Microgrids.	Smart homes.	Microgrid.
P2P energy trading architecture	Localized P2P electricity trading system.	Energy trading among microgrids within a coalition.	Energy transaction between prosumers within the smart home environment.	Prosumers within the microgrid, especially for environments with energy supply constraints.
Objective function	To maximize social welfare.	To maximize the total utility of the coalitions in a coalition structure.	To minimize the total cost of smart homes.	To minimize the consumers' total cost.
Findings	Consortium blockchain improves the security and privacy protection of P2P electricity trading for plug-in hybrid electric vehicles. The iterative double auction mechanism maximizes social welfare.	The coalition formation algorithm could quickly converge and produces improved coalition structure, with higher scalability than a centralized coalition formation algorithm.	Blockchain allows a more cost-efficient P2P trading environment. The P2P energy-transaction unit price is cheaper than the unit price set by utility energy providers.	Demand-side management helps to maintain the supply and demand balance and reduce stress on the grid. It also reduces the utility bill of consumers by P2P energy trading.
	Long et al. [9]	Luo et al. [135]	Zhou et al. [115]	Ghosh et al. [138]
Research context	Developed a two-stage control method to achieve P2P energy sharing in community microgrids. Provided an assessment framework to quantify the advantages of P2P energy sharing. Blockchain not considered.	Proposed a two-layered distributed electricity trading system, with the upper layer, is based on a multi-agent system trading negotiation mechanism and the lower layer is based on a contract settlement system.	Proposed a multiagent framework to simulate the behaviors of prosumers and a novel index system that includes three technical indexes and three economic indexes to evaluate the performance of P2P energy sharing mechanisms. Blockchain not considered.	Formulate the selling or buying strategy selection problem of a prosumer as a game-theoretic problem and proposed a distributed algorithm for each prosumer select its own optimal strategy.

(continued)

Table 3.3 (continued)

	Long et al. [9]	Luo et al. [135]	Zhou et al. [115]	Ghosh et al. [138]
Optimal pricing methods	Proposed constrained non-linear programming.	Proposed a multi-agent coalition formation algorithm.	Proposed a multi-agent simulation framework.	Nash equilibrium.
Price consensus mechanism	The supply and demand variation is used to compute the P2P trading prices.	Negotiation between sellers and buyers.	The coordinator agent facilitates the pricing model to produce the internal trading price for prosumer agents.	There is a platform that sets the prices for exchange among the prosumers.
P2P energy trading domain	Community microgrid.	Active distribution network.	The microgrid, distribution network.	Geographically adjacent prosumers.
P2P energy trading architecture	Prosumers in a community microgrid.	Prosumers in an active distribution network.	Part of the distribution network or prosumers in a microgrid.	Peer-to-peer energy trading or peer-to-grid energy trading.
Objective function	Minimize the total energy cost of the community.	Minimize the total electricity cost within the active distribution network.	Minimize the electricity cost of electrical devices of prosumers or maximize its revenues in the P2P energy sharing.	From an economic perspective, the method maximizes the user's total payoff. From the technical perspective, the method maximizes the transmission of energy between the prosumers or minimizes energy consumption.
Findings	P2P energy trading promotes the benefits of communities and each individual than peer-to-grid energy trading, increase the self-sufficiency, and decrease the bill of consumers.	The proposed distributed electricity trading system can promote energy sharing among the prosumers and overall enhance the energy efficiency of the distribution network.	P2P energy sharing among residential consumers has the potential to bring many economic and technical benefits to Great Britain in the future, compared to the conventional paradigm.	The distributed algorithm determines the Nash equilibrium and optimal transaction price. The total energy consumption the peak load is reduced with the optimal transaction price.

References

1. G. Boyle, *Renewable Energy* (Oxford University Press, Oxford, 2004), p. 456. ISBN-10: 0199261784, ISBN-13: 9780199261789
2. I. Dincer, Renewable energy and sustainable development: A crucial review. Renew. Sust. Energ. Rev. **4**(2), 157–175 (2000)
3. A. Monacchi, W. Elmenreich, Assisted energy management in smart microgrids. J. Ambient. Intell. Humaniz. Comput. **7**(6), 901–913 (2016)
4. R. Ramakumar, Role of distributed generation in reinforcing the critical electric power infrastructure, *2001 IEEE Power Engineering Society Winter Meeting. Conference Proceedings (Cat. No. 01CH37194)*, vol. 1, 2001, p. 139
5. E.J. Ng, R.A. El-Shatshat, Multi-microgrid control systems (MMCS), in *IEEE PES General Meeting*, (IEEE, Piscataway, NJ, 2010), pp. 1–6
6. Z. Xu, P. Yang, C. Zheng, Y. Zhang, J. Peng, Z. Zeng, Analysis on the organization and development of multi-microgrids. Renew. Sust. Energ. Rev. **81**, 2204–2216 (2018)
7. C. Zhang, J. Wu, Y. Zhou, M. Cheng, C. Long, Peer-to-peer energy trading in a microgrid. Appl. Energy **220**, 1–12 (2018)
8. J.W. Chao Long, C. Zhang, L. Thomas, M. Cheng, N. Jenkins, Peer-to-peer energy trading in a community microgrid. *2017 IEEE Power & Energy Society General Meeting*, Chicago, IL, USA, 2017
9. C. Long, J. Wu, Y. Zhou, N. Jenkins, Peer-to-peer energy sharing through a two-stage aggregated battery control in a community microgrid. Appl. Energy **226**, 261–276 (2018)
10. Y. Luo, S. Itaya, S. Nakamura, P. Davis, Autonomous cooperative energy trading between prosumers for microgrid systems, *39th Annual IEEE Conference on Local Computer Networks Workshops*, (IEEE, Piscataway, NJ, 2014), pp. 693–696
11. C. Zhang, J. Wu, C. Long, M. Cheng, Review of existing peer-to-peer energy trading projects. Energy Procedia **105**, 2563–2568 (2017)
12. M. Andoni et al., Blockchain technology in the energy sector: A systematic review of challenges and opportunities. Renew. Sust. Energ. Rev. **100**, 143–174 (2019)
13. A. Goranović, M. Meisel, L. Fotiadis, S. Wilker, A. Treytl, T. Sauter, Blockchain applications in microgrids an overview of current projects and concepts, *IECON 2017—43rd Annual Conference of the IEEE Industrial Electronics Society*, (IEEE, Piscataway, NJ, 2017), pp. 6153–6158
14. S. Nakamoto, Bitcoin: A peer-to-peer electronic cash system, https://bitcoin.org/bitcoin.pdf. Accessed 11 Apr 2019
15. C. Pathak, Blockchain for EVs, 25 June 2018, https://faculty.washington.edu/dwhm/2018/06/25/blockchain-for-evs/. Accessed 21 Feb 2019
16. H. Watanabe, S. Fujimura, A. Nakadaira, Y. Miyazaki, A. Akutsu, J. Kishigami, Blockchain contract: Securing a blockchain applied to smart contracts, *2016 IEEE International Conference on Consumer Electronics (ICCE)*, (IEEE, Piscataway, NJ, 2016), pp. 467–468
17. L. Thomas, C. Long, P. Burnap, J. Wu, N. Jenkins, Automation of the supplier role in the GB power system using blockchain-based smart contracts. CIRED **2017**(1), 2619–2623 (2017)
18. Blockchain white paper (2018), China Academy of Information and Communication Technology, Dec. 2018, http://www.caict.ac.cn/english/research/whitepapers/202003/P020 200327550628685790.pdf (Accessed 7 Aug 2020)
19. G.E. Agency, Blockchain in the energy transition. A survey among decision-makers in the German energy industry, 2016, https://shop.dena.de/fileadmin/denashop/media/Downloads_ Dateien/esd/9165_Blockchain_in_der_Energiewende_englisch.pdf. Accessed 12 May 2019
20. U.G.C.S. Adviser, Distributed ledger technology: Beyond blockchain, 2016, https:// assets.publishing.service.gov.uk/government/uploads/system/uploads/attachment_data/ file/492972/gs-16-1-distributed-ledger-technology.pdf. Accessed 12 May 2019
21. C. Park, T. Yong, Comparative review and discussion on P2P electricity trading. Energy Procedia **128**, 3–9 (2017)

22. IEA, Electricity information 2018 overview, 2018, https://www.iea.org/statistics/electricity/. Accessed 9 Apr 2019
23. IEA, Solar PV tracking clean energy progress, 2018, https://www.iea.org/tcep/power/renewables/solar/. Accessed 9 Apr 2019
24. I. Markit, 90 GW residential solar by 2021, 2017, https://www.pveurope.eu/News/Markets-Money/90-GW-residential-solar-by-2021. Accessed 22 May 2019
25. Z. Zhao, J. Guo, X. Luo, J. Xue, C.S. Lai, Z. Xu, L.L.Lai, Energy transaction for multi-microgrids and internal microgrid based on blockchain, IEEE Access, 2020, https://doi.org/10.1109/ACCESS.2020.3014520
26. IEA, Energy storage tracking clean energy progress, 2019, https://www.iea.org/tcep/energy-integration/energystorage/. Accessed 9 Apr 2019
27. T. Kenning, Residential storage costs will fall 84% globally by 2040—BNEF, 2015, https://www.energy-storage.news/news/residential-storage-system-costs-to-fall-by-84-globally-by-2040-bnef. Accessed 9 Apr 2019
28. C.S. Lai, Y. Jia, L.L. Lai, Z. Xu, M.D. McCulloch, K.P. Wong, A comprehensive review on large-scale photovoltaic system with applications of electrical energy storage. Renew. Sust. Energ. Rev. **78**, 439–451 (2017)
29. A. Bhargava, Smart meters to power the energy and utilities sector, 2019, https://www.wns.com/insights/articles/articledetail/73/smart-meter-data-management-systems-to-power-the-energy-and-utilities-sector. Accessed 9 Apr 2019
30. IEA, Smart grids tracking clean energy progress, 2019, https://www.iea.org/tcep/energyintegration/smartgrids/. Accessed 16 Apr 2019
31. H.R. Chi, K.F. Tsang, K.T. Chui, H.S.-H. Chung, B.W.K. Ling, L.L. Lai, Interference-mitigated ZigBee-based advanced metering infrastructure. IEEE Trans. Ind. Inform. **12**(2), 672–684 (2016)
32. N. Batista, R. Melício, J. Matias, J. Catalão, Photovoltaic and wind energy systems monitoring and building/home energy management using ZigBee devices within a smart grid. Energy **49**, 306–315 (2013)
33. N.M. Kumar, P.K. Mallick, Blockchain technology for security issues and challenges in IoT. Proc. Comput. Sci. **132**, 1815–1823 (2018)
34. D. Minoli, B. Occhiogrosso, Blockchain mechanisms for IoT security. Internet Things **1–2**, 1–13 (2018)
35. Z. Li, J. Kang, R. Yu, D. Ye, Q. Deng, Y. Zhang, Consortium blockchain for secure energy trading in industrial Internet of Things. IEEE Trans. Ind. Inform. **14**(8), 3690–3700 (2018)
36. S. Lanzisera, A.R. Weber, A. Liao, D. Pajak, A.K. Meier, Communicating power supplies: Bringing the internet to the ubiquitous energy gateways of electronic devices. IEEE Internet Things J. **1**(2), 153–160 (2014)
37. D. Shrier, D. Sharma, A. Pentland, *Blockchain & Financial Services: The Fifth Horizon of Networked Innovation* (Massachusetts Institute of Technology, Washington, DC, 2016), https://www.getsmarter.com/blog/wp-content/uploads/2017/07/mit_blockchain_and_fin_services_report.pdf
38. Y. Yuan, F.-Y. Wang, Blockchain and cryptocurrencies: Model, techniques, and applications. IEEE Trans. Syst. Man Cybernet. Syst. **48**(9), 1421–1428 (2018)
39. I.V. Krsul, J.C. Mudge, A.J. Demers, Method of electronic payments that prevents double-spending, Google Patents, 1998
40. D. Puthal, N. Malik, S.P. Mohanty, E. Kougianos, G. Das, Everything you wanted to know about the blockchain: Its promise, components, processes, and problems. IEEE Consum. Electron. Mag. **7**(4), 6–14 (2018)
41. M. Crosby, P. Pattanayak, S. Verma, V. Kalyanaraman, Blockchain technology: Beyond bitcoin. Appl. Innov. **2**(6–10), 71 (2016)
42. Y. Zhu, R. Guo, G. Gan, W. Tsai, Interactive incontestable signature for transactions confirmation in bitcoin blockchain, *2016 IEEE 40th Annual Computer Software and Applications Conference (COMPSAC)*, vol. 1, 2016, pp. 443–448

43. E.B. Hamida, K.L. Brousmiche, H. Levard, E. Thea, Blockchain for enterprise: Overview, opportunities and challenges, *The 13th International Conference on Wireless and Mobile Communications (ICWMC 2017)*, 2017

44. F. Glaser, Pervasive Decentralisation of Digital Infrastructures: A Framework for Blockchain Enabled System and Use Case Analysis, 2017. *Proceedings of the 50th Hawaii International Conference on System Sciences*, https://pdfs.semanticscholar.org/859d/0535e16095f274df4d 69df54954b21258a13.pdf

45. A. Kareem, R. Bin Sulaiman, M. Umer Farooq, Algorithms and Security Concern in Blockchain Technology: A Brief Review, 19 Aug 2018

46. V. Dhillon, D. Metcalf, M. Hooper, Foundations of blockchain, *Blockchain Enabled Applications*, (New York City, Springer, 2017), pp. 15–24

47. Y. Yuan, F.-Y. Wang, Towards blockchain-based intelligent transportation systems, *2016 IEEE 19th International Conference on Intelligent Transportation Systems (ITSC)*, (IEEE, Piscataway, NJ, 2016), pp. 2663–2668

48. A. Baliga, Understanding blockchain consensus models, *Persistent*, 2017, https://www.persistent.com/whitepaper-understanding-blockchain-consensus-models/

49. C. Cachin, M. Vukolić, *Blockchain Consensus Protocols in the Wild*, 2017. arXiv preprint arXiv:1707.01873

50. G. Liang, S.R. Weller, F. Luo, J. Zhao, Z.Y. Dong, Distributed blockchain-based data protection framework for modern power systems against cyber attacks. IEEE Trans. Smart Grid **10**, 3162–3173 (2019)

51. H. Sukhwani, J.M. Martínez, X. Chang, K.S. Trivedi, A. Rindos, Performance modeling of PBFT consensus process for permissioned blockchain network (hyperledger fabric), *2017 IEEE 36th Symposium on Reliable Distributed Systems (SRDS)*, (IEEE, Piscataway, NJ, 2017), pp. 253–255

52. H. Sheng, X. Fan, W. Hu, X. Liu, K. Zhang, Economic incentive structure for blockchain network, *International Conference on Smart Blockchain* (Springer, 2018), pp. 120–128

53. Ethereum, Ethereum, 2019, https://www.ethereum.org/. Accessed 26 Feb 2019

54. N. Szabo, Smart contracts: Building blocks for digital markets, 1996, http://www.fon.hum. uva.nl/rob/Courses/InformationInSpeech/CDROM/Literature/LOTwinterschool2006/szabo. best.vwh.net/smart_contracts_2.html. Accessed 26 Feb 2019

55. M. Pilkington, 11 Blockchain technology: Principles and applications, *Research Handbook on Digital Transformations*, vol. 225, (Edward Elgar Publishing, Cheltenham, 2016)

56. C. Miles, Blockchain security: What keeps your transaction data safe? 2017, https://www. ibm.com/blogs/blockchain/2017/12/blockchain-security-what-keeps-your-transaction-data-safe/. Accessed 18 Mar 2019

57. Statista, Market capitalization of Bitcoin from 1st quarter 2012 to 4th quarter 2018 (in billion U.S. dollars), 2019, https://www.statista.com/statistics/377382/bitcoin-market-capitaliza-tion/. Accessed 18 Mar 2019

58. CoinMarketCap, 2019, https://coinmarketcap.com/charts/. Accessed 22 Feb 2019

59. C. Natoli, V. Gramoli, The blockchain anomaly, *2016 IEEE 15th International Symposium on Network Computing and Applications (NCA)*, (IEEE, Piscataway, NJ, 2016), pp. 310–317

60. P. Fairley, Blockchain world-Feeding the blockchain beast if bitcoin ever does go mainstream, the electricity needed to sustain it will be enormous. IEEE Spectr. **54**(10), 36–59 (2017)

61. Z. Zheng, S. Xie, H.-N. Dai, H. Wang, Blockchain challenges and opportunities: A survey, Int. J. Web and Grid Services, **14**(4), (2018), https://www.henrylab.net/wp-content/uploads/2017/10/blockchain.pdf

62. B. Wiki, SHA-256, 2016, https://en.bitcoin.it/wiki/SHA-256. Accessed 19 Mar 2019

63. B. Curran, What is a Merkle Tree? Beginner's guide to this blockchain component, 2018, https://blockonomi.com/merkle-tree/. Accessed 19 Mar 2019

64. L. Parker, Timestamping on the blockchain, 2015, https://bravenewcoin.com/insights/time-stamping-on-the-blockchain. Accessed 19 Mar 2019

65. B. Whittle, What is a Nonce? A no-nonsense dive into Proof of Work, 2018, https://coincentral.com/what-is-a-nonce-proof-of-work/. Accessed 19 Mar 2019

66. R. Bowden, H.P. Keeler, A.E. Krzesinski, P.G. Taylor, *Block Arrivals in the Bitcoin Blockchain*, 2018. arXiv preprint arXiv:1801.07447

67. S. King, S. Nadal, Ppcoin: Peer-to-peer crypto-currency with proof-of-stake, Aug 19, 2012, https://decred.org/research/king2012.pdf

68. Nxt, Nxt, 2019, https://nxtplatform.org/. Accessed 22 Feb 2019

69. Blackcoin, Blackcoin, 2019, www.blackcoin.co. Accessed 22 Feb 2019

70. Bitshares, Bitshares, 2019, https://bitshares.org/. Accessed 22 Feb 2019

71. Hyperledger, Hyperledger, 2019, https://www.hyperledger.org/. Accessed 22 Feb 2019

72. Tendermint, Tendermint, 2019, https://tendermint.com/. Accessed 22 Feb 2019

73. D. Mingxiao, M. Xiaofeng, Z. Zhe, W. Xiangwei, C. Qijun, A review on consensus algorithm of blockchain, *2017 IEEE International Conference on Systems, Man, and Cybernetics (SMC)*, 2017, pp. 2567–2572

74. O. Dib, K.-L. Brousmiche, A. Durand, E. Thea, E.B. Hamida, Consortium blockchains: Overview, applications and challenges. Int. J. Adv. Telecommun. 11(1&2) (2018)

75. BlockchainHub, Blockchains and distributed ledger technologies, 2019, https://blockchainhub.net/blockchains-and-distributed-ledger-technologies-in-general/. Accessed 19 Mar 2019

76. Github, Types of blockchain, 2019, https://mastanbtc.github.io/blockchainnotes/blockchaintypes/. Accessed 19 Mar 2019

77. IBM, Hyperledger: Blockchain collaboration changing the business world, 2019, https://www.ibm.com/blockchain/hyperledger. Accessed 18 Mar 2019

78. S. Khatwani, What are private blockchains and how are they different from public blockchains? 2018, https://coinsutra.com/private-blockchain-public-blockchain/. Accessed 19 Mar 2019

79. Statista, Distribution of blockchain initiatives in the electric power sector worldwide between March 2017 and March 2018, by application, 2019, https://www.statista.com/statistics/866609/electricity-blockchain-initiatives-globally-by-application/. Accessed 6 Apr 2019

80. E. Mengelkamp, J. Gärttner, K. Rock, S. Kessler, L. Orsini, C. Weinhardt, Designing microgrid energy markets: A case study: The Brooklyn Microgrid. Appl. Energy 210, 870–880 (2018)

81. Power Ledger White paper, 2018, https://whitepaperdatabase.com/power-ledger-powr-whitepaper/. Accessed 21 Feb 2019

82. P. Ledger, Power Ledger, 2019, https://www.powerledger.io/. Accessed 21 Feb 2019

83. J. Sessa, What is Power Ledger (POWR)? The complete guide, 2018, https://coincentral.com/power-ledger-beginner-guide/. Accessed 7 Mar 2019

84. T. Sousa, T. Soares, P. Pinson, F. Moret, T. Baroche, E. Sorin, Peer-to-peer and community-based markets: A comprehensive review. Renew. Sust. Energ. Rev. 104, 367–378 (2019)

85. C. Zhang, J. Wu, M. Cheng, Y. Zhou, C. Long, A bidding system for peer-to-peer energy trading in a grid-connected microgrid. Energy Procedia 103, 147–152 (2016)

86. A. Paudel, K. Chaudhari, C. Long, H.B. Gooi, Peer-to-peer energy trading in a prosumer based community microgrid: A game-theoretic model. IEEE Trans. Ind. Electron., 1–1 (2018)

87. B.P. Majumder, M.N. Faqiry, S. Das, A. Pahwa, An efficient iterative double auction for energy trading in microgrids, *2014 IEEE Symposium on Computational Intelligence Applications in Smart Grid (CIASG)*, (IEEE, Piscataway, NJ, 2014), pp. 1–7

88. M. Sabounchi, J. Wei, Towards resilient networked microgrids: Blockchain-enabled peer-to-peer electricity trading mechanism, *2017 IEEE Conference on Energy Internet and Energy System Integration (EI2)*, 2017, pp. 1–5

89. C. Block, D. Neumann, C. Weinhardt, A market mechanism for energy allocation in micro-chp grids, *Proceedings of the 41st Annual Hawaii International Conference on System Sciences (HICSS 2008)*, (IEEE, Piscataway, NJ, 2008), pp. 172–172

90. T. Liu, X. Tan, B. Sun, Y. Wu, X. Guan, D.H. Tsang, Energy management of cooperative microgrids with p2p energy sharing in distribution networks, *2015 IEEE International*

Conference on Smart Grid Communications (SmartGridComm), (IEEE, Piscataway, NJ, 2015), pp. 410–415

91. M.L. Di Silvestre, P. Gallo, M.G. Ippolito, E.R. Sanseverino, G. Zizzo, A technical approach to the energy blockchain in microgrids. IEEE Trans. Ind. Inform. **14**(11), 4792–4803 (2018)

92. E.R. Sanseverino, M.L. Di Silvestre, P. Gallo, G. Zizzo, M. Ippolito, The blockchain in microgrids for transacting energy and attributing losses, *2017 IEEE International Conference on Internet of Things (iThings) and IEEE Green Computing and Communications (GreenCom) and IEEE Cyber, Physical and Social Computing (CPSCom) and IEEE Smart Data (SmartData)*, (IEEE, Piscataway, NJ, 2017), pp. 925–930

93. R. Alvaro, J. González, C. Gamallo, J. Fraile-Ardanuy, D.L. Knapen, Vehicle to vehicle energy exchange in smart grid applications, *2014 International Conference on Connected Vehicles and Expo (ICCVE)*, 2014, pp. 178–184

94. J. Kang, R. Yu, X. Huang, S. Maharjan, Y. Zhang, E. Hossain, Enabling localized peer-to-peer electricity trading among plug-in hybrid electric vehicles using consortium blockchains. IEEE Trans. Ind. Inform. **13**(6), 3154–3164 (2017)

95. R. Alvaro-Hermana, J. Fraile-Ardanuy, P.J. Zufiria, L. Knapen, D. Janssens, Peer to peer energy trading with electric vehicles. IEEE Intell. Transp. Syst. Mag. **8**(3), 33–44 (2016)

96. X. Huang, Y. Zhang, D. Li, L. Han, An optimal scheduling algorithm for hybrid EV charging scenario using consortium blockchains. Futur. Gener. Comput. Syst. **91**, 555–562 (2019)

97. G. Hua, T. Cheng, S. Wang, Managing carbon footprints in inventory management. Int. J. Prod. Econ. **132**(2), 178–185 (2011)

98. C. Hepburn, Carbon trading: A review of the Kyoto mechanisms. Annu. Rev. Environ. Resour. **32**, 375–393 (2007)

99. E. Al Kawasmi, E. Arnautovic, D. Svetinovic, Bitcoin-based decentralized carbon emissions trading infrastructure model. Syst. Eng. **18**(2), 115–130 (2015)

100. K.-H. Liu, S.-F. Chang, W.-H. Huang, I.-C. Lu, The framework of the integration of carbon footprint and blockchain: Using blockchain as a carbon emission management tool, *Technologies and Eco-innovation Towards Sustainability I*, (Springer, Singapore, 2019), pp. 15–22

101. CarbonX, CarbonX, 2019, https://carbonx.com/. Accessed 9 Mar 2019

102. IBM, Creating a more efficient green energy marketplace with IBM Blockchain technology, 2019, https://www.ibm.com/case-studies/energy-blockchain-labs-inc. Accessed 9 Mar 2019

103. Veridum, Veridum, 2019, https://www.veridium.io/. Accessed 9 Mar 2019

104. IBM, Energy Blockcahin Labs Inc., 2019, https://www.ibm.com/case-studies/energy-block-chain-labs-inc. Accessed 12 Mar 2019

105. ConsenSys, ConsenSys introduces CarbonX, a blockchain initiative to fight climate change, 2017, https://www.prnewswire.com/news-releases/consensys-introduces-carbonx-a-block-chain-initiative-to-fight-climate-change-300524455.html. Accessed 9 Mar 2019

106. M.J. Ashley, M.S. Johnson, Establishing a secure, transparent, and autonomous blockchain of custody for renewable energy credits and carbon credits. IEEE Eng. Manag. Rev. **46**(4), 100–102 (2018)

107. B. Fu, Z. Shu, X. Liu, Blockchain enhanced emission trading framework in fashion apparel manufacturing industry. Sustainability **10**(4), 1105 (2018)

108. K.N. Khaqqi, J.J. Sikorski, K. Hadinoto, M. Kraft, Incorporating seller/buyer reputation-based system in blockchain-enabled emission trading application. Appl. Energy **209**, 8–19 (2018)

109. G.M. Masters, *Renewable and Efficient Electric Power Systems* (Wiley, New York, 2013)

110. G. Strbac, Demand side management: Benefits and challenges. Energy Policy **36**(12), 4419–4426 (2008)

111. M. Mihaylov, S. Jurado, N. Avellana, K. Van Moffaert, I.M. de Abril, A. Nowé, NRGcoin: Virtual currency for trading of renewable energy in smart grids, *11th International Conference on the European Energy Market (EEM14)*, (IEEE, Piscataway, NJ, 2014), pp. 1–6

112. M. Mihaylov, S. Jurado, K. Van Moffaert, N. Avellana, A. Nowe, NRG-X-Change—A novel mechanism for trading of renewable energy in smart grids, *Smartgreens*, 2014, pp. 101–106
113. S. Noor, W. Yang, M. Guo, K.H. van Dam, X. Wang, Energy demand side management within micro-grid networks enhanced by blockchain. Appl. Energy **228**, 1385–1398 (2018)
114. N. Liu, X. Yu, C. Wang, C. Li, L. Ma, J. Lei, Energy-sharing model with price-based demand response for microgrids of peer-to-peer prosumers. IEEE Trans. Power Syst. **32**(5), 3569–3583 (2017)
115. Y. Zhou, J. Wu, C. Long, Evaluation of peer-to-peer energy sharing mechanisms based on a multiagent simulation framework. Appl. Energy **222**, 993–1022 (2018)
116. W. Tushar, C. Yuen, H. Mohsenian-Rad, T. Saha, H.V. Poor, K.L. Wood, Transforming energy networks via peer-to-peer energy trading: The potential of game-theoretic approaches. IEEE Signal Process. Mag. **35**(4), 90–111 (2018)
117. Z. Su, Y. Wang, Q. Xu, M. Fei, Y. Tian, N. Zhang, A secure charging scheme for electric vehicles with smart communities in energy blockchain. IEEE Internet Things J., 1–1 (2019)
118. P. Yuan, X. Xiong, L. Lei, K. Zheng, Design and implementation on hyperledger-based emission trading system. IEEE Access **7**, 6109–6116 (2019)
119. T. Zhang, H. Pota, C.-C. Chu, R. Gadh, Real-time renewable energy incentive system for electric vehicles using prioritization and cryptocurrency. Appl. Energy **226**, 582–594 (2018)
120. C. Pop, T. Cioara, M. Antal, I. Anghel, I. Salomie, M. Bertoncini, Blockchain based decentralized management of demand response programs in smart energy grids. Sensors (Basel, Switzerland) **18**(1), 162 (2018)
121. Z. Dong, F. Luo, G. Liang, Blockchain: A secure, decentralized, trusted cyber infrastructure solution for future energy systems. J. Mod. Power Syst. Clean Energy **6**(5), 958–967 (2018)
122. U. S. E. P. Agency, Renewable Energy Certificates (RECs), 2019, https://www.epa.gov/greenpower/renewable-energy-certificates-recs. Accessed 18 May 2019
123. D. Strugar, R. Hussain, M. Mazzara, V. Rivera, J. Lee, R. Mustafin, *On M2M Micropayments: A Case Study of Electric Autonomous Vehicles*, 2018. arXiv preprint arXiv:1804.08964
124. S. Popov, "The tangle", 2018 https://assets.ctfassets.net/r1dr6vzfxhev/2t4uxvsIqk0EUau6g2sw0g/45eae33637ca92f85dd9f4a3a218e1ec/iota1_4_3.pdf
125. IOTA, IOTA, 2019, https://www.iota.org/. Accessed 22 Feb 2019
126. J.J. Sikorski, J. Haughton, M. Kraft, Blockchain technology in the chemical industry: Machine-to-machine electricity market. Appl. Energy **195**, 234–246 (2017)
127. D. Sembroiz, D. Careglio, S. Ricciardi, U. Fiore, Planning and operational energy optimization solutions for smart buildings. Inf. Sci. **476**, 439–452 (2019)
128. Y. Liu et al., Coordinating the operations of smart buildings in smart grids. Appl. Energy **228**, 2510–2525 (2018)
129. Verdigris, Verdigris, 2019, https://verdigris.co/. Accessed 22 Feb 2019
130. Tesla, Tesla, 2019, https://www.tesla.com/. Accessed 22 Feb 2019
131. BYD, BYD, 2019, http://www.byd.com/. Accessed 22 Feb 2019
132. C. Plus, Chinese automotive company switches focus to clean energy, 2018, http://chinaplus.cri.cn/news/business/12/20180730/163809.html. Accessed 19 Mar 2019
133. Sonnen, Sonnen, 2019, https://sonnengroup.com/. Accessed 22 Feb 2019
134. Wepower, Wepower, 2019, https://wepower.network/. Accessed 22 Feb 2019
135. F. Luo, Z.Y. Dong, G. Liang, J. Murata, Z. Xu, A distributed electricity trading system in active distribution networks based on multi-agent coalition and blockchain. IEEE Trans. Power Syst., 1–1 (2018)
136. S. Thakur, J.G. Breslin, Peer to peer energy trade among microgrids using blockchain based distributed coalition formation method. Technol. Econ. Smart Grids Sustain. Energy **3**(1), 5 (2018)
137. L. Park, S. Lee, H. Chang, A sustainable home energy prosumer-chain methodology with energy tags over the blockchain. Sustainability **10**(3), 658 (2018)
138. A. Ghosh, V. Aggarwal, H. Wan, *Exchange of Renewable Energy Among Prosumers Using Blockchain with Dynamic Pricing*, 2018. arXiv preprint arXiv:1804.08184

Chapter 4
A Time-Synchronized ZigBee Building Network for Smart Water Management

4.1 Introduction

Water is essential and important to improve the health of people. It is an invaluable resource and therefore smart water management is necessary to keep water usage efficiently. To provide the public with a more comfortable and healthier environment, the methods and solutions to enhance water management have attracted a lot of interest and are being developed intensively. The basic idea of enhancing water management is to control the related parameters that are water flow, period, temperature, quality, leakage, etc. Good water management benefits all. For example, indoor air quality (IAQ) is a good indicator to determine whether the public could have a healthy life or not. As such, to achieve good IAQ, two perspectives are proposed, which are greenery concept [1, 2] and intelligent control [3–5]. Greenery concept is to use the natural processes of the plants inside the buildings. The plants are mainly planted at roof, wall, balcony, sky garden, and indoor garden. The natural processes will reduce the heat content, improve the ventilation, and also save the energy consumption from heating, ventilating, and air conditioning (HVAC) system. These improvement schemes are highly related to water control, such as the automatic irrigation system, the water flow in HVAC, water cooling system, etc. Therefore, developing water management will be the first step to achieve a better quality of life.

Smart metering (SM) supports distributed technologies and consumer participation and extracts energy data based on two-way communication [6, 7]. The wireless sensor network (WSN) is a vital component in SM communication [8, 9]. The ZigBee wireless protocol is commonly used in WSN and adopted as one of the standards in SM [10]. Figure 4.1 shows the real-time bi-directional communication between customers and the water industry. Figure 4.2 shows the scheme for Smart Water Management (SWM) and Advanced Metering Infrastructure (AMI).

Fig. 4.1 Real-time bi-directional communication

Fig. 4.2 Smart water management (SWM) and advanced metering infrastructure (AMI)

To realize the intelligent water control, many sensors monitoring water-related parameters are required, and thus, the wireless sensor network is considered due to its high scalability, high flexibility, easy installation, and replacement. To have scalable, reliable, low power consumption, and low production cost, ZigBee is an appropriate wireless protocol to develop the wireless sensor network [11]. ZigBee is a short-range communication implemented on Wireless Personal Area Network (WPAN) and is recognized by the US and Europe as a wireless communication technique for energy management. Figure 4.3 shows the reasons in adopting ZigBee in terms of the bit error rate and signal to noise ratio.

However, the time synchronization of the network has not been addressed for timely data transmission that can lead to degraded network performance such as packet loss, collision, and low latency. In this subchapter, a time-synchronized ZigBee building network (TS-ZBN) will be proposed for water management. The objective of TS-ZBN is to collect the remotely-sensed data of the whole building and transfer them to the backend server. However, some key issues have to be addressed. The first one is the huge data flow of the building and the second one is the time synchronization of each sensor node. By adapting the idea of high-rise building network for advanced metering infrastructure [12], a three-core network that consists of a normal layer network (NLN), transverse layer network (TLN), and unity area network (UAN) will be proposed and integrated to form ZBN. In order to synchronize the nodes properly, a node-to-node synchronization including

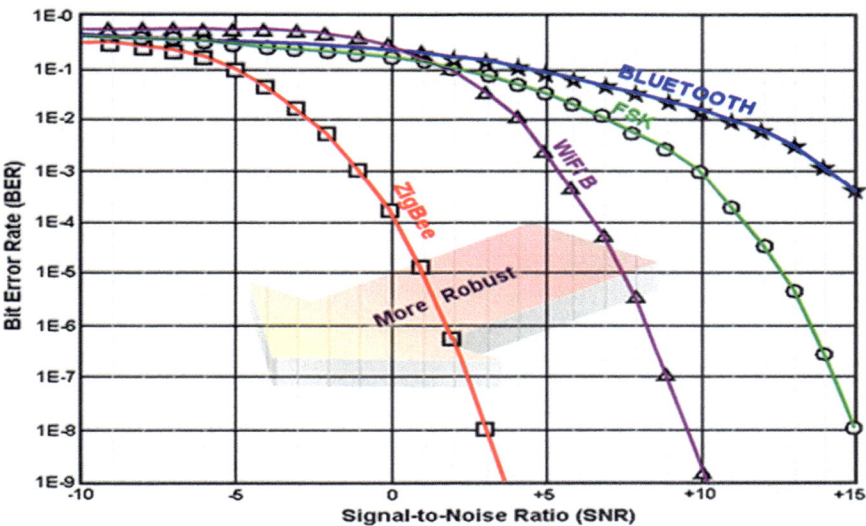

Fig. 4.3 Robustness ZigBee

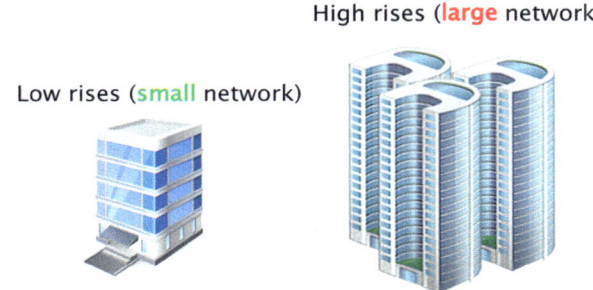

Fig. 4.4 Low rises (small network) versus high rises (large network)

coordinator-to-coordinator synchronization and end-to-end synchronization will be discussed. It is essential to estimate the clock difference and make use of the common clock.

In most smart cities, there are many tall buildings. Figure 4.4 shows a comparison between low-rise and high-rise buildings in terms of communication network requirements.

These high-rises normally present a hostile environment for wireless signals. ZigBee is dedicated to smart energy applications and has been extensively adopted in smart energy applications. By taking advantage of ZigBee smart energy open standard and its mesh capability for scalability, researchers find it superior and well organized to use ZigBee in SM applications. It was pointed out that, in an urban area, a huge aggregation of data creates the need to investigate building-area networks (BANs) [13, 14]. However, high rises are typically comprised of hard reinforced concrete and this leads to signal propagation problems in general. A modern smart city is full of people with a busy life that normally demand communication using WiFi and Bluetooth for wireless delivery in the same frequency band. Thus, the application of ZigBee to advanced metering infrastructure (AMI) in high-traffic areas needs to be dealt with the potential interferences.

A former design of high-traffics AMI (HTAMI) did not consider interference [15]. However, the high attenuation and dispersive characteristics of concrete construction in ZigBee BAN (ZBAN) demand AMI features that mitigate interference. In this investigation, an interference model will be investigated. A new design and implementation of interference-mitigated ZBAN for HTAMI will be proposed and developed.

In this design, there are multiple parameters for consideration, for instance, high power and high throughput for fast data transmission and low latency for good quality of service (QoS). QoS refers to the technology that manages data traffic to reduce packet loss, latency, and jitter on the network. However, the magnitude of these factors may have contradictory requirements, e.g., the high-power transmission that causes the feeling of potential health hazard versus the well-accepted low power, the high throughput demanded by users versus the low throughput generally achieved in a hostile environment, the low latency commonly requested versus the high latency normally occurs in noisy communications. A noticeable and practical solution can be achieved by optimizing these key parameters. In this investigation, experimental work was conducted to acquire the background data applicable to the characteristics of the ZBAN.

In the experiment, measurements of a five-story building were conducted to collect good quality data to make it easier for the large-scale modeling and analysis of the complicated high traffics scenario to happen. The interference mitigation model for ZigBee transmission will also be derived. It will be explained that the nondominated sorting genetic algorithm-II (NSGA-II) is customized to obtain the pareto fronts (PFs) from which the appropriate design will be developed. The optimized network engineering tool which is a packet-level network simulator (OPNET) is employed for the large-scale evaluation and analysis. The measured data are used for optimization and model generation in the OPNET environment. Measurement results show that the developed IMM2ZM satisfies the demand-response requirement of the US standards among the hostile environment of HTAMI.

The contribution of this study is as follows:

1. A measurement was performed to obtain the data for the formulation of objective functions of the optimal solution.
2. An interference mitigation model has been derived.
3. A modification to NSGA-II [16] optimization algorithm has been conducted.
4. OPNET evaluation has been carried out for large-scale analysis.
5. A channel-swapping interference-mitigated multiradio multichannel ZigBee metering (IMM2ZM) system has been implemented for IMM2ZM system for HTAMI.

This chapter is organized as follows. Fundamental background is given in Sect. 4.2. The design of IMM2ZM is presented in Sect. 4.3 and the system IMM2ZM model is given in Sect. 4.4. The multiobjective optimization for the IMM2ZM using NSGA-II is described in Sect. 4.5. The analysis and evaluation of the IMM2ZM are shown in Sect. 4.6. Finally, the conclusion is given in Sect. 4.7.

4.2   Fundamental Background

When considering the communication protocol of the control system, flexibility, scalability, and reliability should be included. To make the control system flexible, the devices should be installed, adjusted, and replaced easily. Therefore, the wireless technique is the right choice to form a communication link. Among various wireless communication protocols, ZigBee is a kind of open standard dedicated to sensor networks due to its scalability and mesh capability [11]. The control system using ZigBee can be realized in a short time and low cost. ZigBee, a WPAN standard of IEEE 802.15.4 and is similar to Bluetooth, a widely employed communication protocol in the mobile devices. Both have features such as short-distance communication, low cost, low data rate, and low energy consumption. Besides the similarity, ZigBee has a longer battery life due to its lower energy consumption than that of Bluetooth. The ZigBee device can be operated for at least 2 years and supports more than 65,000 devices simultaneously. Its scalability and low power consumption are highly suitable for monitoring systems. The ZigBee network mainly has three configuration architectures, namely, star, cluster tree, and mesh.

The star network is basically a radial network composed of the main coordinator and a series of terminal nodes. The core nodes are mainly responsible for data exchange and issuing commands. In a star network topology, all nodes except the central node must establish a wireless transmission connection with the central node, but the central node may become the bottleneck of the entire network. Once the central node fails or transmission is blocked, the reliability of the system network will decrease significantly. Since the star network has three characteristics, for example, less frequently implementing upper-layer protocols, lower hardware costs, and lower upper-layer routing maintenance costs, its implementation is relatively simple. But because of this, its central node needs to perform many data operations, such as giving certificates, remote control, and so on. However, the shortcomings of this kind of network are also obvious. When the terminal nodes are outside the communication radius of the central node, they cannot achieve communication, which makes the system inflexible and greatly limits the coverage and extension of the network. In addition, when all the messages in the network are converged to the central node simultaneously, it will cause problems such as communication congestion, packet loss, and transmission abnormality.

In a star network topology, the network coordinator is usually defined as the central node, and the other nodes in the network are the end nodes. They can only communicate with neighboring network coordinator points. Therefore, the establishment of a network coordinator is the first step in the formation of a star network, and then the upper layer of the network coordinator determines the network coordinator. In practice, any kind of full-function device (FFD) can be used as the central node. When an FFD device is activated for the first time, the first step is to send a broadcast signal to find the network coordinator in the network, that is, to lock the central node. As long as there is a network coordinator in the wireless network, it can certainly receive its response to verify the password and establish a connection link to make this FFD as a normal device in the network. On the other hand, if the FFD device does not receive a reply message, then the network coordinator of the central node cannot be found in the entire network. At this time, this FFD device can make itself a network coordinator and establish a network with itself as the central node. If the network coordinator fails, the entire network system will be affected. In addition to this, it cannot be ruled out that if several network coordinators produce time synchronization errors or blockages when the errors or blockages are eliminated, then there may be multiple central nodes in the network at the same time, which will cause conflicts in the decision.

To solve this problem, the network coordinator is given a unique identifier to distinguish its identity relationship with each other. The mutual communication between two different star networks is completed by the respective network coordinators. The communication within the network is transmitted from the bottom to the network coordinator and then to the network coordinator of the other network.

In fact, a tree network is an integration of a V network in which the subordinate is allowed to communicate with his immediate superior as well as with the superior's superior. However, the communication between the subordinate and the superior's superior is limited. Multiple star networks can be combined into a tree network, and multiple tree structures can be combined into a more complex tree network. Therefore, the tree structure is more suitable for cases with large coverage areas and a wide range of information transmission. However, the realization of the tree structure needs to meet several requirements. First, the address structure must be given a dynamic address to each node; second, there must be an effective route between each node in the tree structure network to ensure information transmission; finally, to indicate the resource situation of the network equipment, a configurable tree range must be given.

The advantages of the tree structure are low cost, large coverage, and scalability, but it is also difficult to avoid problems similar to the star structure, that is, when any node in the network fails or moves outside the coverage, the nodes connected to it and subsequent nodes will be separated from the network, causing a collapse in the entire network. Therefore, improving network stability has become the goal of the tree structure.

When a tree-like topology is used in a ZigBee network, the network coordinator is required to function as a relay route and to implement functions such as new node joining and basic network management. Because FFD nodes can play any role in the network, they can implement data exchange and coordinate control links. As such FFD devices dominate the tree network and the number is the largest. The Reduced Function Device (RFD) can only transmit data and does not process the data. At the same time, FFD can play the role of RFD coordinator, which is network coordinator; only one FFD can replace the network coordinator in the entire network.

The cluster (mesh) network is a topology with a high degree of extension and can be applied in many different scenarios. In a cluster network structure, each node has a routing function and their status is the same. They can directly establish communication links with other nodes within the communication radius. However, its shortcomings are also readily noticeable. The wireless communication module of the node must be online at all times, and the state must be continuously detected and refreshed, which causes the node to consume a lot of energy.

When a cluster network is created, the first step is to determine a cluster head, and for other nodes to receive the broadcast information to join the network; the second step is to collect and detect environmental information, and finally summarize the information to the cluster head. Compared with tree and star networks, cluster networks have stronger self-healing ability, which greatly improves reliability. At the same time, network configuration and maintenance are also easier and faster to be carried out.

As mentioned earlier, ordinary ZigBee networks, such as star networks or tree networks, are more sensitive to single node failures. In addition, problems such as the overall unreliability of the network and short information transmission distances have limited the application of ZigBee. In this case, the cluster network brings new opportunities to ZigBee. The cluster network can provide multiple routes as well as the function of automatic routing, which ensures that multi-level hopping can be achieved under low energy consumption conditions. Compared with point-to-point networks, cluster networks are more suitable for large area coverage wireless networks. It has a higher data throughput rate and better fault self-recovery capabilities, which greatly expand the application range of ZigBee technology. Such a multi-hop network provides the possibility of multi-link selection for data transmission between devices, thereby avoiding the obstacles in nodes selection, and reducing energy consumption. This provides a solution to the operation and management of the entire network and the self-recovery of communication failures. Cluster networks are the most efficient type of all current network structures. They become a focus in wireless sensor network research. The familiar wireless networks such as WiFi (IEEE802.11s), WiMAX, and mobile 4G wireless communication networks are all based on a cluster structure.

In summary, among the three configurations, mesh network ensures the data transmission from source to the destination even when some links fail accidentally. Basically, the communication path will be chosen according to power consumption, latency, and throughput, etc. However, when the path between two nodes is blocked or hidden, the transmission will be dynamically routed to another node with a clear path and targets to the destination. The mesh property provides connectivity between devices and guarantees the network reliability. In summary, the ZigBee sensor network can provide high scalability on device connection, low power consumption, and production cost, high reliability due to its mesh capability. Hence, ZigBee is selected as the core protocol in TS-ZBN. TS-ZBN will be illustrated based on two parts. The first part is the construction of the building network and the second one is the mechanism of the time synchronization for coordinator-to-node and node-to-node.

4.2.1   ZigBee Building Network

Because of the inherent nature of scalability and mesh capability of ZigBee, a ZigBee building network can be set up quickly in most existing buildings at a lower cost. Such an adaptive and scalable wireless structure will certainly help to build up an efficient demand response for various smart water management applications. A good demand and response smart water management system will help the gross domestic product (GDP) grow healthily (less carbon emission) to a great extent.

Attention should be drawn to the fact that traffics in a high-rise building network is a few hundred times more than in a traditional building network used for individual houses or low rises. Since data are normally collected every 15–30 min, the major challenge presented to the time-synchronized ZigBee building network in a high-rise building is the design of high-density traffic for smart water management.

The general model of a high-traffic building network has been presented in [12]. Inspired by this, to facilitate the wireless control system that can be implemented to the whole building, three networks, namely, UAN forming unity communication, TLN forming horizontal communication, and NLN forming vertical communication are proposed. The design of ZBN is shown in Fig. 4.5.

1. UAN is a ZigBee mesh network that connecting the unity coordinator (C_U) and all water sensors within a unity. Typically, the coverage area of UAN is determined geographically. For example, a typical room is the coverage area of

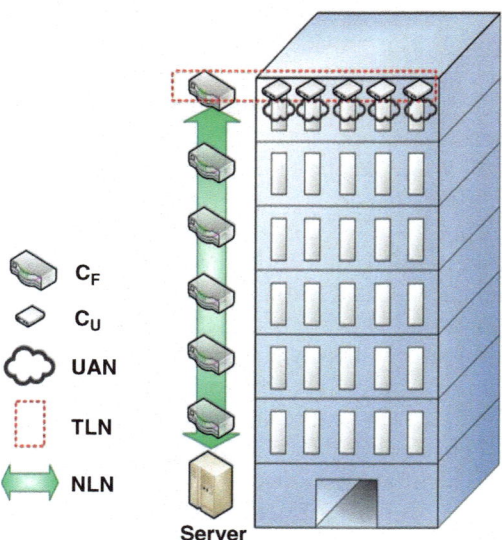

Fig. 4.5 The design of ZigBee building network

UAN. If the size of the room is relatively large, two or more C_U will be implemented on the same network. The responsibility of UAN is to collect and transmit all sensed data to C_U for further processing. By packing well, the sensed data from each node at C_U, the throughput can be increased and the power consumption and latency can be reduced as well. Mesh topology is proposed for the complex environment and also guarantees the communication reliability.

2. TLN is another ZigBee mesh network that connects the floor coordinator (C_F) and all C_U on the same floor. This configuration can remove the external cost of implementing relay to extend the coverage area and can guarantee the stability of the horizontal network. TLN responses to focus all sensed data from each unity and its C_U to C_F on the same floor. Similar to TLN, C_F will gather all the information from C_U and be ready to transmit to the server. TLN can ensure low power consumption and latency as well.

3. NLN is the third network formed by the backend server and all C_F. This network will have a little different from the previous networks. Compared to unity and horizontal communication, vertical communication of NLN is required to deal with the signal penetration of the thick wall between two floors. Since ZigBee is a kind of short-range transmissions that will be attenuated by the thick wall seriously, hence, NLN will be linked by WiFi or powerline as they have much higher penetrating ability compared to ZigBee.

Utilizing UAN, TLN, and NLN, the data from all water sensors can be transmitted to the backend server continuously. The information on the server will be sent via the internet to clouds. The authorized users can access the information for analyzing and make the corresponding control to enhance water management, which is one of the goals of the proposed scheme.

A water sensor can detect the presence of water, often by measuring the electrical conductivity of the water present and completing a circuit to send a signal. Some water sensor systems can be programmed to shut off the water to the house to prevent a small leak from becoming a large one. A water detector is an electronic device that is designed to detect the presence of water to provide an alert in time to allow the prevention of water leakage. These are useful in a normally occupied area near any infrastructure that has the potential to leak water, such as HVAC, water pipes, drain pipes, vending machines, dehumidifiers, or water tanks.

4.2.2 Node-to-Node Time Synchronization

Time synchronization is a critical issue in wireless communication, especially, for distributed measurement networks. All nodes within the network should be equal or close to the reference clock as the coordinator. Practically, the clock time of the devices is generated by crystal oscillator that can be affected by temperature, voltage, tolerance, etc. [17]. The time synchronization issue will be addressed in TS-ZBN. In fact, the synchronization can be considered as a kind of master/slave clock synchronization [18]. The master coordinator will transmit the beacon frame consisting of timestamp data, which can be treated as a reference clock signal, to every slave coordinator periodically for synchronization. The concept of master/slave and beacon frame will be modified. First, the timestamp slot will be implemented to the data frame, which will be followed by the data slot. Because of the embedded timestamp slot, the time difference between the transmitting side and the receiving side can be estimated by comparing the timestamp slot. For TLN, the locations of all C_F and all C_U are known and defined. Before the data transmission is set up, the receiving node will act as a temporal master and send the reference signal to the transmitting node, which acts as a slave, for n times with period T_M as shown in Fig. 4.6.

The slave will receive the reference signals with interval T_s. Then, the clock difference between two nodes can be calculated as follows:

$$\text{Clock difference} = \frac{1}{n}\sum_{i=0}^{n}\left(T_{S+i} - T_{M+i} - \tau_{P+i}\right) \tag{4.1}$$

where τ_p represents the propagation delay during wireless transmission. In the ideal case, τ_p can be simply calculated by the transmission distance divided by the light speed. However, in practice, it cannot be estimated directly and so τ_p will a variable with the following consideration:

1. The length of transmission d: Since the coordinates of the two nodes are known, the length of transmission will provide the most basic information on determining the propagation delay.
2. Multipath propagation $M(t)$: The signal can be reflected or blocked during wireless propagation. The receiving node may receive the bounded signals from all directions and so multipath propagation has to be considered.
3. Path loss model $P_L(d)$: The signal strength will decay during wireless transmission practically which is related to the length of transmission d, path loss exponent ρ, noise $n(t)$, etc. Therefore, analyzing the path loss model will give the

Fig. 4.6 Coordinator-to-coordinator synchronization (CCS) between two known-location nodes where the receiving side acts as a master and transmitting side act as a slave

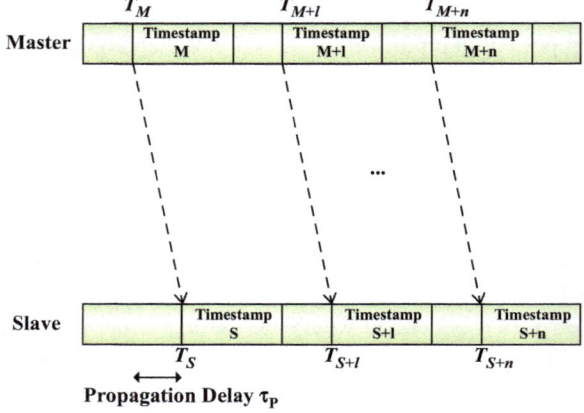

more practical propagation model and it is useful to distinguish the interested signal from the multipath signal as well. Besides, the geodesic-blinded nodes within the mesh network will be much more difficult to process synchronization. It is hard to find and define the dedicated reference clock signal in a mesh network because each node can communicate with another node without a master/slave concept.

It is expected that the sensor can be adjusted freely, which means that the sensor can be placed everywhere at any time. Therefore, the clock offset computation is required in the end-to-end synchronization.

The demand for HTAMI in modernized cities has significantly increased. Wireless data delivery basically meets the "versatility" need of HTAMI. Due to the open-standard nature and mesh capability, ZigBee is the populated candidate adopted by the industry [15]. It is evidenced that ZigBee has been applied to SM [6].

Derived for practical needs, generic design for HTAMI, namely, multiinterface ZBAN (MIZBAN), was developed by partitioning the network into two parts, namely, the Backbone Network and the Floor Network, and multiple interfaces were developed [15]. In the MIZBAN, interference was not particularly treated. It is well evidenced that WiFi, Bluetooth, and ZigBee operate in the same frequency band [19]. In addition, mobile signals such as 3G and LTE also operate in the vicinity which may cause adjacent channel or cross-channel interference. In order to provide a good quality of service, interference mitigation for HTAMI must be developed.

Limited former work was devoted to interference in ZigBee [20–25]. For instance, ZigBee deployment guidelines that include the safe distance and the safe offset frequency for smart grid applications were developed in an attempt to mitigate the potential WiFi interference [20]. However, the WiFi interference in the high-rises environment is much more complex, since the apartments are close to one another and WiFi signals scatter around the environment. Therefore, deployment guidelines alone as given in [20] are not sufficient. In general, an optimal solution to mitigate interference is difficult to be obtained.

A generic cross-layer optimization for caching was also discussed for multiinterface multiradio (M2) WSN [25].

However, only limited discussions were focused on IEEE 802.15.4. A comparative study of WiFi and IEEE 802.15.4 for M2 was provided in [26]. A M2 MAC layer design for IEEE 802.15.4 was also presented in [27], but the discussion was only based on the MAC layer of ZigBee, and the network layer and application layer were not considered. It can be seen that there is still much room for further development. In this study, based on IEEE 802.15.4, a cross-layer design into the network layer and application layer will be investigated. Particular interest will be devoted to the interference mitigation design for HTAMI. An interference mitigation solution, namely IMM2ZM, has been developed.

4.3   Design of IMM2ZM

4.3.1   IMM2ZM Basic Structure

Similar to MIZBAN [15], the proposed architecture of IMM2ZM is also divided into the backbone network and the floor network. The architecture of IMM2ZM is shown in Fig. 4.7.

The backbone network refers to a multiradio ZigBee mesh network that is formed by a reading centralizer (RC) with multiple reading meter terminals (RMTs) deployed into the meter room on each floor (this is a common configuration in Asia). Multiple radios were used in the IMM2ZM backbone network to share the traffic loadings to facilitate fast data delivery. The backbone network interacts with the meter data management system (MDMS) to provide the utility services such as meter management (MM), meter record order (MRO), and load profile (LP). Apart from the backbone network, RMTs are connected wirelessly with in-home displays (IHDs) to form another ZigBee single-radio network, namely, floor network, to facilitate end-users to obtain real-time meter readings. The functions of each component are summarized as follows:

The IMM2ZM incorporates multiple channels to achieve good latency [15]. Also, channel-swapping is incorporated to facilitate interference mitigation.

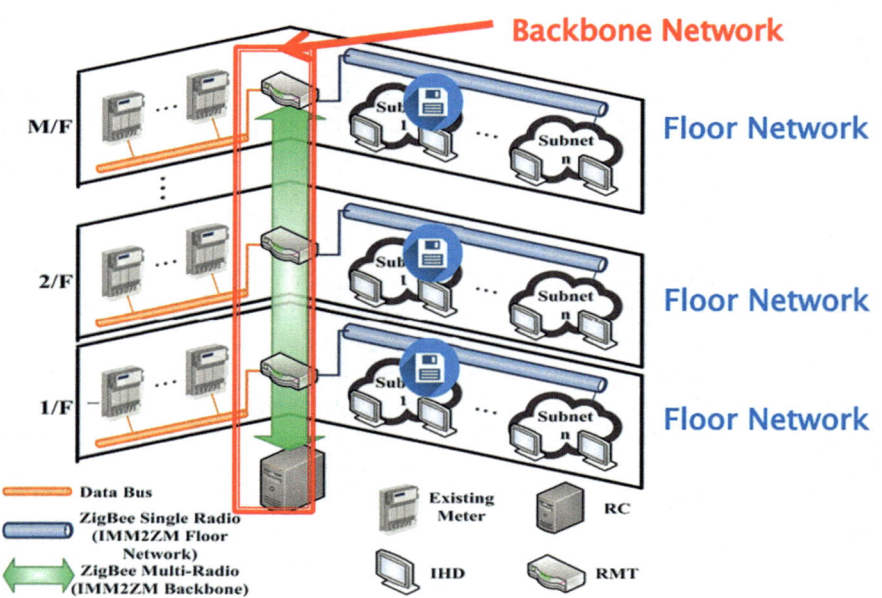

Fig. 4.7 Architecture for IMM2ZM

4.3.2 Multilayer Design of IMM2ZM Backbone Communication

The network layer and the application layer of the M2 backbone network have been designed to interoperate with the current ZigBee standard. ZigBee implements two layers on top of the 802.15.4 MAC layer, namely, the Network layer and the Application security layer. The IMM2ZM design consists of network initialization, swappable channel registration, address distribution, routing control, and application security. The process tasks and protocol architecture will be described later.

The Network layer is situated above the IEEE 802.15.4 MAC. One of the missions of the network layer is to empower IEEE 802.15.4 devices to deal with a variable network size application. There are three main tasks for the network layer such as (1) network initialization; (2) address distribution; and (3) routing control. The network initialization includes the management of network formation and devices. Address distribution aims to arrange a unique network address for each device in the ZigBee network. Routing control is a mechanism to maintain the end-to-end reliability and transfer packets through the network.

1. *Network Initialization:* Basically, this design is mainly applied to multiradio devices, e.g., the RC and RMTs. Generally, RMT is the backbone infrastructure that aims to relay the information across different floors to the RC.

When an interference source is detected at an occupied channel, the channel-swapping process will be activated to ensure the reliability of the IMM2ZM system. For example, if the ZigBee radio 1 of RMT *A* at channel *B* is jammed by strong interference and experiences continuous transmission failure, the ZigBee radio 1 of RMT *A* will issue the Channel_Jam_Report to the RC with the jammed channel ID. Then, the RC will broadcast the Channel_Scan_Req (channel ID) to all RMTs through channel *A*. After the channel scanning, the RC will send Channel_Result_Req to each RMT to collect the scan results and then select a new channel and broadcast Channel_Update_Req to all RMTs.

The selection of the new channel is mainly based on the principle that channels with larger frequency separation will intercept with less cochannel interference. Normally, there are 16 frequency channels available in IEEE 802.15.4, namely, channel 1–channel 16.

Initially, channel 1 will be assigned as the operating channel. If a traffic jam is detected, the channel swapping will be incurred based on Eq. (4.2)

$$CH_{new} = \begin{cases} 17 - CH_{old}, & CH_{old} \left\{ x | x = 2k - 1|, k \in Z^+ \right\} \\ 19 - CH_{old}, & CH_{old} \in \left\{ x | x = 2k|, k \in Z^+ \ \& \ k > 1 \right\} \\ 1 & CH_{old} = 2 \end{cases} \quad (4.2)$$

where CH_{new} refers to the channel to be selected and CH_{old} is the previous channel with jam before channel swapping.

The channel-jamming issue will be detected on the new channel until no Channel_Jam_report is received.

Figure 4.8 shows the initialization process for the channel swapping.

After the initialization process, the RMT has been assigned multiple channels and also registered to the RC to identify the data exchange definition of each channel. Generally, there are two categories of channels defined in CSA, namely control channel and operation channel. The control channel carries not only the data but also transmits system management commands. While the operation channel only carries the meter reading data. The major role of the control channel is to distribute the meter reading collection schedule from the utilities and coordinate the channel swapping.

In order to avoid broadcast storming, RC assigns the RMTs into various groups and each group shares the same control channel and operation channels. As a result, the control broadcast message is sent to RMTs using a multicast transmission. Typically, the best channel is selected to be the control channel and if the control channel is jammed, the second-best channel will take up the role of the control channel.

Figure 4.9 illustrates the registration process for the channel swapping.

Initially, the RC will send Device_Info_Req to each RMT in turn to collect the detailed information of the RMT. The RMT will reply Device_Info_Rsp to the RC. RMT will assign the remaining occupied channels as the operation channel according to the control channel ID. When an interference source is detected at an occupied channel, the channel-swapping process will commence ensuring the reli-

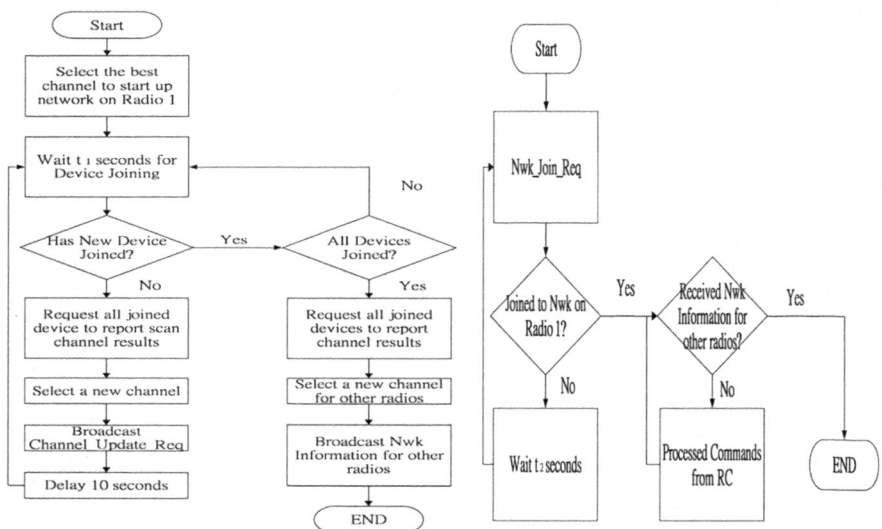

Fig. 4.8 Channel swapping (initialization process)

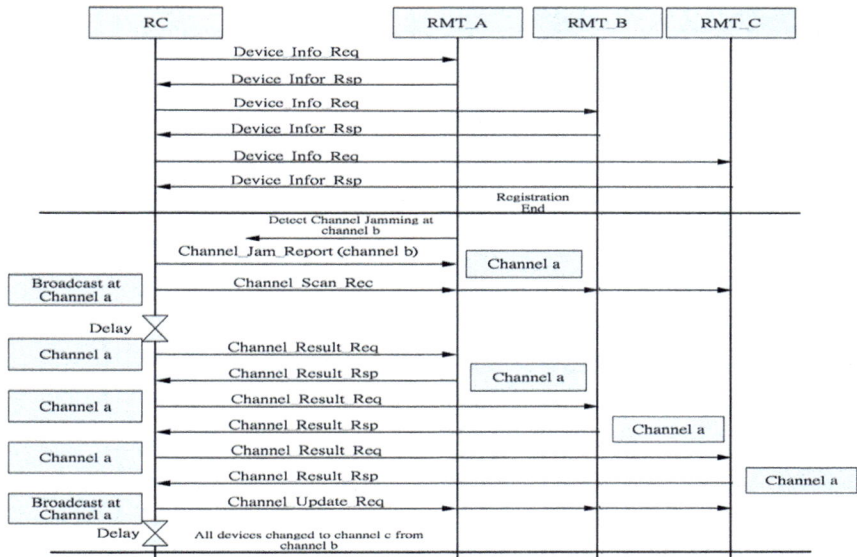

Fig. 4.9 Channel swapping (registration process)

ability of the proposed system. For example, if the ZigBee radio 1 of RMT A at channel b is jammed by strong interference and has continuous transmission failure, the ZigBee radio 1 of RMT A will issue the Channel_Jam_Report to the RC with the jammed channel ID. Then the RC will broadcast the Channel_Scan_Req (channel ID) to all RMTs through channel a. After the channel scanning, the RC will send Channel_Result_Req to each RMT, in turn, to collect the scan result and then select a new channel and broadcast Channel_Update_Req to all RMTs.

2. *Address Distribution:* When a device joins the network, it is given a 16-bit short address (network address). Such an address is a unique address in the ZigBee network. Two distributed addressing schemes are available in the ZigBee network, they are the tree address assignment scheme and the stochastic address assignment scheme.

3. *Routing Control:* Basically, ZigBee supports two routing mechanisms, that is, hierarchical (also known as a tree) and table-driven (also known as mesh) routing. In particular, mesh network routing (table-driven routing) is basically similar to the ad hoc on-demand distance vector (AODV) routing protocol [28, 29] for general multihop ad hoc networks. For the design of IMM2ZM, the address distribution and routing mechanism should be considered together since these two schemes affect each other.

4.4 IMM2ZM Model

In this section, a system model of IMM2ZM is presented. The purpose is to help a system designer to estimate the performance of IMM2ZM. An IMM2ZM is considered with k channels in an n-floor building experiencing the interference from x WiFi devices, y ZigBee devices, z Bluetooth devices, and m other wireless devices such as 3G and LTE devices from both adjacent channels of IMM2ZM and non-IMM2ZM network. The total interference power, $P_{in}(x, y, z, m)$, receipted by a single IMM2ZM ZigBee receiver is calculated as [28].

$$P_{in}\left(x,y,z,m\right) = P_{N0} + \sum_{i=1}^{x}P_{RX,WiFi}^{i} + \sum_{i=1}^{y}P_{RX,ZB} \\ + \sum_{i=1}^{z}P_{RX,BT}^{z} + \sum_{i=1}^{m}P_{RX,others}^{i}$$

(4.3)

where P_{N0}, $P_{RX,WiFi}$, $P_{RX,ZB}$, $P_{RX,BT}$, and $P_{RX,\,others}$ are the noise power, WiFi interferer power, ZigBee interferer power, Bluetooth interferer power, and interferer power from other sources, respectively.

The bit error rate (BER) of a single IMM2ZM ZigBee receiver interfered by x WiFi devices, y ZigBee devices, z Bluetooth devices, and m other wireless devices including from both adjacent channels of IMM2ZM and non-IMM2ZM network, $B_{x,y,z,m}$, is evaluated as

$$B_{x,y,z,m} = Q\left(\sqrt{2\gamma\left[10\log_{10}\frac{P_{RX,ZB}}{P_{in}\left(x,y,z,m\right)} + PG - P_{fading}\right]}\right)$$

(4.4)

where [30].

$$Q\left(x\right) = \frac{1}{\sqrt{2\pi}}\int_{x}^{\infty}\exp\left(-\frac{u^2}{2}\right)du.$$

(4.5)

P_{fading} is the fading loss, PG is the process gain, and $\gamma \approx 0.85$ [31].

The derivation of BER of ZigBee packets among the interference of all potential sources is studied. The extreme cases are considered in which the packets are transmitted successfully (P_{succ}) and all IMM2ZM devices are busy (P_{bs}). It is assumed that the packet length is L bits and h IMM2ZM devices are competing with each other. P_{succ} is the probability of a correct packet successfully transmitted (with every bit in the packet correctly transmitted) and P_{bs} is the probability that all IMM2ZM devices are busy when a packet is sent to a specific ZigBee transceiver of IMM2ZM devices. P_{succ} and P_{bs} are evaluated as

$$P_{\text{succ}} = \left(1 - B_{x,y,z,m}\right)^{L} \tag{4.6}$$

$$P_{\text{bs}} = \left(1 - \tau\right)^{h-1} \tag{4.7}$$

ZigBee performs clear channel assessment (CCA) four times before reporting failure; thus, the transmission probability τ is evaluated from the channel busy probability α. In this study, four channels are used; hence, α is defined as follows:

$$\tau = 1 - \alpha^{4} \tag{4.8}$$

For the purpose of performance evaluation, the packet error rate P_{err} is evaluated by incorporating P_{bs} into consideration.

Hence, P_{err} is now defined as

$$P_{\text{err}} = 1 - P_{\text{succ}} / P_{\text{bs}} \tag{4.9}$$

In IMM2ZM, the channel busy probability α is then derived as

$$\alpha = 1 - \left(1 - \alpha_{\text{IMM2ZM}}^{\text{BT}}\right)\left(1 - \alpha_{\text{IMM2ZM}}^{\text{WiFi}}\right)\left(1 - \alpha_{\text{IMM2ZM}}^{ZB}\right) \times \left(1 - \alpha_{\text{IMM2ZM}}^{\text{others}}\right) \tag{4.10}$$

where $\alpha_{\text{IMM2ZM}}^{\text{BT}}$, $\alpha_{\text{IMM2ZM}}^{\text{WiFi}}$ and $\alpha_{\text{IMM2ZM}}^{ZB}$ denote the CCA busy probability of a given IMM2ZM device due to Bluetooth devices, WiFi devices, and ZigBee, respectively. $\alpha_{\text{IMM2ZM}}^{\text{others}}$ refers to other interferers such as 3G and LTE devices.

The tagged IMM2ZM device is modeled as M/G/1 queuing system. It is assumed that (1) h IMM2ZM devices are competing; (2) each IMM2ZM device generates packet conforming to the Poisson process of packet generation rate λ_{M}; and (3) data packet size is constant with b_{M} seconds. By incorporating T_{BO}, T_{turn}, T_{SW}, T_{ACK}, and following [32], $\alpha_{\text{IMM2ZM}}^{ZB}$ is expressed as

$$\alpha_{\text{IMM2ZM}}^{ZB} = \frac{(h-1)\left(1 - \left[\alpha_{\text{IMM1ZM}}^{ZB}\right]^{4}\right) E[\Gamma]\left(T_{\text{BO}} + b_{\text{M}} + 2T_{\text{turn}} + T_{\text{ACK}} + T_{\text{SW}}\right)}{\dfrac{1}{\lambda_{\text{M}}} + E[\Gamma] \cdot E[D_{\text{q}}]} - 1 \tag{4.11}$$

where T_{BO}, T_{turn}, T_{SW}, and T_{ACK} are the time for back off, turn around, switching, and transmit acknowledgment, respectively. In Eq. (4.10), channel-swapping has specifically addressed to ensure the busy probability of IMM2ZM devices been taken into consideration of the interference. $E[\Gamma]$ is the average number of packets served by the tagged IMM2ZM device in a busy period and is defined as $E[\Gamma] = 1/(1 - \rho)$, where traffic intensity $\rho = \lambda_{\text{M}}(E[D_{\text{q}}] + b_{\text{M}} + 2T_{\text{turn}} + T_{\text{ACK}})$. $E[D_{\text{q}}]$

denotes the queueing delay which refers to the duration that the packet in the system queues before transmission or discarded. Substituting $E[\Gamma]$ into Eq. (4.11), α_{IMM2ZM}^{ZB} is given by

$$\alpha_{IMM2ZM}^{ZB} =$$

$$\frac{\lambda_M (h-1)\left(1-\left[\alpha_{IMM1ZM}^{ZB}\right]^4\right)\left(T_{BO} + b_M + 2T_{turn} + T_{ACK} + T_{SW}\right)}{1-\lambda_M\left(b_M + 2T_{turn} + T_{ACK}\right)} - 1 \quad (4.12)$$

With the newly defined P_{err} in Eq. (4.9), the single-hop transmission channel throughput S for an IMM2ZM device with single radio is expressed as below.

$$S = \frac{8L_p P_s}{\delta P_i + T_s P_s + T_c P_c + T_f P_f} \quad (4.13)$$

$$P_i = (1-\tau)^h \quad (4.14)$$

$$P_s = h\tau (1-\tau)^{h-1}(1-P_{err}) \quad (4.15)$$

$$P_f = h\tau (1-\tau)^{h-1} P_{err} \quad (4.16)$$

$$P_c = 1 - P_i - P_s - P_f \quad (4.17)$$

where L_p is the payload of a packet in bytes; P_i is the probability that the time slot is idle; P_s is the probability of successful transmission without channel error and collision in a time slot; P_f is the probability of channel error occurs in a time slot; P_c is the probability that collision occurs in a time slot; δ is the duration of idle time slot; T_s is the average channel busy time due to successful transmission; T_c is the average channel busy time due to collision; and T_f is the transmission failure time due to channel error. T_s, T_c, and T_f follow the meanings from [33], and the relationship between T_s, T_c, and T_f is given by

$$T_s = b_M + T_{ACK} + 2T_{IFS} \quad (4.18)$$

$$T_c / T_f = b_M + T_{ACK} + T_{IFS} \quad (4.19)$$

The overall transmission of IMM2ZM with k radios is now investigated. Consider a high-rise building with n floors and each floor has N_a apartments. By assuming that a smart meter stores N_r records for data recovery and the record length is N_b bits. The sleep-to-join time for each node is T_{s2j}. Therefore, the meter reading collection duration for a specific floor demanding c hops from transceivers $T(c)$ is newly derived according to the detailed construction of the building as

$$T(c) = \left(\frac{N_a N_r N_b}{S} + T_{s2j} \right) \times \frac{c}{k} \tag{4.20}$$

$T(c)$ gives an account of multiple hops and multi-channels. The general knowledge of the average delay D is the amount of time required to transmit all the packet's bits successfully. D is the primary parameter for wireless communication network design. For SM, a large D largely impacts the effectiveness of the system [32]. To facilitate more advanced applications such as real-time pricing, a low value of D is demanded. In IMM2ZM, D is also defined as the time of collection of the meter readings of the entire building

$$D = \frac{\sum_{i=1}^{n} \left[\left[\left(\frac{N_{a,i} N_r N_b}{S} + T_{s2j} \right) \times \frac{c}{k} \right] + T_{cs,i} \right]}{n} \tag{4.21}$$

where $T_{cs,i}$ is the channel-swapping time of the respective $N_{a,i}$, $T_{cs,i}$ will be defined in Sect. 4.5.

In general, the transmission rate and the number of bits transmitted successfully in a unit time, are important performance indicator for wireless communication. In essence, data overlay the entire network on the application layer from which they are processed. With high traffics in high-rises, the quantity of data transmitted in a time slot is bulky. Thus, the transmission rate on the application layer affects network performance significantly.

Therefore, the application-layer transmission rate σ a pertinent descriptor of IMM2ZM is defined as

$$\sigma = \frac{N_a N_r N_b}{D} \tag{4.22}$$

From the analysis, the descriptors provide a holistic view of the latency performance that takes the total number of hops and the interference mitigation into account. Thus, D and σ are indicative figures to quantify the performance of the IMM2ZM in a BAN.

4.5   Multiobjective Optimization Based on NSGA-II

To investigate the performance of IMM2ZM, the system requirement will be formulated and optimization is needed. It is well known that the genetic algorithm (GA) is a powerful optimization technique. GA is commonly used to generate high-quality solutions to optimization and search problems by relying on biologically inspired operators such as mutation, crossover, and selection. In most practical engineering problems including wireless network design, global optimum can hardly be found. Therefore, the problem cannot be formulated into a single-objective optimization problem. Also, most of the problems in engineering demand the consideration of multiple conflicting objectives to give a comprehensive and excellent performance. Compared to single-objective optimization, multiobjective optimization has super advantages such as the diversity of multiobjective optimization is much wider than single-objective optimization [9]. As a result, multiobjective problems lead to the launch of multiobjective evolutionary algorithms (MOEAs). The MOEA could be a kind of GA that always searches for a set of non-dominated optimal solutions, which is referred to as PF [34].

MOEAs were successfully applied to the optimization of wireless local area network (WLAN) [35]. It is well evidenced that NSGA-II is proven to outperform other MOEAs in terms of convergence and diversity functional analysis [36]. It is envisaged that NSGA-II is powerful and will provide a wider distribution of the solutions during the search for optimal solutions. Thus, NSGA-II [16] is employed in this study for a custom design for an optimal IMM2ZM. The developed model will minimize the influence of potential interference with optimal throughput and minimal latency.

The following tasks illustrate the main design concept behind.

4.5.1   Initialization

During the initialization, the population size, constraints, objective functions, and a number of parameters are determined. The crowding distance, the average distance of the two nearest points representing optimal solutions, are calculated to estimate the number of optimal solutions.

4.5.2   Multiobjective Searching Process

The main scope in the multiobjective searching process aims to generate a new population for further optimization to reach optimal solutions. Selection, crossover, and mutation imitate the process of natural evolution [16]. The objective values of each objective function of the individuals in the new population are estimated based on the designed objective functions. The ranking of the individuals in the same population is based on domination. Recall from [34] that solution u dominates solution v, if and only if two conditions are true, that is (1) all the objectives in u should perform no worse than v; and (2) at least one objective in u should perform better than v. Solution u does not dominate solution v if either of the conditions is violated. Solutions that are not dominated by other solutions in the population have the highest ranking.

The iteration process will be completed when the maximum generation is reached or the output converges, and thus, the PF is obtained. Every solution in the PF is an optimal solution and does not dominate each other.

Owing to the simplicity of computation in optimization, prioritized objective functions are sometimes used, and weighting factors are assigned to the objective functions. In contrast, multiobjective optimization has a wider diversity to search for optimal solutions in a wider range. An investigation is made to explore the effectiveness of these two schemes. The comparison will be shown later.

4.5.3 Network Representation

To model the network, important information such as the number of floors and the maximum number of channels will firstly be obtained. The NSGA-II optimization will then be customized and incorporated to evaluate the optimal solution.

4.5.4 Design Constraints

To facilitate the search, it is necessary to assign reasonable upper and lower limits of the parameters, which conform to the unique design of the network. Reasonable limits may effectively reduce the number of undesirable individuals during the operation, thus reducing the computing time significantly.

4.5.5 Design of Fitness Values

In general, for multiobjective optimization, the objective functions are expressed as [16]

$$Minimize\ F(x) = \left(f_1(x),\dots,f_m(x) \right)^T$$

$$\text{Subject to } x \in \Omega \tag{4.23}$$

where $f_m(x)$ is the objective values for each individual in the whole population, and Ω is the variable range.

A feasibility study was carried out. However, it is impracticable, if not impossible, to perform a full-scale measurement in high-rises. Therefore, a prior measurement was performed for the provision of realistic data to support the model construction of IMM2ZM. For the same reason described in [15], the performance of the large-scale IMM2ZM is analyzed using the OPNET model and simulation [37]. Interference mitigation model developed in Sect. 4.4 will be incorporated into the OPNET to achieve a full-scale performance evaluation of IMM2ZM.

There are mainly two parts of the feasibility study, namely (1) A small-scale IMM2ZM prior measurement using four ZigBee physical channels; and (2) a large-scale simulation of the IMM2ZM using OPNET model. The feasibility study is mandatory since it analyzes the performance of the developed IMM2ZM. Besides, the measured data in the prior measurement also play an important role in the initialization of the parameters in objective functions for the optimization. For example, in Eq. (4.3), P_{N0}, $P_{RX,WiFi}$, $P_{RX,ZB}$, $P_{RX,BT}$, and $P_{RX,others}$, each varies at numerous wireless environment within the floors. These parameters will be estimated based on the measured data in the prior measurement to give a more accurate formulation for ZBAN at large-scale. In essence, α in Eq. (4.10), T_{BO}, T_{turn}, T_{SW}, and T_{ACK} in Eq. (4.12), and T_{s2j} in Eq. (4.20) were evaluated in the prior measurement in the feasibility study and, provided good estimates in the large-scale model.

To facilitate testing, an IMM2ZM was set up in a residential building. In the prior measurement, a five-floor IMM2ZM using four ZigBee physical channels was developed and measured. The five-floor IMM2ZM consists of five four-radio RMTs and one four-radio RC. The experimental setup is shown in Fig. 4.10.

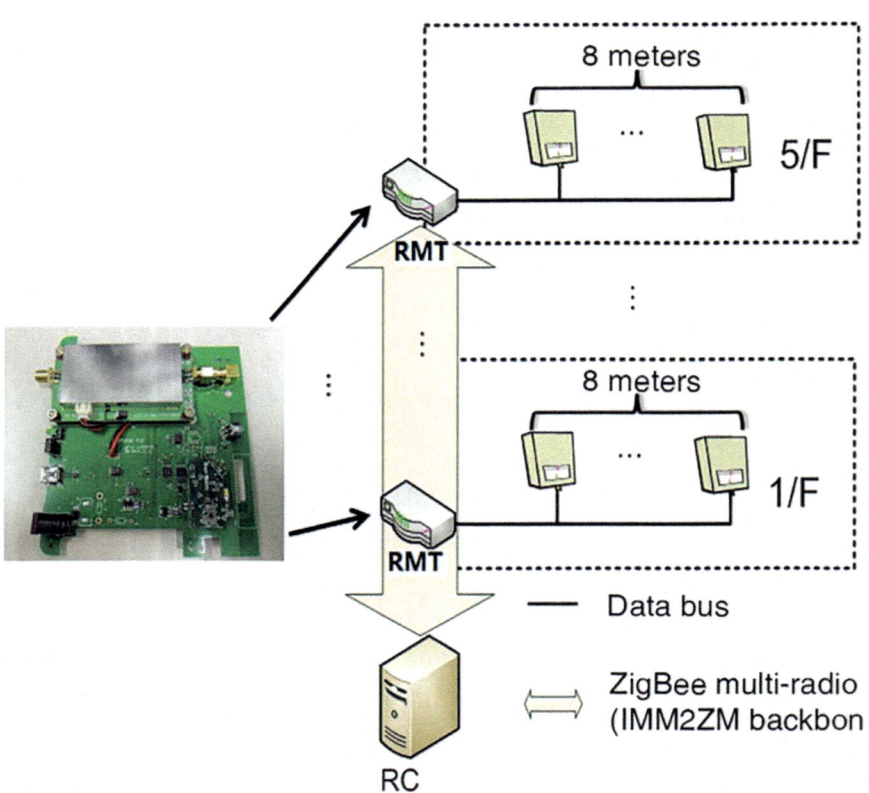

Fig. 4.10 Experimental setup for feasibility test

Table 4.1 System specification of IMM2ZM

Description	Experimental data	Simulation data
Number of floor, n (n-floor)	5	30
Number of apartment per floor, N_a	8	8
Number of record stored by smart meter, N_r	10	10
Record length, N_b, (bits)	32	32
AES 128bit enabled payload length, L_p, (bytes)	60	60
Packet length, L, (bytes)	127	127
Transmission power P_{ZB}(dBm)	19.6	[−20,20]
Receiver antenna gain G_{RX}(dBi)	0	0
Transmitter antenna gain G_{TX}(dBi)	0	0

The data measured in the prior measurement were employed in the formulation of objective functions of the optimal solution at large-scale ($n = 10, 20, 30$). As such, a thirty-floor building with eight apartments on each floor, i.e., $n = 30$ and $N_a = 8$, is considered at large-scale. The RC collected the meters' data once every 30 min, and the smart meter stored the latest ten records, i.e., $N_r = 10$. The system specifications of IMM2ZM for both the experiment and simulation are summarized in Table 4.1.

In the prior measurement, testing was carried out in the meter room from the first to the fifth floor to identify the potential WiFi, Bluetooth, ZigBee, LTE, 3G, and other interference sources. The measured data from the prior measurement form the important trustworthy parameters for objective function analysis. Based on the measured data, important parameters such as the transmitter and receiver gains, the packet generation rate, and the transmission power are optimized ("genes" in the algorithm) for the network and device design. On the other hand, D, BER, and σ are designed as objective functions.

The objective functions are designed to minimize average D ($F1$), average BER ($F2$), and maximize average σ ($F3$). The three objective functions are formulated as

$$Minimize\ F_1 = \sum_{i=1}^{num} D\ /\ num \tag{4.24}$$

$$Minimize\ F_2 = \sum_{i=1}^{num} B_{x,y,z,m}\ /\ num \tag{4.25}$$

$$Minimize\ F_3 = \sum_{i=1}^{num} \sigma\ /\ num \tag{4.26}$$

$$Subject\ to\ G_{RX} \in [0,2] dBi, G_{TX} \in [0,2] dBi, P_{ZB} \in [-20,20] dBm$$

where *num* is the number of replication of the experiment.
Constraints for each objective function:

Table 4.2 Parameters setting of NSGA-II

Population size	100
Maximum number of generations	200
Crossover type	Uniform
Crossover rate	1
Mutation rate	0.2

To fulfill the demand response (DR) requirement for SM, $D \leq 0.5$ s [38].

$$BER \leq 5 \times 10^{-4}; \sigma \geq 20 \, kb/s$$

In the Hong Kong environment, a data rate of $\sim$10–20 kb/s is normally adopted; hence, $\sigma \sim$ 20 kb/s is employed for evaluation. The NSGA-II scheme is then customized to optimize the network. The key parameters are listed in Table 4.2.

With the inclusion of the number of floors (n) and the number of channels (k), the performance of the IMM2ZM is optimized for n = 5, 10, 20, 30 and k = 1, 2, 3, 4, and simulated values for each objective are obtained to search for optimal solutions. As an illustration, the PF for n = 5, k = 4 is shown in Fig. 4.11.

It is reiterated that every solution in PF does not dominate each other. As a representative value for SM wireless communication network, BER is chosen as 5×10^{-4} [7]. From Fig. 4.11(a), D = 0.04 s and σ = 2.1 × 104 b/s. Coupled with the objective functions (4.23), (4.24), and (4.25), P_{ZB} = 100 mW.

The comparison between prioritized objective functions and multiobjective optimization is now investigated.

Objective functions with prioritized weighting factors are formulated as in [34].

$$Minimize\,(4.27)\,F(x) = \sum_{m=1}^{M} \omega_m f_m(x) \tag{4.27}$$

$$Subject\ to\ g_j \geq 0, j = 1, 2, \ldots, J$$
$$h_k(x) = 0, k = 1, 2, \ldots, K$$
$$x_i^{(L)} \leq x_i \leq x_i^{(U)}, i = 1, 2, \ldots, n$$

where ω_m is the weight of the mth objective function. $f_m(x)$ is the normalized objective function. g_j, h_k, and x_i are constraints.

The prioritized objective function is now investigated, and weighting factors are assigned to explore the effectiveness to obtain the optimum solution. As an illustration, indicative designs of assigning weighting factors ω_m to the corresponding objective functions $F_m(x)$ are analyzed, and the corresponding results are shown in Table 4.3.

For $\omega_1 = \omega_2 = 0.1$, $\omega_3 = 0.8$, the priority of transmission rate σ is the highest among D, BER, and σ, the BER exceeds the limitation of SM, i.e., 5×10^{-4}. Similarly, when $\omega_2 = \omega_3 = 0.1$, $\omega_1 = 0.8$ (i.e., the priority of delay is more important).

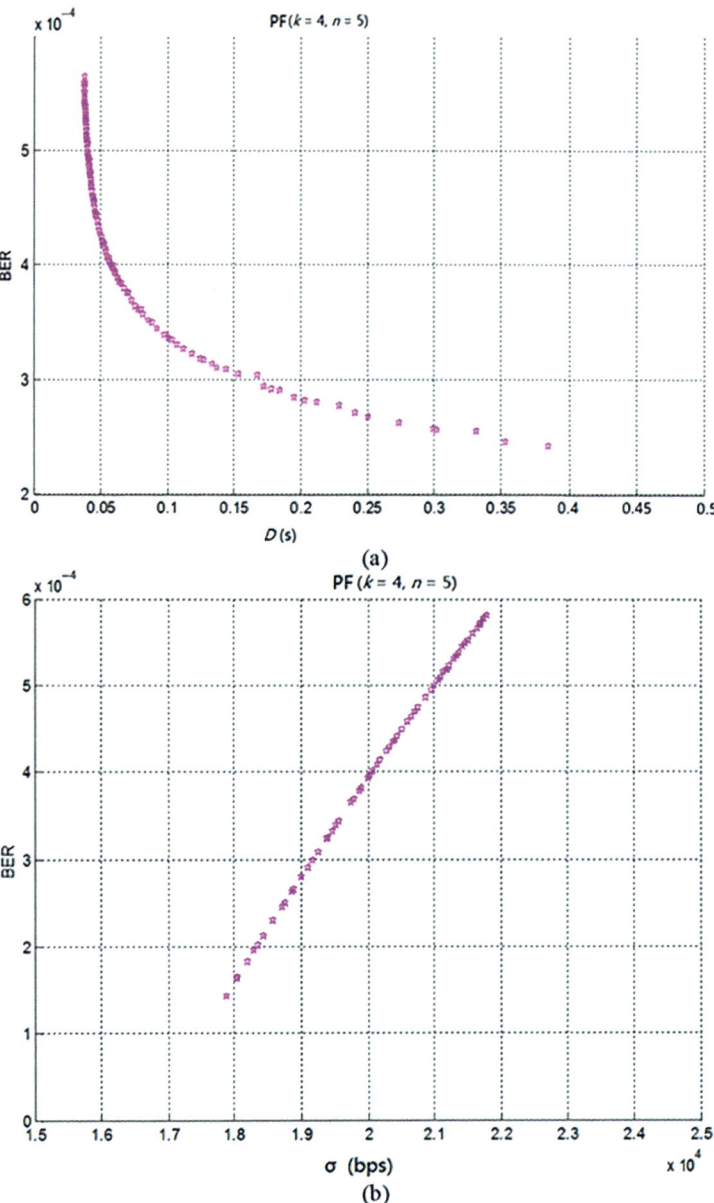

Fig. 4.11 (a) PF of BER versus D for five-floor. (b) PF of BER versus σ for five-floor

Table 4.3 Design of weighting factors for the objective functions and corresponding results

ω_m			$F_m(x)$			Description
ω_1.	ω_2	ω_3	$F_1(x)$ (D,s)	$F_2(x)$ (BER)	$F_3(x)$ $(\sigma$,kbps)	
1/3	1/3	1/3	0.07	3.4×10^{-4}	19.2	D is 42.8% worse than obtained by IMM2ZM; σ is 9.5% less than obtained by 1MM2ZM
0.1	0.1	0.8	0.03	6.2×10^{-4}	22.8	BER > limitation
0.1	0.8	0.1	0.71	1.2×10^{-4}	17.5	D > limitation
0.8	0.1	0.1	0.03	6.3×10^{-4}	23.2	BER > limitation
	N/A		0.04	5.0×10^{-4}	21.0	Optimal result

Besides, when $\omega_1 = \omega_3 = 0.1$, $\omega_2 = 0.8$, BER can be guaranteed within the SM requirement; in contrast, the delay D will be increased and exceeds the limitation of 0.5 s. For cases with an average priority of the three objectives, BER is confined to an acceptable level. From Table 4.3, it is concluded that if the priority of the objectives is assigned, there are negative impacts as follows:

1. The limitation of BER, D may not be guaranteed.
2. The diversity of PF will be reduced.
3. D (multiobjecitve) $- D$(prioritized) > 43%.
4. σ (multiobjective) $- \sigma$(prioritized) > 9%.

Thus, the performance of multiple objective optimizations surpasses priority-based optimization.

The same optimization process was applied to IMM2ZM and reiterated for $n = 6$, . . ., 30, and the corresponding PFs were obtained. The respective optimized values, namely, D, BER, σ, and P_{ZB} are evaluated and plotted in Fig. 4.12.

Figure 4.12(a) shows the variation of P_{ZB} and λ_M versus the network size n. When n increases, a higher received power P_r is needed to overcome the complex interference environment and significant fading.

Figure 4.12(b) shows the variation of D and σ versus n. It is seen that D increases and σ decrease when n increases. It is important to point out that $D < 0.5$ s in all cases and, thus, fulfills the U.S. standard for SM. As an illustration, from Figs. 4.12(a), (b), when $n = 10$ and $N_a = 8$, $P_{ZB} = 91$ mW and $\sigma = 2.1 \times 10^4$ bps, $D = 0.2$ s which falls within specifications. Alternatively, when $n = 10$ and $N_a = 8$, $P_{ZB} = 93$ mW and $\sigma = 2 \times 10^4$ bps, $D = 0.4$ s which also falls within specifications. Thus, the optimization analysis here provides the design platform for the scalable and versatile development of IMM2ZM model essential to HTAMI.

The large-scale analysis of the IMM2ZM is investigated with OPNET based on the characterized five-floor model data for HTAMI in the Hong Kong environment. The large-scale wireless environments are then simulated by incorporating a comprehensive consideration of interference sources in a practical situation. In the IMM2ZM model, the consideration of the interference sources is based on the common specifications of real products as well as the HTAMI nature in the densely populated area as measured in the five-floor experiment. Typically, in Hong Kong,

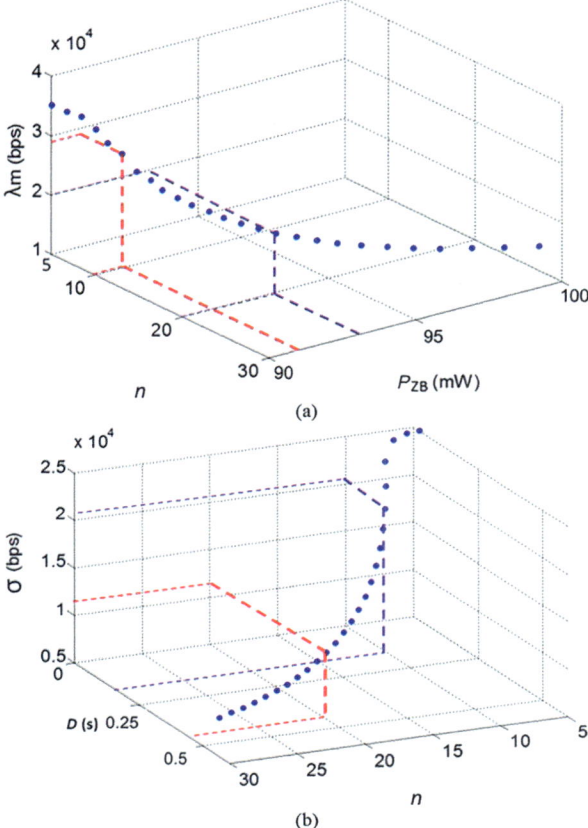

Fig. 4.12 (**a**) Optimized parameters P_{ZB} and λ_M versus n. (**b**) Optimized parameters D and σ versus n

there are eight apartments in a high-rise. Normally, one WiFi router is sufficient to represent the WiFi coverage of one apartment, thus one WiFi interference source per apartment is considered. Moreover, from the analysis of population census by the Hong Kong government [39], the average domestic household size in Hong Kong is 2.9; hence, three cellular phones and three Bluetooth sources per apartment are assigned. The simulation condition incorporating the interference sources is listed in Table 4.4.

The interference will cause delay overshoot, and thus, the IMM2ZM will activate "channel-swapping." Define T_{cs} as the "channel-swapping time" for the duration of channel-swapping. Figure 4.13 shows the simulated results from OPNET regarding D against time for $n = 5, 10, 20, 30$ under the wireless environment as shown in Table 4.4.

Figure 4.13 reveals that, at the turn of the IMM2ZM, there is an unstable period of delay overshoot due to T_{cs}. It can be seen that $T_{cs} = 21, 25, 30$, and 80 s for $n = 5$,

Table 4.4 Wireless environment design in OPNET simulation

Description	Assigned values according to findings from feasibility testing in a prior experiment		
	WiFi	Bluetooth	Cellular phone signal
Number of interference sources per floor	8	24	24
Interference sources	One router per apartment	Three nodes per apartment	Three devices per apartment (3G:LTE = 2:1)
Power level of each interference source	20 dBm	4 dBm	33 dBm (3G) 27 dBm (LTE)
Wireless standards	IEEE 802.11n	IEEE 802.15.1	3G & LTE
Modulation	QPSK	GFSK	EDGE
Frequency channels	Randomly assigned channels (2.4–2.4835 GHz from CH 1 to CH 13)	Randomly assigned channels (2.4–2.4835 GHz from CHI to CH 79)	Randomly assigned in UMTS frequency bands (2.1 GHz as central frequency with CH 1 to CH 26)

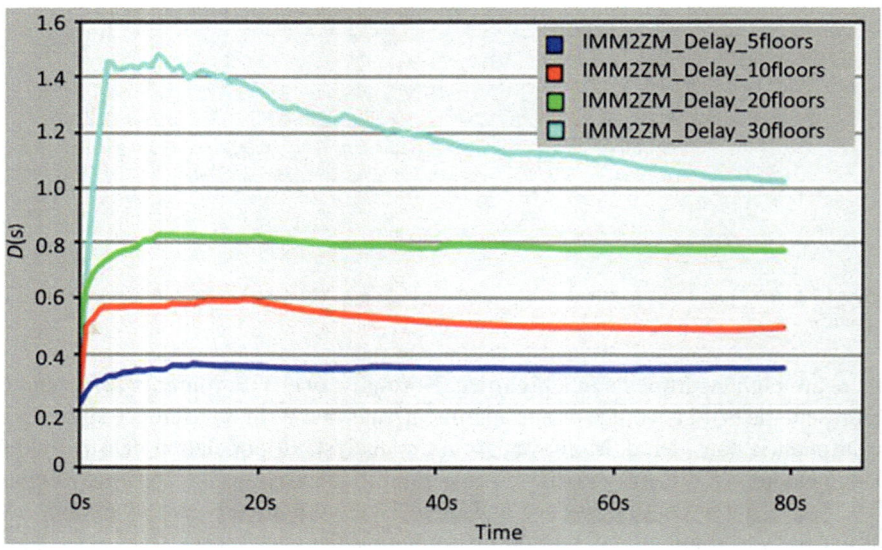

Fig. 4.13 Simulated D for $n = 5, 10, 20, 30$ by OPNET

10, 20, and 30, respectively. The delay overshoot aims to reduce interference and is mainly caused by channel-swapping. After the lapse of delay overshoot (T_{cs}), the transmission remains stable, hence signifying that the channel-swapping process has been completed. It is seen that T_{cs} increases significantly with an increasing n due to the large network-cluster size in HTAMI, and this requires long transmission time between nodes. It can also be observed that when the number of interference sources increases or when P_{ZB} is smaller, T_{cs} increases.

Define P_{RX} as the receiving sensitivity of the IMM2ZM. P_{RX} is related to the gains and losses incurred in the link budget, the transmitting power of interference sources, and its associated distance, as well as the distance away from the interference sources.

P_{RX} is expressed as

$$P_{RX}\left(\text{dBm}\right) = P_{ZB} + G_{TX} + G_{RX} - L_{FS} - L_{I} - L_{TX} - L_{RX} \tag{4.28}$$

where P_{ZB}, G_{TX}, and G_{RX} are given in Table 4.1. L_{FS} (dBm) is the path loss and fading, which is related to the transmission distance and wavelength. L_{I} (dBm) refers to the loss due to interference, and L_{TX} (dBm) and L_{RX} (dBm) are the transmitter loss and receiver loss, respectively. It can be concluded from Eq. (4.28) that P_{RX} increases with an increasing P_{ZB} or a hardware design of larger G_{TX} and G_{RX}. However, with fixed L_{TX} and L_{RX}, as well as L_{FS}, P_{RX} certainly decreases tremendously due to serious interference.

The relationship of T_{cs} versus n and P_{RX} is plotted in Fig. 4.14, when $P_{ZB} = -20$ to 20 dBm and $n = 1$ to 30.

From Fig. 4.14, it is seen that P_{RX} and n affect T_{cs} significantly. When n increases, T_{cs} increases significantly, because the channel-swapping process requires time to detect channel condition and reiterates network-traffics information between RC and RMT in high-traffics networks in HTAMI. The improvement of P_{RX} will reduce T_{cs}. It is evaluated that when $P_{RX} = -12$ dBm, T_{cs} will be increased tremendously, because the link budget reaches the bottom margin of the sensitivity of IMM2ZM.

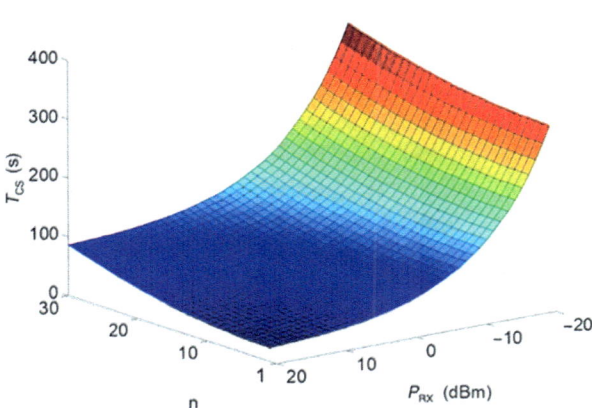

Fig. 4.14 Relationship between T_{cs} and n, P_{RX}

4.6   Analysis and Evaluation

To investigate the performance of IMM2ZM, an interference mitigation study and a latency study were conducted. It was shown that the latency study accounted for the IMM2ZM system performance.

4.6.1   Interference Mitigation Study

Interference under high-traffic conditions weakens signal reception. However, the potential interference cannot be ignored for high-rises as a result of the ever-increasing number of wireless users. As a result, interference mitigation is important for high performance, and thus, a study is necessary.

With the experimental setup as shown in Fig. 4.10, the interferers were established in the vicinity of the RMT. The RMT was located in the meter room, and the access point operated at the same frequency channel as the operating channel of IMM2ZM.

During the experiment, D and T_{cs} were measured from the meter reading collection. In order to investigate a comprehensive performance of IMM2ZM, five buildings with $n = 3$, 4, and 5, were measured. The results are shown in Fig. 4.15.

Figures 4.15(a)–(c) show the real-time performance of D with the introduction of interferers into the buildings for $n = 3$, 4, and 5, respectively. On each floor, the real-time delays of a maximum of ten individual hops (referred to as "Hop_<floor_No.>_<hop_No.>") are recorded and analyzed. It is seen that D increases by 60–70% between $t = 0$ and $t = 5$ s for $n = 3$, 4, and 5, since IMM2ZM collects meter readings using a single channel. The channel-swapping period ends at $t = 15$, 20, and 25 s for $n = 3$, 4, and 5, respectively, and D becomes relatively constant afterward. This phenomenon is attributed to the fact that the IMM2ZM has successfully found a channel with insignificant interference for transmission. When $t \leq T_{cs}$, the delay is high since data delivery enters the overshoot period. When $t > T_{cs}$, D returns to a stable lower value. As an illustration, $T_{cs}(n = 5) = 25$ s is longer than $T_{cs}(n = 3)$ and $T_{cs}(n = 4)$ by 10 and 5 s, respectively. Thus, a larger network obviously occupies a longer swapping period and leading to a higher delay. Nevertheless, for all scenarios, IMM2ZM recovers its normal transmission after channel-swapping is completed. It is noted that T_{cs} is relatively small with respect to the data-collection period, i.e., 15–30 min. Therefore, an IMM2ZM generally with a small T_{cs} is a figure to reflect a robust HTAMI.

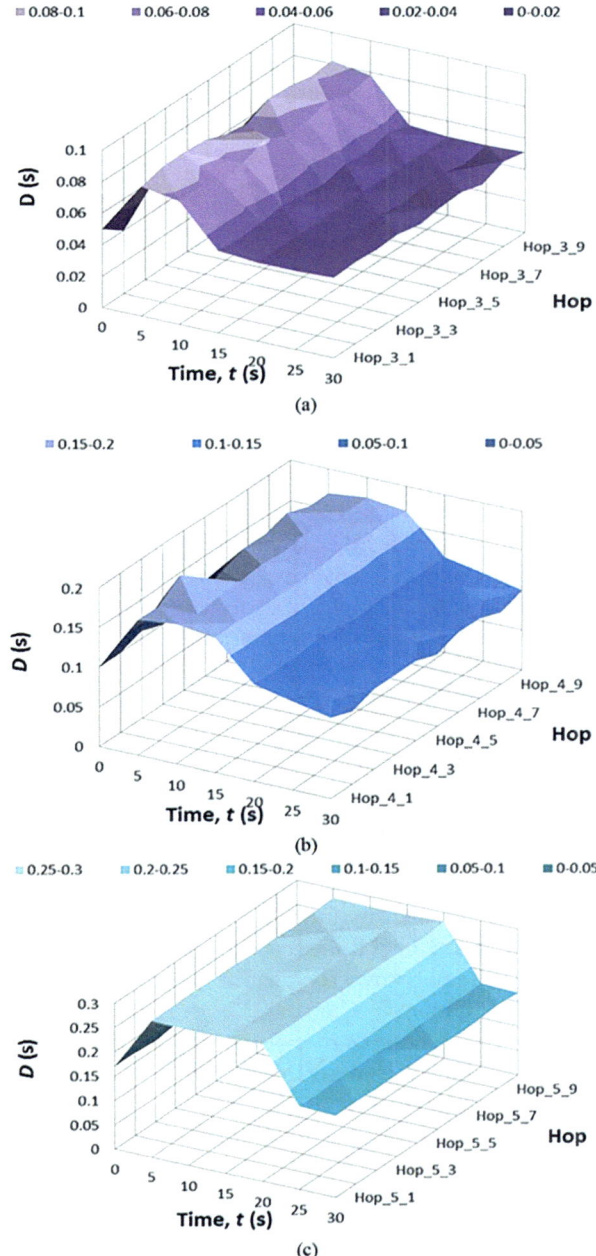

Fig. 4.15 (**a**) Real-time D under interference for $n = 3$ (**b**) $n = 4$ (**c**) $n = 5$ when the maximum of hop = 10

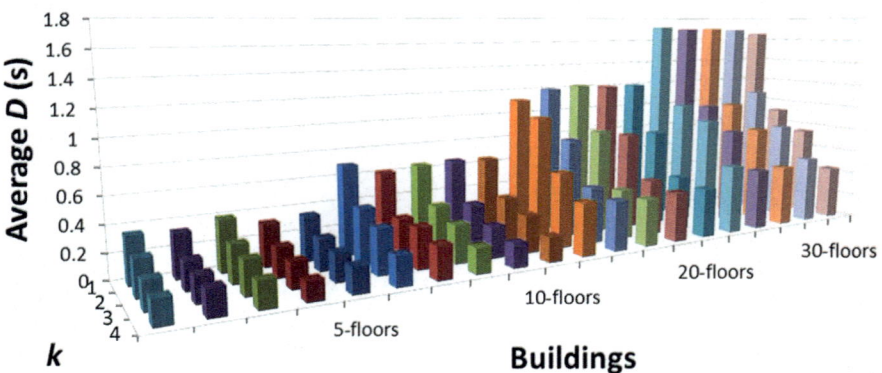

Fig. 4.16 Investigation of D when $k = 1, 2, 3, 4$

4.6.2   Latency Study

In this investigation, analysis of IMM2ZM with $n = 5$, 10, 20, and 30 has been studied to give a holistic view of the effectiveness. The results of D and σ versus k ($k = 1$, 2, 3, 4) are plotted in Figs. 4.16 and 4.17, respectively. For $k = 4$, the performance improvement of IMM2ZM, with and without interference mitigation over MIZBAN [15] is shown in Fig. 4.18.

Figure 4.16 shows the variation of D against n ($n = 1$–30) and k ($k = 1$–4). In general, D increases as n increases, since the average number of hops for the routing path as well as the traffic loading increases. In contrast, D decreases as k increases because the traffic loadings can be shared by the multiple operation channels. It is seen from Fig. 4.16 that the improvement of D for 5-floor ($n = 5$) buildings is approaching saturation when $k = 2$. At $k = 2$, the improvement of D for 5-floor buildings is not significant as compared to 10-floor and 20-floor buildings. These findings are attributed to the low-density traffic characteristics at $n = 5$. Besides, when k increases, in particular at $k = 4$, it is seen that the probability of finding a busy channel for RMTs is extremely low. The channel-access delay will be minimized, and thus, D reaches a minimum.

Figure 4.17 shows the variation of σ against n ($n = 5$, 10, 20, 30) and k ($k = 1$, 2, 3, 4). In general, σ increases as k increases since IMM2ZM transmits data in parallel via multiple channels simultaneously.

The strength of IMM2ZM versus MIZBAN is now analyzed. The maximum capacity should be examined, and thus, $k = 4$ is investigated. Figure 4.18 shows the performance improvement of IMM2ZM ($k = 4$), with and without interference mitigation over MIZBAN [15]. It can be seen that as n increases ($n = 5$, 10, 20, 30), σ increases from 174% when $n = 5$ to 329% when $n = 10$; 280% when $n = 20$; 274% when $n = 30$. It is seen that the gradient increase of σ is tremendous from $n = 5$ to 10. Thus, it is concluded that IMM2ZM performs very well at increasing network size (say, $n = 30$). The performance of D is also investigated. The improvement of D

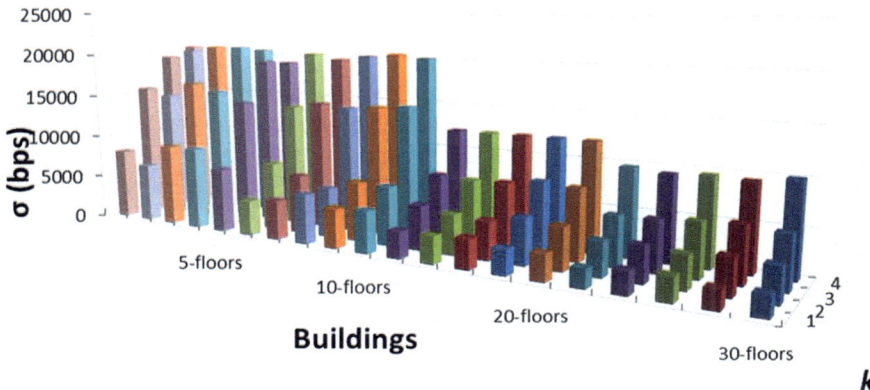

Fig. 4.17 Investigation of σ when $k = 1, 2, 3, 4$

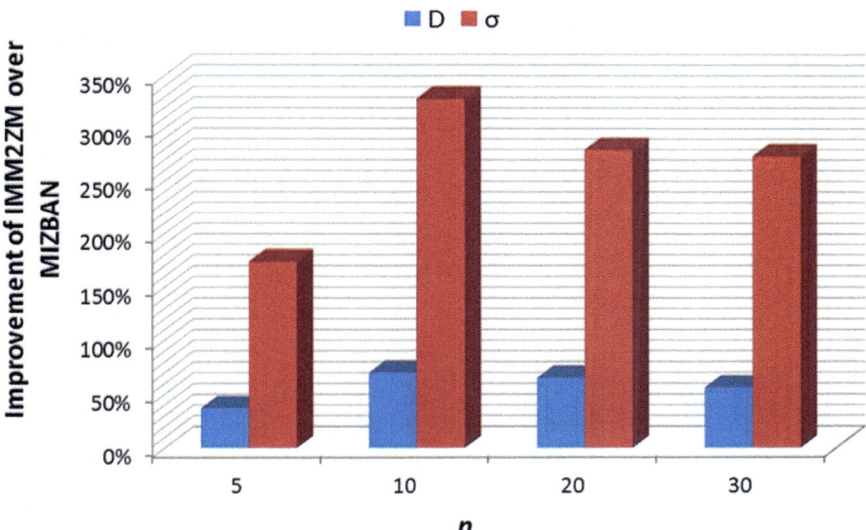

Fig. 4.18 Performance improvement of IMM2ZM ($k = 4$), with/without interference mitigation over MIZBAN [15]

increases rapidly from 37% at $n = 5$ to 72% when $n = 10$; 65% when $n = 20$; 56% when $n = 30$. Hence, it is concluded that the performance of the optimized IMM2ZM well surpasses MIZBAN. In Hong Kong, the Hong Kong Housing Authority of the Census and Statistics Department of the Government of Hong Kong [40] revealed that $n \sim 12$ in 2014. Apparently, n will increase significantly with urban modernization in the future. From the analysis here, it is evidenced that the IMM2ZM should be adopted for high-performance HTAMI.

4.6.3 Performance of TS-ZBN

To evaluate the performance of TS-ZBN, ten floors building with 100 m*100 m per floor is designed with OPNET to simulate the building environment. Each floor contains ten unities and ten water sensors were employed to each utility randomly. ZigBee is chosen for the wireless protocol for UAN and TLN. WiFi is chosen for the wireless protocol for NLN.

During coordinator-to-coordinator synchronization, each C_U will send the information to C_F for 100 times. Since that TLN is a mesh network, a number of CCS may be processed which depends on the path from source to destination. For example, if the signal from the source is required to pass through m C_U before it reaches C_F, the number CCS will be $m + 1$.

Based on the simulation result, the maximum number of CCS is found as 9. For end-to-end synchronization, each sensor node sends the sensed data to C_U for 100 times. Similar to the previous simulation, the number of node-to-node synchronization (NNS) depends on the transmission path from the source node to destination C_U. The maximum number of NNS is also found as 9 after simulation. Figure 4.19 shows the delay in interference coexistence study while Fig. 4.20 gives the latency performance.

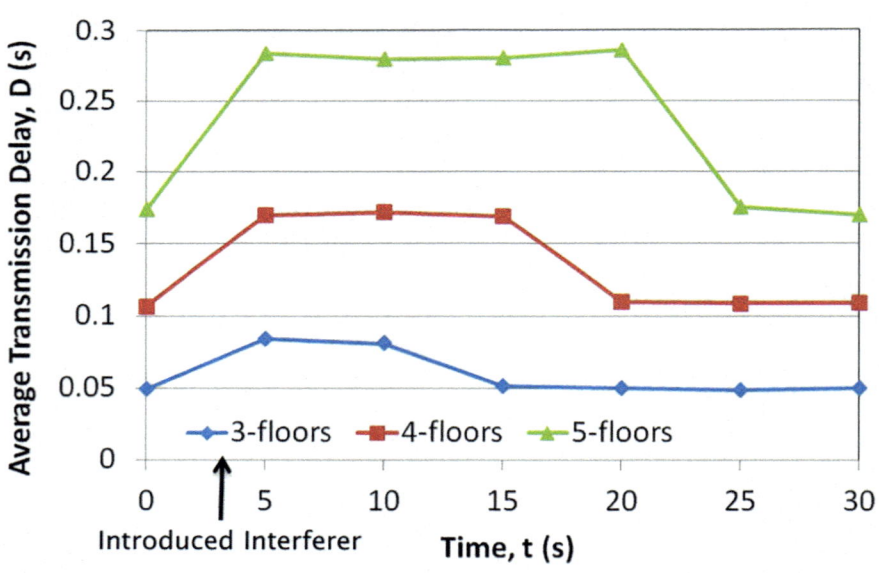

Fig. 4.19 Interference coexistence study results

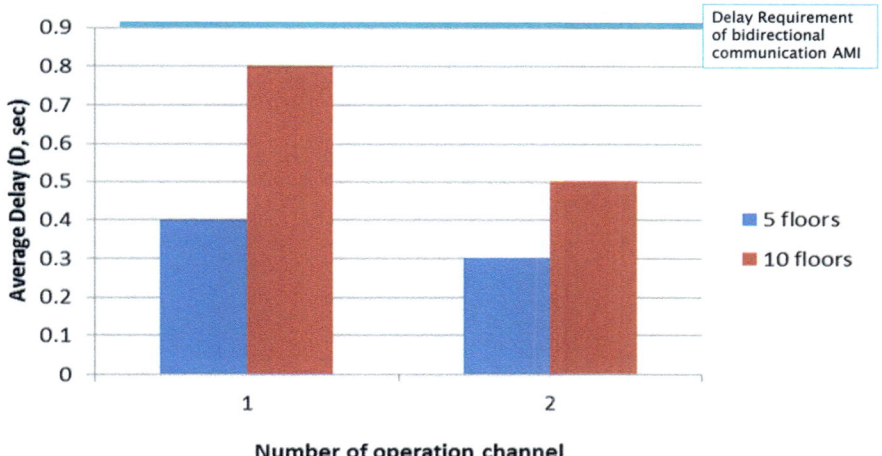

Fig. 4.20 Latency performance

4.7   Conclusion

The simulation result shows that TS-ZBN achieves low mean synchronization error and variance. Current SM solutions focus on low traffic for individual houses. SM traffics are ever growing, in particular, for buildings in Asia. This study proposes the IMM2ZM, a new multiobjective optimization interference-mitigated ZigBee-based AMI as a SM solution for high-traffics data. The contribution of this study is five-folded. First, a prior measurement was performed to obtain the essential data for the formulation of objective functions for the optimal solution at large-scale. Second, an interference mitigation model has been derived. Third, customization to NSGA-II optimization has been developed. Fourth, the OPNET evaluation has been implemented for large-scale analysis. Fifth, a channel-swapping IMM2ZM system has been implemented and analyzed for HTAMI.

T_{cs} evaluates the efficiency of channel-swapping, hence giving an account of the latency performance of the network due to interference. It is concluded that when the IMM2ZM sensitivity (P_{RX}) is less than -12 dBm, T_{cs} increases tremendously. It is important to highlight that the IMM2ZM achieves an effective performance in a HTAMI and results in a significant improvement in the performance of the application-layer transmission rate (σ) and the average delay (D). The improvement figures are $\sigma > \sim300\%$ and $D > 70\%$ in a 10-floor building; $\sigma > \sim280\%$ and $D > 65\%$ in a 20-floor building; and $\sigma > \sim270\%$ and $D > 56\%$ in a 30-floor building. In conclusion, this confirms the feasibility to adopt time-synchronized ZigBee building network for water management.

Acknowledgments The permission given to use the materials from the following papers is very much appreciated.

A. Hao Ran Chi, K F Tsang, K T Chui, Henry Chung, Bingo Wing Kuen Ling, Loi Lei Lai, Interference-mitigated ZigBee based advanced metering infrastructure. IEEE Trans. Ind. Inform. **12**(2), 672–684 (2016)
B. Chung Kit Wu, Hongxu Zhu, Loi Lei Lai, Anna S. F. Chang, Fengjun Li, Kim Fung Tsang, Roy Kalawsky, A time-synchronized ZigBee building network for smart water management, IEEE INDIN2017, Germany, July 2017

References

1. D. Newkirk, J.S. Evans, O.S. Alraddadi, C.G. Kelemen, R. Mietusch, X. Yu, B. Rajkhowa, Plant-assisted air-conditioning systems for a better tomorrow. IEEE Potentials **34**(1), 11–17 (2015)
2. B. Raji, M.J. Tenpierik, A. van den Dobbelsteen, The impact of greening systems on building energy performance: a literature review. Renew. Sust. Energ. Rev. **45**, 611–621 (2015)
3. H. Liu, S.C. Lee, M.J. Kim, H. Shi, J.T. Kim, K.L. Wasewar, C.K. Yoo, Multi-objective optimization of indoor air quality control and energy consumption minimization in a subway ventilation system. Energ. Buildings **66**, 553–561 (2013)
4. Z. Wang, L. Wang, Intelligent control of ventilation system for energy-efficient buildings with CO_2 predictive model. IEEE Trans. Smart Grid **4**(2), 686–693 (2013)
5. A. Afram, F. Janabi-Sharifi, Theory and applications of HVAC control systems—a review of model predictive control (MPC). Build. Environ. **72**, 343–355 (2014)
6. N. Liu, J. Chen, L. Zhu, J. Zhang, Y. He, A key management scheme for secure communications of advanced metering infrastructure in smart grid. IEEE Trans. Ind. Electron. **60**(10), 4746–4756 (2012)
7. V.C. Gungor et al., Smart grid technologies: communication technologies and standards. IEEE Trans. Ind. Inf. **7**(4), 529–539 (2011)
8. M.E. Kantarci, H.T. Mouftah, Wireless sensor networks for cost efficient residential energy management in the smart grid. IEEE Trans. Smart Grid **2**(2), 314–325 (2011)
9. V.C. Gungor, G.P. Hancke, Industrial wireless sensor networks: challenges, design principles, and technical approaches. IEEE Trans. Ind. Electron. **56**(10), 4258–4265 (2009)
10. U.S. Department of Energy, *Locke, Chu Announce Significant Steps in Smart Grid Development* [Online] (2009, May 18), http://www.energy.gov/news2009/7408.htm
11. K.F. Tsang, H.Y. Tung, K.L. Lam, *ZigBee: From Basics to Designs and Applications* (Prentice Hall, Upper Saddle River, 2009)
12. H.Y. Tung, K.F. Tsang, K.T. Chui, H.C. Tung, H.R. Chi, G.P. Hancke, K.F. Man, The generic design of a high-traffic advanced metering infrastructure using ZigBee. IEEE Trans. Ind. Inf. **10**(1), 836–844 (2014)
13. B. Heile, Smart grids for green communications. IEEE Wireless Commun. **17**(3), 4–6 (2010)
14. P. Varahram, B. Ali, A crest factor reduction scheme based on recursive optimum frequency domain matrix. IEEE Trans. Consum. Electron. **60**(2), 179–183 (2014)
15. H.Y. Tung et al., The generic design of a high-traffic advanced metering infrastructure using ZigBee. IEEE Trans. Ind. Inf. **10**(1), 836–844 (2014)
16. K.S. Tang, T.M. Chan, R.J. Yin, K.F. Man, *Multiobjective Optimization Methodology—A Jumping Gene Approach* (CPC Press, Boca Raton, 2012), Chapter 2
17. D. Stanislowski, X. Vilajosana, Q. Wang, T. Watteyne, K.S.J. Pister, Adaptive synchronization in IEEE802.15.4e networks. IEEE Trans. Ind. Inf. **10**(1), 795–802 (2014)
18. O. Tipmongkolsilp, S. Zaghloul, A. Jukan, The evolution of cellular backhaul technologies: current issues and future trends. IEEE Commun. Surv. Tutorials **13**(1), 97–113 (2011)

19. X. Zhang, K.G. Shin, Cooperative carrier signaling: harmonizing coexisting WPAN and WLAN devices. IEEE/ACM Trans. Netw. 21(2), 426–439 (2012)
20. P. Yi, A. Iwayemi, C. Zhou, Developing ZigBee deployment guideline under WiFi interference for smart grid applications. IEEE Trans. Smart Grid 2(1), 110–120 (2011)
21. A. Mukherjee, A.L. Swindlehurst, Robust beamforming for security in MIMO wiretap channels with imperfect CSI. IEEE Trans. Signal Process. 59(1), 351–361 (2010)
22. E. Toscano, L.L. Bello, Multichannel superframe scheduling for IEEE 802.15.4 industrial wireless sensor networks. IEEE Trans. Ind. Inf. 8(2), 337–350 (2012)
23. I. Ho, P. Lam, P. Chong, S. Liew, Harnessing the high bandwidth of multi-radio multi-channel 802.11n mesh networks. IEEE Trans. Mobile Comput. 13(2), 448–456 (2013)
24. M.K. Denko, T. Jun, T. Nkwe, M.S. Obaidat, Cluster-based cross-layer design for cooperative caching in mobile ad hoc networks. IEEE Syst. J. 3(4), 499–508 (2009)
25. C.E.A. Campbell, K.K. Loo, H.A. Kurdi, S. Khan, Comparison of IEEE802.11 and IEEE 802.15.4 for future green multichannel multiradio wireless sensor networks. Int. J. Commun. Netw. Inf. Sec. 3(1), 96–103 (2011)
26. Z. Liu, W. Wu, A dynamic multi-radio multi-channel MAC protocol for wireless sensor networks, in *Proc. 2nd Int. Conf. Commun. Softw. Netw. (ICCSN'10)* (2010), pp. 105–109
27. S.Y. Shin, H.S. Park, S.H. Choi, W.H. Kwon, Packet error rate analysis of ZigBee under WLAN and bluetooth interferences. IEEE Trans. Wirel. Commun. 6(8), 2825–2830 (2007)
28. S.Y. Shin, S. Choi, H.S. Park, W.H. Kwon, *Lecture Notes in Computer Science: Packet Error Rate Analysis of IEEE 802.15.4 Under IEEE 802.11b Interference* (Springer, New York, 2005), Chapter 4
29. G. Anastasi, M. Conti, M.D. Francesco, A comprehensive analysis of the MAC unreliability problem in IEEE 802.15.4 wireless sensor networks. IEEE Trans. Ind. Inf. 7(1), 52–65 (2010)
30. G. Casella, R.L. Berger, *Statistical Inference* (Duxbury Press, Singapore, 2001), p. 159
31. K.F. Tsang, *Wireless Communication* (Pearson, Upper Saddle River, 2007), Chapter 4
32. J.W. Chong, H.Y. Hwang, C.Y. Jung, D.K. Sung, Analysis of throughput in a ZigBee network under the presence of WLAN interference, in *Proc. Int. Symp. Commun. Inf. Technol. (ISCIT'07)* (2007), pp. 1166–1170
33. K.F. Tsang, H.Y. Tung, K.L. Lam, *ZigBee: From Basics to Designs and Applications* (Prentice Hall, Englewood Cliffs, 2009), Chapter 3
34. K. Deb, *Search Methodologies* (Springer, Berlin, 2014), Chapter 15
35. T.M. Chan, K.F. Man, K.S. Tang, S. Kwong, A jumping-genes paradigm for optimizing factory WLAN network. IEEE Trans. Ind. Inf. 3(1), 33–43 (2007)
36. K. Deb, A. Pratap, S. Agarwal, T. Meyarivan, A fast and elitist multiobjective genetic algorithm: NSGA-II. IEEE Trans. Evol. Comput. 6(2), 182–197 (2002)
37. Riverbed Technology, Inc., *OPNET University Program* [Online] (2015), http://www.opnet.com/services/university/
38. Communications requirements of smart grid technologies, Dept. Energy, USA, Oct. 5 (2010)
39. Hong Kong Digest of Statistics, *Trends in Population and Domestic Households in Hong Kong* [Online] (2012, Apr), http://www.census2011.gov.hk/pdf/Feature_articles/Trends_Pop_DH
40. Hong Kong Housing Authority, *Sustainability Report 2013/14* [Online] (2015, Jan 28), http://www.housingauthority.gov.hk/minisite/hasr1314/en/common/index.html

Chapter 5
A Narrowband Internet of Thing-Based Temperature Prediction for Valve-regulated Lead Acid Battery

5.1 Introduction

5.1.1 NB-IoT

The Third-Generation Partnership Project (3GPP) introduced the first IoT-specific user equipment (UE) in Long-Term Evolution (LTE) Release 12, known as LTE-M with features including peak data rate at 1 Mb/s over 1.08 MHz bandwidth and support UEs with the half duplex operation and power-saving mode. Recently, in the LTE Release 13, 3GPP has standardized a new radio access network (RAN) technology called narrowband IoT (NB-IoT) [1]. The narrow-band internet of things (NB-IoT) is a massive low power wide area (LPWA) technology proposed by 3GPP for data perception and acquisition particularly for intelligent low-data rate applications [2].

It inherits basic functionalities from the LTE system, while it operates in a narrowband. With a software upgrade, the existing LTE network can be enabled to support NB-IoT. This is essential for reducing deployment cost and time.

Technically, NB-IoT is developed under the specification of LTE. The bandwidth of NB-IoT is about 180 kHz, and the coverage is less than 10 km in practice. NB-IoT protocol can be deployed in not just LTE, but also GSM or UMTS, whose downlink speed is from 160 to 230 kbps and uplink speed is from 160 to 250 kbps. Moreover, the NB-IoT communication protocol is half-duplex. The maximum transport block size in the downlink is 680 bits, and the uplink is 1000 bits.

NB-IoT has three deployment methods, namely, Independent Deployment, Guard-band Deployment, and In-band Deployment. In Independent Deployment, the 180 kHz frequency band is located out of the LTE carrier. While for Guard-band Deployment, the 180 kHz frequency is on the edge of LTE carrier. Regarding In-band Deployment, the frequency band is located in the LTE carrier. Communication operators define the deployment method. Mobile network operators

C. S. Lai et al., *Smart Grids and Big Data Analytics for Smart Cities*, https://doi.org/10.1007/978-3-030-52155-4_5

generally tend to operate the NB-IoT technology in the in-band mode due to the low cost and lower deployment complexity.

5.1.1.1   NB-IoT Features

Super Coverage

In IoT application scenarios such as water meter reading, smart parking, the requirement for wide coverage exceeds the performance of the traditional 2G/3G/4G network. Therefore, 3GPP proposed that NB-IoT should have 20 dB coverage enhancement compared to GSM [2, 3].

The 3GPP standardization adopted two solutions to enhance the coverage for NB-IoT. The first method is using reduced bandwidth to promote the user equipment's (UE) transmission Power Spectral Density (PSD) and bring additional gains of coverage enhancement. The second method is the repeat transmission. Putting the two technologies together, NB-IoT could have 20 dB gains of coverage enhancement, as compared to GPRS. However, there are side effects of these two technologies. Reduced bandwidth degrades the data rate and repeat transmission could lead to more severe latency [4, 5].

Low Power Consumption

There are a large number of internet of things terminals. Some terminals are in the environment where replacing battery or battery charging is impossible, therefore, low power consumption is an essential feature of internet of things terminals. In TS 45.820, in combination with the industrial demand, for periodical report services, 3GPP demands a low power consumption requirement for IoT terminals. The normal working time should be about 10 years.

There are two main features, that is, power-saving mode (PSM) and extended discontinuous reception (eDRX), which are used to extend the battery life of NB-IoT devices for up to 10 years. Both technologies leverage the advantage of a low frequency of data transmission.

There are three operation states in NB-IoT devices, CONNECT, IDLE, and PSM. PSM state is added as a sub-state of the original IDLE state. When working in PSM, a device will go into deep sleep mode and could not receive any signal since the radio frequency unit is completely shut off and the downlink is inaccessible. It could only wake up when the UE needs to transmit mobile-originated uplink data or when it is triggered by an exterior RTC wakeup signal. PSM is similar to power-off, but the UE remains registered with the network. Because PSM resembles UE powered off, and the maximum PSM duration time in NB-IoT is 310 h, UE's energy consumption is slashed considerably, making it feasible to adopt battery as UE's only power source during its lifetime.

Discontinuous reception (DRX) indicates that the UE turns on the receiver and works in CONNECT state only when necessary and works in IDLE state, turns off the receiver, stops receiving downlink data during the rest of the time. The extended version aims at increasing the paging monitoring interval. With eDRX adopted, in each eDRX cycle, the UE only needs to monitor the paging frame at the prescribed Paging Time Window (PTW) to check whether there is a paging-radio network temporary identity on the physical downlink control channel.

Power consumption is closely related to every electronic component and every module in every terminal equipment. And the power consumption optimization is an accumulative process since it is affected by numerous factors, ranging from the operating system to the hardware devices. Communication protocol optimization also contributes to power saving.

Low Cost

Low Power Wide Area Network (LPWAN) IoT market has the highest potential in the whole IoT field. Low cost is an outstanding feature of LPWAN IoT. As a mobile IoT technology designed for LPWAN, NB-IoT inevitably needs to realize a low cost.

In order to realize the ultra low cost of NB-IoT devices, it is necessary to consider lowering the complexity of protocols and products, because NB-IoT originates from the complicated LTE specification. To simplifying protocol volume, it removes many features of LTE, including physical uplink control channel, Physical Hybrid ARQ Indicator Channel, and Measurement Report. NB-IoT only supports Frequency Division Duplex and Half Duplex, and only requires one antenna. All these modifications together lead to the low cost of the NB-IoT module and NB-IoT chipset. This will definitely drive a huge boost in the NB-IoT business applications.

Apart from the cost of the chipset and module, there is another cost factor worth considering, which is the network installation and maintenance cost. Fortunately, NB-IoT does not need to construct the network from scratch because it can be deployed in three different operation modes within the existing LTE carrier. A complete industrial chain has been formed for LTE and LTE is still prospering. NB-IoT can take advantage of LTE technology to effectively lower the cost.

Massive Connections

Traditional human-based telecommunication has almost reached the ceiling, due to the limited number of terminals held by each person. It is forecasted that by 2025, international IoT connection number will reach 27 billion, and most of the IoT connections are sensors, monitoring, and control use cases, rather than human-based connection. They are widely distributed, insensitive to delay, sensitive to cost, and energy consumption, having a low data transmission rate.

According to [6], the supported connection number is 52,547 per cell site sector, based on the calculation and assumptions for London, where the area of cell site sector is 0.866 km^2, the household density per km^2 is 1517, and the number of devices in a household is 40. However, such capability does not allow for high user-concurrency.

In actual use cases, the deployment scheme must avoid different UEs requesting for data upload at the same time. The massive connection capacity requires reassessment based on the actual service scheme, which will give different data packet sizes.

NB-IoT users transmit a small amount of data and they can tolerate latency. Therefore, over 50,000 users can camp on the same cell. In addition, NB-IoT supports two schemes multi-tone and single-tone transmission simplifies signaling overhead, to further sustain a large capacity.

Since 2017, many IC manufacturers like Qualcomm, MediaTek have put NB-IoT chips into production, like MT2625, and correspondingly, NB-IoT module manufacturers are manufacturing NB-IoT modules for the market, such as BC26, BC95, and BC28. Three communication carriers in China have all been deploying NB-IoT base stations nationwide. The NB-IoT industries including chips, modules, and platforms are prospering. There are also lots of PaaS, SaaS IoT platforms, such as Microsoft Azure, Cisco Jasper, Telit, and China Mobile OneNet.

5.1.1.2   Comparisons with LoRa and eMTC

Reference [7] made a comparison between LoRa and NB-IoT. Among Low Power Wide Area Network, LoRa and NB-IoT are the two leading emergent technologies. LoRa, designed by Semtech Company, is built on proprietary spread spectrum techniques and Gaussian frequency shift keying (GFSK). LoRa is a non-cellular network while NB-IoT is a cellular network. LoRa operates in a non-licensed band below 1 GHz for long-range communication link operation whereas NB-IoT uses the licensed frequency bands, which are the same frequency bands in LTE. Though NB-IoT is integrated into the LTE standard, it is kept more simply simple than LTE in order to reduce device costs and minimize battery energy consumption.

In general, NB-IoT has better Quality of Service (QoS) because of its licensed band spectrum, but its advantage in low cost is no longer prominent compared with LoRa. In terms of battery life, because NB-IoT needs to upload and synchronize data regularly, so it is also a little less advantageous compared to LoRa. In summary, LoRa and NB-IoT have their respective advantages in different aspects of IoT.

Each application has its specific requirements, which lead to a specific technology selection. Both LoRa and NB-IoT can cater to more use cases spreading from the low-end to the high-end scenarios in a variety of fields and play important roles in the LPWAN IoT market. LoRa focuses on low cost applications, whereas NB-IoT is dedicated to applications that require higher QoS, wider coverage, and lower latency. Most importantly, NB-IoT can get support from the operators due to the licensed attribute [4]. In contrast, LoRa just lacks such powerful business drivers. Especially, NB-IoT still manifests some advantages in both peak rate and coverage range, which can also help it to win more market share in the future competitions with LoRa.

Reference [8] studied two technologies built from LTE, namely Enhanced Machine Type Communication (eMTC) and NB-IoT. eMTC targets the applications such as VoLTE, mobility with tracking devices, tasks that need high data rate and low power consumption with wide area coverage. NB-IoT targets the applications such as wireless sensors and meters with low complexity and low power consumption with wide area coverage. Both technologies have power saving features.

Later, simulation experiments were conducted to compare the two technologies in three aspects. Considering the energy consumption, NB-IoT is a good choice for simple sensors and low data transmission rate applications in medium to poor coverage cases, while eMTC is for applications transmitting a large amount of data in good to medium coverage cases. A battery lifetime of 8 years can be achieved by both technologies in a poor coverage case with daily reporting interval. As far as the end-to-end latency and scalability, the delay of transmitting a packet in eMTC is lower than the delay of transmitting a packet in NB-IoT, and eMTC can serve more devices in a network than NB-IoT.

In general, eMTC has the highest speed due to the widest bandwidth but also suffers from the highest cost, so it is only fit for high-end applications without sensitivities to prices [4].

5.1.1.3   NB-IoT Application Scenarios

Advanced information technology will be used to consolidate and renovate the infrastructure of the smart city. Intelligent collaboration, resource sharing, interconnection, and comprehensive perception, will provide the intelligent service management of the city to better solve the problems of urban development and realize the sustainable development of the city. Comprehensive perception of the smart city requires the infrastructure to collect various data and information for further data analysis and integration. These massive cross-region and cross-industry data and information are potential resources for urban services and management decisions. To be more specific, each city needs numerous cameras, sensors, detectors, and other devices to form the perception layer, to help city administrators to better understand the city and carry out the corresponding actions.

NB-IoT fits for services that are not sensitive to latency, requiring little data transmission, located in places inconvenient for power supply, with high installation density and number, and strong signal shielding. Smart cities will digitize the infrastructure such as street lamps, manhole covers, underground pipelines, parking lots, and make full use of the network, database, and other technical means to make the information technologies more widely and comprehensively applied in the field of city management and operation. These applications rely on using NB-IoT combined with sensors to transmit a huge amount of structured and unstructured data that can be used for automation, decision-making, and analysis. Smart city not only implements smart applications, but it also fosters a data-driven and low carbon economy.

Smart city benefits not only its residents but also its citizens, tourists, investors, and government.

Designing a smart city is a top-down intelligently engineered process, whereas city evolution is unpredictable, not following the way it was once designed. However, evolution is highly integrated with the activities of humans. This section introduces several typical smart cities and municipal applications of NB-IoT, including smart street lights, smart parking, smart meter reading, and other businesses. Current status, existing problems and challenges, solutions based on NB-IoT, and the advantages are discussed.

Smart Manhole Cover

The problem of stolen and poorly managed urban well covers has become increasingly prominent. It is frequently reported in the news that manhole cover missing, and damage can cause accidents. Also, the communication or electric cables, sewage tubes under the manhole covers will face threats.

Smart manhole cover system can monitor the manhole cover, detecting its movement and integrity, as well as the situation under the cover, then report the data periodically via the NB-IoT network to the cloud platform. As a result, the smart manhole cover system could reduce a fair amount of accidents caused by a broken or stolen manhole, contributing to the smart city development [9, 10].

The cover detector is attached to the inner side of the cover, fixed with screws. It can detect the inclination angle as well as the displacement of the cover. If there is any irregularity, it will immediately send the emergency message to the back end. Opening the cover is permitted only if the administrator applies to the cloud platform for unlocking and receives permission. After inspection and maintenance with manhole cover closed, the cover detector will send messages to the cloud platform, notifying that the cover is locked.

Apart from these basic functions, the detector could be further equipped with various sensors that collect information such as pressure, water level, flammable gas like CH_4, CO concentration, and temperature. And transmit the data back to the operation and maintenance cloud platform via the NB-IoT network. The cloud platform displays the status, location of the covers, as well as the situation beneath the cover. With the information, the municipal administration staff can assign workers to the field to deal with the situation.

NB-IoT is extremely suitable for smart manhole cover for the following reasons. Usually, a specific urban area requires a large number of manhole covers, and the data generated by each cover for transmission is relatively limited. In addition, the cover state perception can tolerate the latency of several seconds. Figure 5.1 shows an application of NB-IoT to smart manhole management.

Fig. 5.1 Application of NB-IoT to smart manhole management

5.1.1.4   Smart Meters

Traditionally, water meter, gas meter, and electricity meter are read manually, usually on a monthly or seasonal basis. Manual meter reading demands intense laboring work, causing extremely low efficiency and low accuracy due to human error. In general, the metering data could not be collected and analyzed in real-time. As a result, differentiated and tiered pricing could not be implemented easily. Furthermore, electricity and water thefts could hardly be prevented. The integration of NB-IoT technology and conventional meters is an effective solution.

NB-IoT smart metering is a system where the measured value of water, electricity, and gas usage is sent to NB-IoT module embedded in the meter, and transmitted to the cloud platform through the wide-covering NB-IoT network. Usually, water meters, gas meters, and electricity meters are numerous and placed in a hidden environment such as the basement, and connection to a power source is not always available or easy. Due to the huge number of meters in urban areas, the cost must be low enough. In such conditions, NB-IoT effectively solves all these problems. NB-IoT is especially advantageous in terms of massive connections, wide coverage, low power consumption, and long battery life. Since the meters only send a small amount of data and transmission delay is bearable, the relatively long latency and low transmission rate of NB-IoT is acceptable in meter reading applications [11].

It is possible to modify the conventional meters by embedding NB-IoT modules and sensors, connecting the meters to the NB-IoT network, and, further, to the core network and IoT cloud platform. Therefore, the users can acquire information about water, gas, and electricity usage and pay their bills by simply scrolling the screen on their phones. The water and gas supply companies and the property management companies can read the meters more accurately, efficiently at a higher frequency, and receive payment in a much more efficient and convenient way. The system allows the administrators to promptly discover any error or anomaly value or even leakage by data analysis to ensure safety. Figure 5.2 shows an application of NB-IoT to smart meters operation.

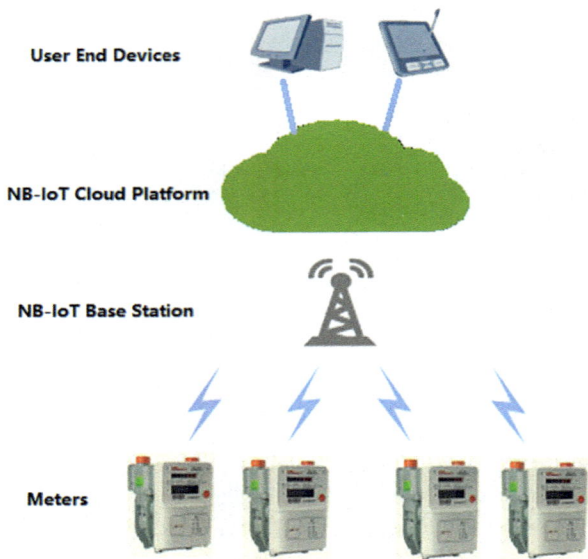

Fig. 5.2 Application of NB-IoT to smart meters operation

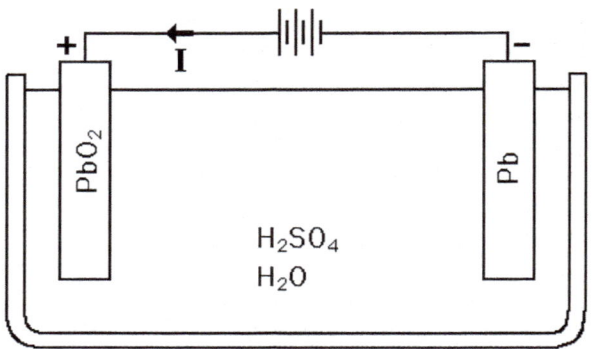

Fig. 5.3 Basic structure of a lead-acid battery

5.1.2 *Battery*

Traditional lead-acid battery is widely used due to the low cost, large power-to-weight ratio, high current density, and easy maintenance. There are various applications, such as providing surge current for motor starter for vehicles, acting as energy storage for backup power supply for an emergency, and power buffer for unstable energy sources, for example, solar power plants.

A basic structure of a lead-acid battery is shown in Fig. 5.3:

There are various heat sources in the battery and they lead to heating problems. In reversible electrochemistry reaction, there is heat absorption when discharging and heat is released during charging. Due to the Joule heating effect, heat is releasing in both the charging and discharging process. This affects the temperature as compared to the electrochemistry reaction.

As a result, the operation temperature is limited. Performance drops in low temperature and internal resistance increases. There is permanent damage when the electrolyte is frozen, which is a destruction of internal structure as the solution expands. Thermal runaway exists in high temperature and battery could explode. High temperatures could cause unexpected electrical short circuits, release hydrogen gas to build up internal pressure, and cause fire and explosion.

Figures 5.4 and 5.5 show an exploded battery and car fire due to raising of temperature and pressure inside the battery.

Valve-regulated lead acid (VRLA) battery owns a huge market which reached USD 51.2 billion in 2017 [12]. By comparing to other kinds of batteries, VRLA has a distinct advantage that it does not require water to replenish the electrolyte [13]. VRLA reduces the amount of labor and time required for energy storage. Also, VRLA can work at the extreme temperature, and this makes VRLA acquire a big portion in the battery industry. Currently, VRLA is widely used in various industry scenarios such as data center and airplane. Also, the deep-cycle absorbent glass mat (AGM) is commonly used in off-grid solar power and wind power installations as an energy storage bank, and so on [13–15]. The battery is used in the uninterruptible power supply (UPS) as a backup when the electrical power goes off too.

Unfortunately, with the higher float current inside the VRLA, it is more prone to produce heat due to the chemical reactions [16, 17], which generate serious consequences. On the one hand, the performance of VALR is seriously affected by the temperature. Potential hazards will be caused in applications due to an ineffective battery. On the other hand, the battery life of VRLA will easily be decreased because of the excessive internal temperature [18]. Thus, extra cost will be needed for

Fig. 5.4 Battery explosion

Fig. 5.5 Battery fire

maintenance or replacement [19]. Besides, thermal runaway, an unstoppable self-heating reaction, will happen if the battery keeps working at an inappropriate internal temperature. This self-heating reaction always causes serious consequences such as fires, electrolyte leakage, and venting due to lack of risk assessment [20]. To prevent potential hazards from internal temperature, two main types of algorithms are developed to measure internal temperature. One of the methods is to measure the internal temperature directly by electrical devices. The other is to predict the internal temperature based on several external parameters using an established battery thermal model. The thermal model needs to have the following characteristics:

- Thermal parameters should be self-adaptive
- Temperature range must cover battery's operation temperature
- The model can be used for various size battery
- Suitable for longer time prediction

One of the traditional battery internal temperature estimation methods is direct measurement. To measure the internal temperature directly, researchers installed thermocouples inside the battery [21]. An obvious disadvantage of this method is that the cost of manufacturing a battery will be increased a lot. Also, a potential safety problem may exist if a redundant component is added and, thus, destroy the original structure of the battery. In conclusion, it is not an appropriate and practical method to directly measure internal temperature unless the mentioned issues could be solved. Another familiar method is to estimate internal temperature by a thermal model.

Figure 5.6 shows a thermal model of the battery.

As shown in the figure, the three sources represent the reversible electrochemistry reaction, oxygen cycle, and Joule heating effect.

A comparison between thermal and electrical model is given in Table 5.1:

Based on the thermal model as shown in Fig. 5.6, the following equations could be derived.

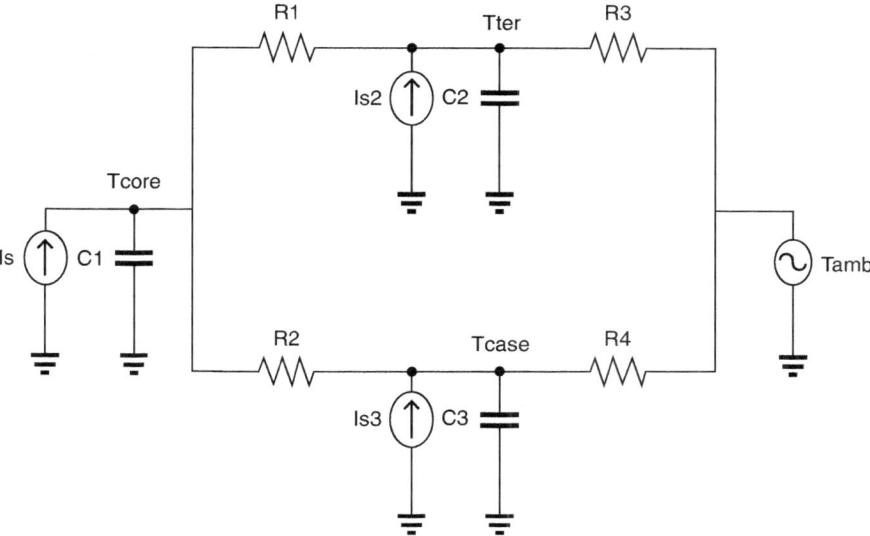

Fig. 5.6 Battery thermal model

Table 5.1 Parameters comparison between thermal and electrical model

Thermal model	Electrical model
Temperature	Voltage
Heat transfer rate	Current
Thermal resistance	Electrical resistance
Heat capacitance	Electrical capacitance
Heat source	Electrical source

$$C_1 \times \frac{dT_{core}(t)}{dt} = \frac{T_{ter}(t) - T_{core}(t)}{R_1} + \frac{T_{case}(t) - T_{core}(t)}{R_2} + I_s(t) \tag{5.1}$$

$$C_2 \times \frac{dT_{ter}(t)}{dt} = \frac{T_{amb}(t) - T_{ter}(t)}{R_3} + \frac{T_{core}(t) - T_{ter}(t)}{R_1} + I_{S2}(t) \tag{5.2}$$

$$C_3 \times \frac{dT_{case}(t)}{dt} = \frac{T_{amb}(t) - T_{case}(t)}{R_4} + \frac{T_{core}(t) - T_{case}(t)}{R_2} + I_{S3}(t) \tag{5.3}$$

The Joule heating effect model is shown in Fig. 5.7:

$$R_{int}(SOC, T) \propto \frac{1}{T} \tag{5.4}$$

Fig. 5.7 Joule heating
effect modelling

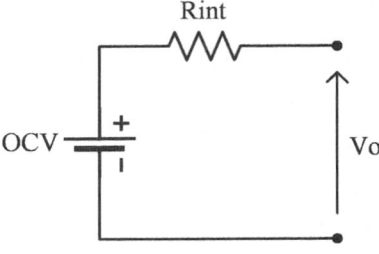

$$P_{JH} = I^2_{current} \times R_{int} \qquad (5.5)$$

SOC is state of charge. Previously, many researches have worked on this topic. However, most of them concentrate on the temperature prediction models of Lithium-ion but hardly on the VRLA battery [22]. The methodology of Lithium-ion battery using an external parameter for prediction provides inspirations for the internal temperature prediction of VRLA battery. However, instead of just a thermal model, a neural network-based model is also involved in the algorithm.

In this research, a NB-IoT-based VRLA battery internal temperature prediction (VBITP) algorithm is developed. As mentioned earlier, NB-IoT is one of the most suitable mobile network technologies for IoT applications, which need exceptionally deep coverage and extremely low power consumption. These applications, such as smart metering, usually require low data rates and moderate reaction times of a few seconds. The first network deployment began in late 2017 and global commercial deployment started in 2018. NB-IoT can be deployed inside Long-term Evolution (LTE) carrier, in the LTE guard band and as a standalone solution. In March 2019, the Global Mobile Suppliers Association announced that more than 100 operators have deployed/launched either NB-IoT or LTE-M networks [23]. The VBITP, utilizing neural network and NB-IoT, does not fall into either of the mentioned categories. It consists of three parts, that is, measurement of input parameters, data transmission by NB-IoT network, and establishment of a prediction model. Two external parameters, namely ambient temperature (AT) and input current (IC), are measured and saved as the input. The output of the prediction model, internal temperature (IT) is measured at the same time. A dataset including the two inputs and the output is thus established.

Based on computational intelligence techniques, a recurrent neural network, called nonlinear autoregressive exogenous (NARX) neural network, is applied to find the potential relationship between the input and output. As one kind of the recurrent neural network, NARX considers the continuous change in time series of the input. In this work, the value changes of ET and IC are both associated with the time. Thus, by taking advantage of this, NARX is applied to train the prediction model. In the application, the measured data will be transmitted to the backend server for decision making. The details of the NB-IoT infrastructure will be explained later.

The contributions can be summarized as follows:

1. A VRLA battery internal temperature prediction (VBITP) algorithm is developed to effectively monitor the internal temperature of the VRLA battery.
2. NARX neural network is involved in the proposed model as a novel method to find the relationship between the input parameters and internal temperature.
3. NB-IoT system has been implemented.

The rest of this sub-chapter is organized as follows. In Sect. 5.2, related work on VRLA temperature prediction is reviewed and the proposed VBITP algorithm is introduced. In Sect. 5.3, the NB-IoT system is described. Section 5.4 describes the implemented experiments and the results are analyzed and discussed. In Sect. 5.5, a conclusion is made.

5.2   Overview on Intelligent-Based Approach

The diagram of the proposed methodology in this study is shown in Fig. 5.8.

This section consists of three parts, namely, NARX neural network, model establishment, and model validation. In the first part, the basic structure of the NARX neural network will be described. In the second part, the extraction of the input feature and the establishment of the prediction model will be introduced. Finally, in the third part, the validation method will be defined to verify the effectiveness of the developed model.

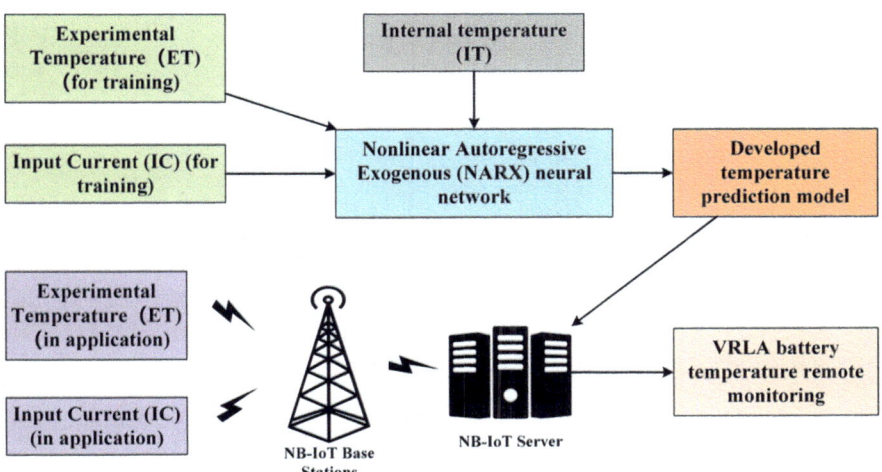

Fig. 5.8 The diagram of VBITP methodology

5.2.1 Nonlinear Autoregressive Exogenous (NARX) Neural Network

NARX is an extension of Autoregressive Exogenous (ARX), which is widely used in a linear system for analyzing [24] and generally modeling a variety of non-linear systems. It is a kind of nonlinear dynamic neural network used for prediction purposes. Neural networks are mathematical tools stimulated by the biological neural system, which have a powerful capacity in learning, storing, and recalling information. They are black-box modeling tools that map the low dimensional input space to the high dimensional output space for nonlinear mapping when the relationship between the input space and the output space is unknown. The nonlinear problems in low dimensions become linear in high dimension, which decreases the complexity and difficulty. To be specific, NARX belongs to a recurrent neural network (RNN). RNN considers the relationship between the current input and previous one and thus good at dealing with a prediction problem of time series; composed of an input layer, output layer, feedback layer, and multiple hidden layers. In the NARX neural network model, the function of the feedback layer node is to store the output value of the output layer node at the previous moment.

The structure of NARX neural network is shown in Fig. 5.9:

The mathematical model is formulated as Eqs. (5.6)–(5.9) [25].

$$x_1(k) = f\left[w^1 u(k) + w^c x_c(k)\right] \tag{5.6}$$

$$x_i(k) = f\left[w^i x_{i-1}(k)\right], \quad i = 1, 2, \ldots, s \tag{5.7}$$

$$x_c(k) = y(k-1) \tag{5.8}$$

$$y(k) = g\left[w^{s+1} x_s(k)\right] \tag{5.9}$$

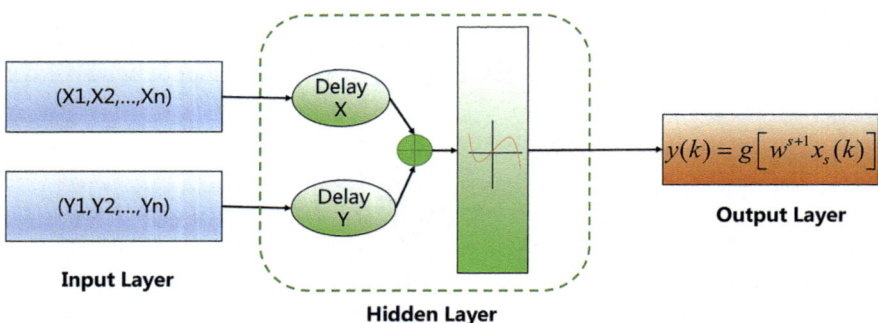

Fig. 5.9 Structure of nonlinear autoregressive exogenous (NARX) neural network

In Eq. (5.6), w^i is the weight matrix of connections between hidden layers. w^c is the weight matrix connecting the feedback layer and the hidden layer $u(k)$ is the input of the neural network at time k. In Eqs. (5.6) and (5.7), $x_i(k)$ and $x_c(k)$ are the outputs of the feedback layer and the hidden layer. In Eq. (5.8), $y(k)$ is the output of the output layer. In Eqs. (5.7) and (5.9), s is the hidden layer.

5.2.2 Model Establishment and Validation

To validate the performance of the developed prediction model, the mean absolute percentage error (MAPE) is selected as an indicator of the model. It can be formulated as in Eq. (5.10):

$$\text{MAPE} = \frac{1}{n} \sum_{i=1}^{n} \frac{\left| y_{t,i} - y_{p,i} \right|}{y_{t,i}} \times 100\% \tag{5.10}$$

In Eq. (5.10), $y_{t,i}$ represents the true value of output (internal temperature) and $y_{p,i}$ represents the predictive value given by the model. The MAPE reflects the relative error in the prediction process to measure the performance of the model.

In the model training phase, tenfold cross-validation is applied. The dataset is divided into ten equal groups randomly. In each turn, nine groups are used as the training data and the other one is used for validation. The choice of training and validation data in each turn should not be repeated. The average MAPE in these ten turns will be identified as the final result and reflect the performance of the model.

5.3 NB-IoT System

As mentioned before, NB-IoT is developed under the specification of LTE [26]. Information flow in the NB-IoT-based intelligent system is shown in Fig. 5.10.

Fig. 5.10 Data transmission based on NB-IoT

 **Fig. 5.11** NB-IoT data characteristics

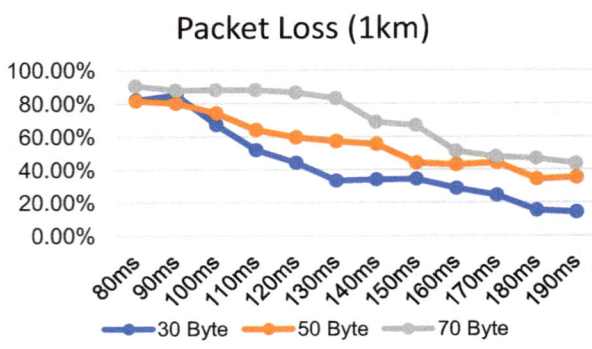

In this investigation, the battery temperature data will be carried back to the backend server through NB-IoT. NB-IoT data delivery is shown in Fig. 5.11.

5.4   Experiments and Results

In the experiments, the scope of input data, AT, and IC are limited to simulate the real situation in the VRLA application. The steps to carry out the task to evaluate the battery internal temperature remotely based on NB-IoT are summarized as below:

- Step 1: Obtaining VRLA Internal Temperature (IT)
- Step 2: Measuring Ambient Temperature (AT) and Input Current (IC)
- Step 3: Data transmission by NB-IoT
- Step 4: Data analysis by NARX algorithm
- Step 5: Establishing the VRLA evaluation model

Devices used in the experiment are shown in Fig. 5.12:

The apparatus consists of resistance temperature detectors (RTD), data logger, and battery test system.

Considering the normal range of VRLA temperature, four typical ambient temperature value, 15, 25, 35, and 45 °C are considered in the experiments. The experiment will be conducted at these four ATs. In each AT phase, a repeatable current pulse set is introduced to act as pseudo-random frequency pulses. Three typical values of IC are also settled as 12.5 A, 15 A, 20 A. The output of IT is measured synchronously. The range of IT is corresponding to the ET of the battery. In each phase, IC increases as the IC is floating and charging the battery. These three sets of data are used for training the prediction model. Tenfold cross-validation is applied to validate the performance. As shown in Fig. 5.13, the curve of target value and output is very close, which can barely be distinguished. The curve below shows the error rate of the established model. The average error rate is about 0.04. Regarding the final example, some real experiments are carried out to verify the proposed narrowband internet of thing based VRLA battery internal temperature prediction

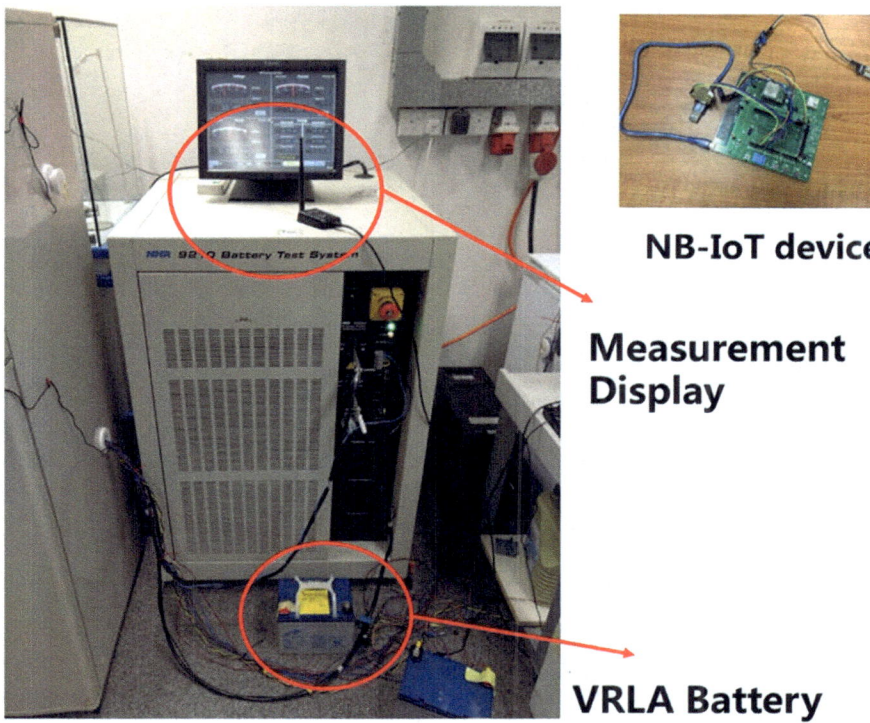

Fig. 5.12 Measurement apparatus used

algorithm to conclude all thermal reactions such as Joule heating effect, entropy changes due to H_2SO_4 reaction, and the water cycle in VRLA battery. The uncontrollable self-heating problem, thermal runaway, can be easily avoided by using the prediction as to the precaution. The NB-IoT VBIPT can predict the battery's temperature without any thermal knowledge but simulate the thermal runaway. Regarding the temperature range, it is agreed that 45 °C ambient temperature is the highest temperature that should go on to prevent any danger from happening. Also, the battery's temperature that rises to more than 55 °C after the experiment should be considered as the threshold of becoming a thermal runaway. So that the NB-IoT VBIPT is capable to prevent self-heating problem.

5.5 Conclusions and Future Work

In this chapter, a narrowband internet of thing (NB-IoT)-based valve-regulated lead acid (VRLA) battery internal temperature prediction (VBITP) algorithm is developed to monitor the internal temperature of VRLA battery. VBITP could provide

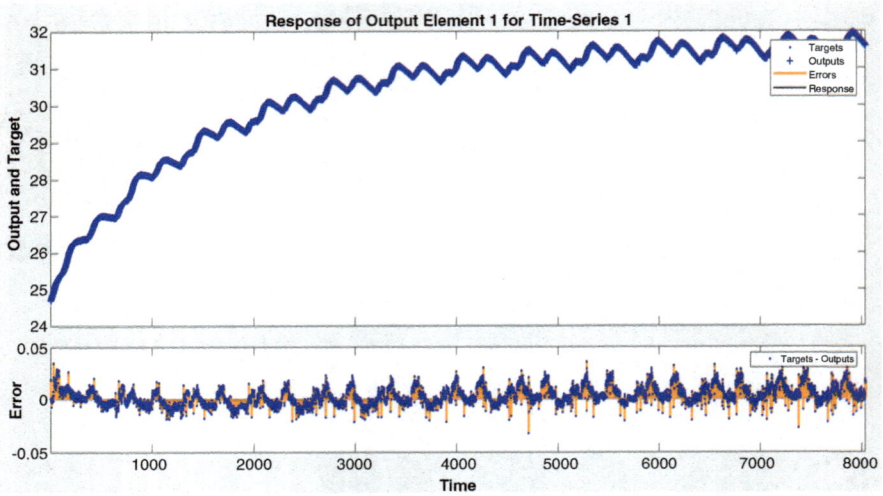

Fig. 5.13 The comparison between output and target (top) and the MAPE result (bottom)

early warning of VRLA battery temperature and thus prevent potential hazards in applications. Different from traditional methods, IoT networks, and neural networks are involved in VBITP to predict the internal battery temperature of VRLA battery based on two external parameters, namely, input current and ambient temperature. By taking the advantages of the NARX neural network, the prediction model in VBITP could predict internal battery temperature with excellent performance. The NB-IoT system sent the measured data back to the server and battery temperature is monitored and an alert could be activated when overheating occurs. The experimental results show that the error rate of the temperature prediction model is no more than 0.04. In the future, this algorithm will be applied to the study of lithium battery internal temperature. At the same time, a new training structure could be developed to improve the performance further.

Acknowledgments   The permission in using the materials from the following paper is very much appreciated.

A. Hao Wang, Lai Tsz Chun, Yang Wei, Yucheng Liu, Kim Fung Tsang, Chun Sing Lai and Loi Lei Lai, A narrowband internet of thing connected temperature prediction for valve regulated lead acid, in *International Conference on Applied Energy 2019*, 12–15 August 2019, Västerås, Sweden, Paper ID: 204

References

1. Y.D. Beyene et al., NB-IoT technology overview and experience from cloud-RAN implementation. IEEE Wirel. Commun. **24**(3), 26–32 (2017)
2. M. Chen, Y. Miao, Y. Hao, K. Hwang, Narrow band internet of things. IEEE Access **5**, 20557–20577 (2017)

3. S. Popli, R.K. Jha, S. Jain, A survey on energy efficient narrowband internet of things (NBIoT): architecture, application and challenges. IEEE Access **7**, 16739–16776 (2019)
4. J. Xu, J. Yao, L. Wang, Z. Ming, K. Wu, L. Chen, Narrowband internet of things: evolutions, technologies, and open issues. IEEE Internet Things J. **5**(3), 1449–1462 (2018)
5. B. Martinez, F. Adelantado, A. Bartoli, X. Vilajosana, Exploring the performance boundaries of NB-IoT. IEEE Internet Things J. **6**(3), 5702–5712 (2019)
6. [SPEC] 3GPP TR 45.820—cellular system support for ultra-low complexity and low throughput Internet of Things (CIoT)
7. R.S. Sinha, Y. Wei, S.-H. Hwang, A survey on LPWA technology: LoRa and NB-IoT. ICT Express **3**(1), 14–21 (2017)
8. M. El Soussi, P. Zand, F. Pasveer, G. Dolmans, Evaluating the performance of eMTC and NB-IoT for smart city applications, in *2018 IEEE International Conference on Communications (ICC)* (Kansas City, 2018), pp. 1–7
9. X. Guo, B. Liu, L. Wang, Design and implementation of intelligent manhole cover monitoring system based on NB-IoT, in *2019 International Conference on Robots & Intelligent System (ICRIS)* (Haikou, 2019), pp. 207–210
10. G. Jia, G. Han, H. Rao, L. Shu, Edge computing-based intelligent manhole cover management system for smart cities. IEEE Internet Things J. **5**(3), 1648–1656 (2018)
11. N.S. Živic, O. Ur-Rehman, C. Ruland, Evolution of smart metering systems, in *2015 23rd Telecommunications Forum Telfor (TELFOR)*, Belgrade, 2015, pp. 635–638
12. Global stationary lead-acid (SLA) battery market research report 2019, MarketResearchNest, March 2019 [Online], https://www.marketresearchnest.com/Global-Stationary-Lead-Acid-SLA-Battery-Market-Research-Report-2019.html. (Visited on 1 May 2019)
13. E. McKenna, M. McManus, S. Cooper, M. Thomson, Economic and environmental impact of lead-acid batteries in grid-connected domestic PV systems. Appl. Energy **104**, 239–249 (2013)
14. C. Glaize, S. Genies, *Lead-Nickel Electrochemical Batteries* (Wiley, New York, 2012)
15. C.S. Lai, Y. Jia, L.L. Lai, Z. Xu, M.D. McCulloch, K.P. Wong, A comprehensive review on large-scale photovoltaic system with applications of electrical energy storage. Renew. Sust. Energ. Rev. **78**, 439–451 (2017)
16. J. Yang, C. Hu, H. Wang, K. Yang, J.B. Liu, H. Yan, Review on the research of failure modes and mechanism for lead–acid batteries. Int. J. Energy Res. **41**(3), 336–352 (2017)
17. G. Kujundžić, Š. Ileš, J. Matuško, M. Vašak, Optimal charging of valve-regulated lead-acid batteries based on model predictive control. Appl. Energy **187**, 189–202 (2017)
18. D. Berndt, E. Meissner, W. Rusch, Aging effects in valve-regulated lead-acid batteries, in *Proceedings of Intelec 93: 15th International Telecommunications Energy Conference*, vol 2 (IEEE, 1993), pp. 139–145
19. C.S. Lai, Y.W. Jia, Z. Xu, L.L. Lai, X.C. Li, J. Cao, M.D. McCulloch, Levelized cost of electricity for photovoltaic/biogas power plant hybrid system with electrical energy storage degradation costs. Energy Convers. Manag. **153**, 34–47 (2017)
20. D. Valkovska, M. Dimitrov, T. Todorov, D. Pavlov, Thermal behavior of VRLA battery during closed oxygen cycle operation. J. Power Sources **191**(1), 119–126 (2009)
21. Z. Li, J. Zhang, B. Wu, J. Huang, Examining temporal and spatial variations of internal temperature in large-format laminated battery with embedded thermocouples. J. Power Sources **241**, 536–553 (2013)
22. Z. Liu, H.-X. Li, A spatiotemporal estimation method for temperature distribution in lithium-ion batteries. IEEE Trans. Ind. Inf. **10**(4), 2300–2307 (2014)
23. https://en.wikipedia.org/wiki/Narrowband_IoT. (Visited on 1 May 2019)
24. M.A. Arain, H.V.H. Ayala, M.A. Ansari, Nonlinear system identification using neural network, in *International Multi Topic Conference* (Springer, 2012), pp. 122–131
25. T. Lin, B.G. Horne, P. Tino, C.L. Giles, Learning long-term dependencies in NARX recurrent neural networks. IEEE Trans. Neural Netw. **7**(6), 1329–1338 (1996)
26. H. Malik, H. Pervaiz, M. Mahtab Alam, Y. Le Moullec, A. Kuusik, M. Ali Imran, Radio resource management scheme in NB-IoT systems. IEEE Access **6**, 15051–15064 (2018). https://doi.org/10.1109/ACCESS.2018.2812299

Chapter 6
Health Detection Scheme for Drunk Drivers

6.1 Introduction

Concluded by the World Health Organization (WHO), 70% of the population over the world can be protected by giving drunk drivers harsh punishments [1]. Annually, more than 1 million people and nearly 50 million are dead and injured, respectively, because of traffic accidents [2]. This leads to a heavy burden to the medical services and it also leads to an expenditure of around $500 billion [3]. As reported by the WHO, traffic accidents potentially turn into the fifth primary cause of death [2]. One of the main reasons leading to traffic accidents is drunk driving. Drunk drivers can be found in 40% of the total traffic accidents and drunk-related traffic accidents cost 22% of the total expenditures [4]. Therefore, it is worth developing drunk driver detection (DDD) to reduce the losses from drunk-related traffic accidents. DDD could contribute to protecting the public and decreasing the related costs. Around 0.02 of Blood Alcohol Concentration (BAC) level will probably lead to the loss of judgment. DDD should be able to provide early warning once the drivers are classified as in drunk status. These early warnings can protect the public including drivers and pedestrians. Basically, there are three types of DDD approaches, namely, direct approach, vehicle-based approach [5, 6], and bio signal-based approach [7]. Direct approaches are widely adopted. These approaches require to collect drivers' breaths, blood, or urines and then detect the drivers' BACs from the collected samples. The vehicle-based approaches mainly detect the differences in the drivers' behavior between normal cases and drunk cases. If a large variation is detected, the driver will be classified as drunk. However, these two approaches can hardly provide real-time, automatic detection, and early warning at the same time. However, the bio-signal based approaches can achieve this. The plethysmogram signal was used to detect the variations in organ volume and the corresponding status of the drivers [7]. But the drawback is that the method requires a long processing time. Among the bio-signals, electrocardiogram (ECG) and electroencephalography (EEG) have

C. S. Lai et al., *Smart Grids and Big Data Analytics for Smart Cities*, https://doi.org/10.1007/978-3-030-52155-4_6

been proven to provide the status of the human timely. As compared to EEG, ECG is easier for wearable applications implementation.

Five main waves can be provided from the ECG signals, namely, P, Q, R, S, and T waves and these waves relate to the dedicated electrical activities of the human heart. In the future, more and more wearable ECG sensors will be available in the market and this will facilitate the development of ECG-based DDD.

6.2   Cardiovascular Diseases Classifier

Electrocardiogram (ECG) signals are important information for cardiovascular disease diagnosis conducted by cardiologists. Such a diagnosis requires the development of cardiovascular disease classifier (CDC). Generally, a CDC mainly comprises feature vectors extraction and machine learning algorithms like an Artificial Neural Network or Support Vector Machine. Features can be divided into three categories, that is non-fiducial features, fiducial features, and hybrid features. Non-fiducial features normally refer to features that do not characterize the ECG signals using P waves, Q waves, S waves, QRS complexes, and T waves [8–12], and vice versa for fiducial features [13, 14]. Hybrid features refer to feature vectors constructed by both non-fiducial and fiducial features [15–17]. In this investigation, a Support Vector Machine (SVM) is used to construct the CDC for the four most common types of cardiovascular diseases, namely bundle branch block, myocardial infarction, heart failure, and dysrhythmia. Seven criteria, including overall accuracy (OA), sensitivity (S_e), specificity (S_p), area under the curve (AUC), training time (T_r), testing time (T_e), and number of features (N_f), which are features to indicate the speed and accuracy of detection, are used as the essential parameters to compute the analytic hierarchy process (AHP) score to aid the multiple criteria decision analysis (MCDA) for the evaluation of the optimal CDC. Traditional work usually aims at the highest overall accuracy and/or lowest testing time. In reality, every end-user has to specify the weights between criteria. It is not uncommon to find a ratio setting by quick understanding or simply adopting a direct 1:1 assignment. It is noted that the needs of volunteers are neglected or not required. In the new method, assignments of criteria are devised for AHP analysis. The incorporation of AHP analysis in the classifier enables the consideration of the needs of the volunteers.

6.2.1   Design of the Optimal CDC

Figure 6.1 summarizes the block diagram of the new method. After the retrieval of ECG data, feature vectors are extracted. The SVM classifiers are then designed based on the feature combinations. Therefore, N configurations can be obtained. The best model is selected among configuration f_1 to configuration f_N based on seven criteria, namely overall accuracy, sensitivity, specificity, area under the curve,

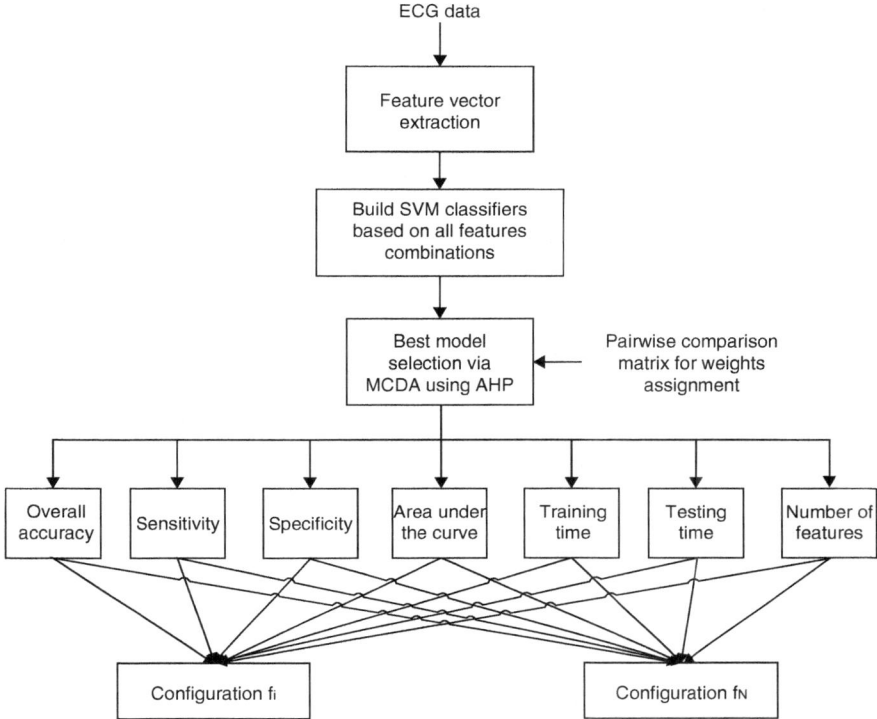

Fig. 6.1 Block diagram of the new method

training time, testing time, and number of features, with the aid of MCDA via AHP. The details of the new method are illustrated in the following figure.

6.2.2 Data Pre-processing and Features Construction

The data is obtained from an online and open-access database [18, 19]. A group of healthy candidates, as well as candidates with the four most common types of cardiovascular diseases, are selected. They are 52 candidates from health control, 15 bundle branch block candidates, 148 myocardial infarction candidates, 18 heart failure candidates, and 14 dysrhythmia candidates. The unequal sample size in each class will lead to a bias of the SVM classifier [20]. The ECG signal is further partitioned into 30 s sub-signals to obtain 500 samples of healthy candidates and 125 samples of unhealthy candidates for each type of cardiovascular disease. This process aims at equalizing the number of samples in each healthy/unhealthy class. Before the introduction of these four diseases, the notations are briefed. Denote RR-interval to be the consecutive R points between consecutive ECG signals, the

QRS complex is the time between Q wave and S wave where point R is between Q wave and S wave. Similarly, the QT interval refers to the time between point Q wave and T wave. The background of these four diseases is presented as follows:

1. Myocardial infarction: Irregular heartbeat and thus irregular RR-interval may occur in the ECG signal of the patients [21];
2. Bundle branch block: Patients have QRS complex with a value exceeding 0.12 ms [22];
3. Dysrhythmia: The heartbeat can be more than 100 beats per minute or less than 60 beats per minute. Thus, RR-interval is different from the normal ECG signal. Also, the QT interval may increase if the type of cardiovascular disease is ventricular arrhythmias [23];
4. Heart failure: A finding of prolonged QT interval in the ECG signals of the patients [24].

As a result, Q wave, R wave and S wave, QRS complex, and RR-interval are representative features to identify between healthy persons versus cardiovascular patients. The feature vector consists of ten features using the average and standard deviation of these five parameters. Before detecting and computing the features, the ECG signals will undergo data pre-processing [25]. The maximum frequency of an ECG signal is typically less than 60 Hz, thus a bandpass filter with cut-off frequencies at 1 and 60 Hz is implemented. A derivative filter is then applied to sharpen the Q, R, and S wave. Finally, signal squaring and sliding window integration are utilized for the location of Q, R, and S wave.

6.2.3   Cardiovascular Diseases Classifier Construction

The CDC is constructed by employing SVM with a ten-dimensional feature vector. This algorithm uses a Lagrange Multiplier with a set of support vectors, a set of weighting, and an offset bias [26, 27]. This section focuses on the design of the CDC. The performance of the CDC is dictated by OA, S_e, S_p, AUC, T_r, T_e, and N_f. It directly classifies the ECG signal into healthy (negative response) candidates and unhealthy (positive response) candidates. OA, S_e, S_p, and AUC are related to the accuracy of the CDC. T_r is the time required to train the CDC and T_e is the time needed to detect the ECG signal. In this investigation, CDC will be trained up and validated with the ECG datasets. For the analysis of positive response—Class 0, 500 healthy patients are used. For the analysis of positive response—Class 1, 125 bundle branch block patients, 125 myocardial infarction patients, 125 heart failure patients, and 125 dysrhythmia patients are retrieved from the database. Table 6.1 lists the datasets for the CDC with a binary classifier.

The CDC will use tenfold cross-validation for performance evaluation [28] and the polynomial kernel function (third-order) is used for SVM analysis. There is a total of 1023 combinations $\left(\sum_{n=110}^{10} C_n \right)$, thus 1023 configurations can be formulated

Table 6.1 Database specification of ECG data for CDC

Class 0 (healthy/negative response)	Number of samples	Class 1 (unhealthy/positive response)	Number of samples
PTB diagnostic (healthy)	500	Bundle branch block	125
		Myocardial infarction	125
		Heart failure	125
		Dysrhythmia	125

Table 6.2 CDC of each configuration

f_j	OA	S_e	S_p	AUC	T_r (S)	T_e (S)	N_f
f_1	0.324	0.350	0.298	0.321	3.5	2.3	1
f_2	0.310	0.324	0.296	0.303	3.4	2.5	1
f_3	0.298	0.288	0.308	0.287	3.6	2.4	1
...	...	...	...	...	...	...	...
f_{1021}	0.986	0.988	0.984	0.972	4.9	3.4	10
f_{1022}	0.964	0.970	0.958	0.946	**5.1**	3.4	10
f_{1022}	0.970	0.974	0.966	0.949	4.3	3.5	10

from a selection (from 1 to 10) of the ten features. For the jth configuration where $j = 1,\ldots,1023$, namely f_j, its corresponding criteria, OA, S_e, S_p, AUC, T_r, T_e, and N_f are recorded. The main settings of SVM are summarized as follows, in general, the default setting is adopted in the MATLAB toolbox:

1. Number of classes: Two;
2. Class 0: 500 healthy candidates;
   Class 1: 125 bundle branch block candidates, 125 myocardial infarction candidates, 125 heart failure candidates, and 125 dysrhythmia candidates;
3. Feature vector: The maximum dimensionality is 10, which consists of Q wave average, Q wave standard deviation, R wave average, R wave standard deviation, S wave average, S wave standard deviation, QRS complex average, QRS complex standard deviation, RR-interval mean, and RR-interval standard deviation;
4. Kernel function: third-order polynomial;
5. Fold of cross-validation: Tenfold 1023 classifiers are constructed in 1023 configurations; the results are tabulated in Table 6.2.

6.3   Multiple Criteria Decision Analysis of the Optimal CDC

In Table 6.2, seven criteria, namely OA, S_e, S_p, AUC, T_r, T_e, and N_f, are employed for performance evaluation of the 1023 scenarios. Multiple criteria decision making (MCDM) has been utilized in many areas since the 1990s [29]. It entails using the particular characteristics of cardiovascular diseases. By allocating appropriate weightings, the analytic hierarchy process (AHP) is adopted to evaluate and analyze the best scenarios among the 1023 scenarios investigated. The allocation of weightings confronts the feedback from an AHP analysis of 200 volunteers from which a pairwise comparison 7×7 matrix Am ($m = 1, \ldots, 200$) is formulated. It is intuitively understood that T_e should be as low as possible and that the accuracy should be kept to an acceptable level. Since the speed of detection is essential, the analysis on MCDA reveals that high weightings should be assigned to OA, S_e, S_p, AUC, T_e. These five parameters are referred to as the primary parameters. While N_f is typically preferred to be small for speedy detection, it is noted that T_r will not affect the detection time. Hence N_f and T_r are classified as the secondary parameters.

The volunteers are required to fill in the $a_{m,ij}$, where i and j are between 1 and 7, in Table 6.3. The AHP based MCDA CDC is referred to as the new classifier (NC). Traditional classifiers (TC) in [10, 14, 15] are also evaluated. Both the NC and the TC are applied to the three feature groups, namely, non-fiducially features, fiducially features, and hybrid features in [10, 14, 15]. The performance comparison between the NC and the TC is tabulated in Table 6.4. Based on the discussion for AHP formulation, the assignment of values of $a_{m,ij}$ are based on the following guidelines:

1. Write 1 if there is equal importance of i and j
2. Write 3 if i is slightly more important than j
3. Write 5 if i is more important than j
4. Write 7 if i is strongly more important than j
5. Write 9 if i is absolutely more important than j

The pairwise comparison 7×7 matrix A_m is then normalized, and an Anorm$_m$ can be obtained by modifying the matrix entries $a_{m,ij}$ in A_m into matrix entries anorm$_{m,ij}$ in Anorm$_m$:

$$\text{anorm}_{m,ij} = \frac{a_{m,ij}}{\sum_{l=1}^{7} a_{m,lj}} \tag{6.1}$$

By averaging each row of Eq. (6.1), the corresponding 7×1 priority matrix w_m with entries $w_{m,k}$ for $k = 1,\ldots,7$ is given by:

Table 6.3 Pairwise comparison 7×7 matrix A_m

	OA	S_e	S_p	AUC	T_r	T_e	N_f
OA	1	$a_m,12$	$a_m,13$	$a_m,14$	$a_m,15$	$a_m,16$	$a_m,17$
S_e	$a_m,21$	1	$a_m,2$	$a_m,24$	$a_m,25$	$a_m,26$	$a_m,27$
S_p	$a_m,31$	$a_m,32$	1	$a_m,34$	$a_m,35$	$a_m,36$	$a_m,37$
AUC	$a_m,41$	$a_m,42$	$a_m,43$	1	$a_m,45$	$a_m,46$	$a_m,47$
T_r	$a_m,51$	$a_m,52$	$a_m,53$	$a_m,54$	1	$a_m,56$	$a_m,57$
T_e	$a_m,61$	$a_m,62$	$a_m,63$	$a_m,64$	$a_m,65$	1	$a_m,67$
Nr	$a_m,71$	$a_m,72$	$a_m,73$	$a_m,74$	$a_m,75$	$a_m,76$	1

Table 6.4 Performance of NC versus TC

Method	Datasets (number of samples)	Features	Results (related work TC)	Results (new work NC)
Two-layered Hidden Markov Model [3]	MIT-BIH database (34,799 samples from 16 Arrhythmia candidates)	P-R interval, QRS complex interval and T sub-wave interval	OA = 0.992	OA = 0.987
			S_e = 0.993	S_e = 0.99
			S_p = 0.992	S_p = 0.984
			AUC = 0.971	AUC = 0.966
			T_r = 3.7 s	T_r = 3.4 s
			T_e = 2.7 s	T_e = 1.9 s
			N_f = 3	N_f = 2
Cross wavelet transform with a threshold based classifier [7]	The PTB Diagnostic ECG database (18,489 samples from 52 healthy control candidates and 148 myocardial infarction candidates)	Total sum of wavelet cross spectrum value and total sum of wavelet coherence	OA = 0.976	OA = 0.966
			S_e = 0.973	S_e = 0.978
			S_p = 0.988	S_p = 0.958
			AUC = 0.949	AUC = 0.933
			T_r = 6.2 s	T_r = 5.6 s
			T_e = 4.1 s	T_e = 2.8 s
			N_f = 6	N_f = 4
SVM [8]	CU database, VF database, and AHA database (40,956 samples from 67 Ventricular fibrillation and rapid ventricular tachycardia candidates)	Leakage, count 1, count 2, count 3, A1, A2, A3, time delay, FSMN, cover bin, frequency bin, kurtosis, and complexity	OA = 0.952	OA = 0.947
			S_e = 0.951	S_e = 0.952
			S_p = 0.951	S_p = 0.942
			AUC = 0.943	AUC = 0.937
			T_r = 4.8 s	T_r = 4.5 s
			T_e = 2.7 s	T_e = 1.6 s
			N_f = 13	N_f = 10

$$w_{m,k} = \frac{1}{7}\sum_{l=1}^{7} \text{anorm}_{m,kl} \tag{6.2}$$

Let $C_{p,q}$, ($p = 1,\dots,7$ and $q = 1,\dots,1023$) be the pth criteria, and qth scenario of CDC. $C_{p,q}$ is normalized to become $C_{p,q,\text{norm}}$. The final score for each scenario, AHP_q, is evaluated by:

$$\text{AHP}_q = \sum_{l=1}^{7} C_{p,q,norm}\left(\frac{1}{200}\sum_{m=1}^{200} w_{m,l}\right) \tag{6.3}$$

To avoid inconsistency in the construction of pairwise comparison matrices, the optimal CDC is concluded from the highest value of AHP_q [30]. It is evaluated that the optimal CDC is obtained from scenario f_{652}, with feature vector composes of average of Q, standard deviation of Q, standard deviation of S, average of QRS mean, standard deviation of QRS, average of RR-interval, and standard deviation of RR-interval, with AHP_{652} as follows: OA = 0.988, S_e = 0.992, S_p = 0.985, AUC = 0.982, T_r = 4.5 s, T_e = 2.8 s, N_f = 7.

6.4   AHP Scores and Analysis

The performance scores between the NC and the TC [10, 14, 15] are evaluated and tabulated in Table 6.4. In this investigation, the algorithms in related work have been evaluated, with the addition of MCDA using AHP to obtain the best scenario by assigning weights to the seven criteria. As the new work and related works are in the same application area, the classification of cardiovascular diseases, the weight assignment can be reused to facilitate performance comparisons. From Table 6.4, the percentage changes are evaluated as follows:

1. Percentage change compared with AHP scores from [10]: OA = −0.504%, S_e = −0.302%, S_p = −0.807%, AUC = −0.515%, T_r = −8.109%, T_e = −29.630%, and N_f = −33.333%. It is concluded that there is an improvement of 30% in speed of detection of cardiovascular diseases @ ~99.5% accuracy.
2. Percentage change compared with AHP scores from [14]: OA = −1.025%, S_e = 0.514%, S_p = −3.036%, AUC = −1.686%, T_r = −9.677%, T_e = −31.707%, and N_f = −33.333%. It is concluded that there is an improvement of 30% in speed of detection of cardiovascular diseases @ ~99% accuracy.
3. Percentage change compared with AHP scores from [15]: OA = −0.525%, S_e = 0.105%, S_p = −0.946%, AUC = −0.636%, T_r = −6.250%, T_e = −40.741%, and N_f = −23.077%. It is concluded that there is an improvement of 40% in speed of detection of cardiovascular diseases @ ~99.5% accuracy.

The analysis reveals that in NC, the speed of detection has been increased by 30–40% while the accuracy is retained at ~99–99.5% of the TC. It is seen that there the reduction of OA, S_e, and S_p are less than 1%. Thus, the AHP based MCDA CDC is a reliable and speedy detection scheme for cardiovascular diseases.

To collect ECG data, the ECG sensors will be implemented to convert the raw data into meaningful representation. There are four stages in the development of ECG wearable sensors, which are pre-amplification stage, filtering stage, tertiary amplification and DC high-pass filtering, and voltage level shifter. Then, the support vector machine (SVM) will be used to classify two classes, normal situation and drunk situation. The ECG samples of the normal situation and drunk situation will be first collected and then pre-processed. After that, feature extraction will be performed and the extracted features will then be utilized to construct the kernel function for the classifier. The kernel function responses to transform the data into high-dimensional space.

The result demonstrates that the accuracy of proposed DDD achieves a satisfying accuracy compared to those conventional methods. Besides, using ECG-based detection can realize early detection and fully automated detection.

6.5   Development of EDG-Based Drunk Driver Detection

To collect ECG data, the ECG sensors will be implemented to convert the raw data into meaningful representation. There are four stages in the development of ECG wearable sensors, which are pre-amplification stage, filtering stage, tertiary amplification and DC high-pass filtering, and voltage level shifter. Then, the support vector machine (SVM) will be used to classify two classes, normal situation and drunk situation. The ECG samples of the normal situation and drunk situation will be first collected and then pre-processed. After that, feature extraction will be performed and the extracted features will then be utilized to construct the kernel function for the classifier. The kernel function responses to transform the data into high-dimensional space.

The result demonstrates that the accuracy of proposed DDD achieves a satisfying accuracy compared to those conventional methods. Besides, using ECG-based detection can realize early detection and fully automated detection.

The ECG sensor front-end consists of four stages, preamplification, filtering, tertiary amplification and DC-offset high-pass filtering, and voltage level shifter. The front-end is responsible to convert the raw ECG data into meaningful representation and attenuate noises and interferences. After ECG pre-processing, the classifier will be developed using the collected data. Feature extraction will then be performed and utilized in the kernel functions for building classifiers. To test the DDD classifiers' performance, a widely adopted method, K fold cross-validation, is performed [31]. The overall accuracy is recorded.

6.5.1   ECG Sensors Implementations

Instrumentation amplifier is necessary for the ECG sensor design. The instrumentation amplifier is the combination of the differential amplifier with two non-inverting amplifiers as buffering input. Therefore, the input impedance matching can be neglected. The instrumentation amplifier has a high common-mode rejection ratio which means that the whole amplifier will just amplify the difference of the input without amplifying the input noise. This is an important feature especially for the bio-signal with tinny amplitude such as ECG signal. Also, the low DC input and voltage noise can be achieved by comparing it with the typical operational amplifier. There are four stages [32] that can be found in the front-end of ECG sensors. The operation is explained below:

Table 6.5 Energy spectral of
ECG signal

Useful components	Noises and interferences
P-wave: 0.6–5 Hz	Muscle noise: 5–50 Hz
T-wave: 1–7 Hz	Respiratory noise: 0.12–0.5 Hz
QRS complex: 10–15 Hz	Line frequency: 50 or 60 Hz
	Human DC offset: 0.01–0.4 Hz

6.5.1.1 Stage 1: Pre-amplification Stage

The gain of the pre-amplify stage is chosen to be 50, which is based on the consideration of making a suitable range of the input ECG signal. The input should be within the power supply range of ±9 V without any collapsing and allows for the next stage filtering.

6.5.1.2 Stage 2: Filtering Stage

The relative energy spectra of the ECG signal after the Fourier transform is summarized in Table 6.5 [31].

There are a variety of components with various frequency ranges that coexist in the ECG signals. Therefore, it is required to determine the cut-off frequencies of the band-pass filter in order to filter out the unnecessary frequency components. The frequency range was chosen from 0.6 to 15.9 Hz which contains P, Q, R, S, T waves of the ECG signals. This frequency range can exclude most of the unwanted noises in the ECG signals. Then, the first order passive type low-pass filter will be cascaded with the first order active type low pass filter to form the second-order filter. The higher order of filter can be more likely to be formed by cascading another passive filter since the active filter is constructed from the non-inverting operational amplifier with the defined input impedance.

6.5.1.3 Stage 3: Tertiary Amplification and DC-Offset High Pass Filtering

The gain is designed to be controllable to adapt to various situations and provide a convenient measurement. Based on the dedicated frequency range selection from the ECG signal power spectrum evaluation, the first-order high-pass filter will be constructed in a passive way with the controllable gain and behind the cut-off frequency 0.7 Hz.

6.5.1.4 Stage 4: Voltage Level Shifter

The level shifter is to make positive feedback to the voltage level of the positive input and shift the voltage level up when the DC-offset is found at the negative input. The design of the voltage level shifter should be adapted to the variable input of the DC-offset input so that suitable DC-offset can be implied to shift the entire ECG signal level up.

6.5.2   Drunk Driving Detection Algorithm

To classify whether the driver is drunk or not, two sets of data, normal ECG signals, and drunk ECG signals, are collected using the ECG sensor. The collected data will be used to develop the drunk driving classifier using a support vector machine (SVM). SVM is a famous learning machine for data analysis and classification. A high dimensional space can be obtained after training a set of data. In this case, there are two sets of data. Hence, after training, these two sets of data will be separated in the resultant space as far as possible. They are normally separated by hyperplane which is used to classify the input data to the class it belongs to. For the input information with low dimensionality, kernel function has been adopted to transform the input information to higher dimensionality for classification. Sampling will be performed on the ECG signals. The sampling data points will be the feature and they will be utilized to customize the kernel function and classifier. A large margin between the two classes (normal and drunk) would be required. SVM is a kind of machine learning algorithm to recognize patterns and classify the unknown input data to the appropriate category. Therefore, Lagrangian dual optimization is considered for maximizing the margin distance using SVM to solve the maximum margin problem [32]:

$$L(\alpha) = \arg\max_{\alpha} \left\{ \sum_{i=1}^{N} \alpha_i - \frac{1}{2} \sum_{i=1}^{N} \sum_{j=1}^{N} \alpha_i \alpha_j S_i S_j K(X_j, X_j) \right\} \qquad (6.4)$$

Subject to

$$\alpha_i \geq 0, \quad i = 1, \ldots, N$$

$$\sum_{i=1}^{N} \alpha_i S_i = 0$$

where α is defined as the Lagrange multiplier, s belongs to $\{1,-1\}$ which is the class label of input ECG signals, $K(x_i, x_j)$ denotes the kernel function which is used to transform the input data to the desired high-dimensional feature dimension.

6.6 Result Comparisons

The performance of the ECG-based DDD classifier was evaluated by using the K-fold validation. First, all ECG samples will be divided into K groups. During a fold, one group will be selected to evaluate the classifier and the rest of the groups will be used to train the classifier. The process will repeat for $K - 1$ times and all samples will be evaluated. Also, the accuracy of the classifier will be calculated at each fold and the overall accuracy will be obtained by averaging them. Normally, the accuracy can be calculated by taking an average of sensitivity and specificity where sensitivity depends on true positive ratio, and specificity is determined by the true negative ratio [33].

The comparisons between the proposed work and other methods are summarized in Fig. 6.2.

In real-life applications, early detection and fully automated detection are important considerations. For the direct methods, they cannot meet the requirement of real-time protection and fully automated detection as they involve manpower to monitor the data collection process. For the vehicle-based method, it cannot provide early detection since it is determined by the changes of the vehicle motions. As such, only a bio-signal-based method can fulfill the requirement of DDD since it can achieve simultaneous monitoring on the drivers and so early detection and fully automated detection can be provided.

6.7 Human Status Detection Scheme

In 2015, the United Nations (UN) announced 17 Global Goals to achieve a better world [35]. The Global Goals were adopted by world leaders. As such, a sub-goal 3.6 aims at halving the number of road traffic injuries and death by 2020 [35]. It is found that the total number of road traffic injuries and death is more than 50 million annually [36]. The annual expenditure of the injuries is more than $500 billion [36].

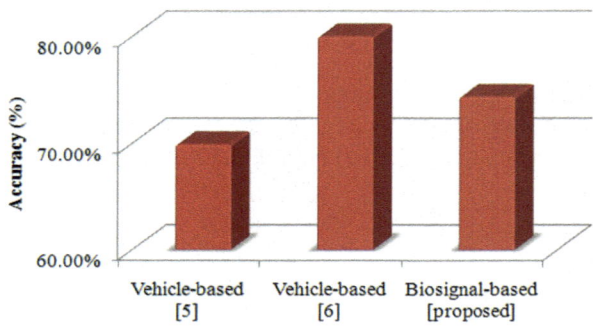

Fig. 6.2 Comparing DDD with other methods [34]

Besides, traffic accidents are the top leading cause of death in the age group of 15–29 years. It is predicted that traffic accidents will become the seventh leading cause of death by 2030 if there is no prevention scheme taking place urgently [37]. Figures 6.3 and 6.4 give the statistics of drowsy driving and statistics of drunk driving.

Real-time monitoring of the status of the human and giving early alert could be considered as the most effective method of preventing traffic accidents. For example, if a candidate is identified as abnormal (such as the candidate is drunk) before driving, prohibiting an engine start protects all other drivers and pedestrians. The reviews on traffic accidents indicated that drowsy driving and drunk driving are two major causes of traffic accidents. More than half of professional drivers felt sleepy and more than 30% of drivers fell asleep while driving [38]. The situation of drunk driving is more serious such that 30% of total traffic accidents involve drunk drivers [39]. Nearly one person is killed by drunk driving every hour. Therefore, a real-time detection scheme on both drowsy driving and drunk driving renders a significant reduction in traffic injuries and deaths. Roughly estimated $50 billion can be saved from the expenditure of traffic accidents.

Figure 6.5 gives a traditional drunk driving test. The conventional detection schemes are divided into three types, that is, (1) image-based detection [40], (2) behavior-based detection [5, 41], and (3) bio-signal-based detection [42, 43].

It is worth noting that image-based detection and behavior-based detection cannot achieve the purposes of providing pre-warning before driving and high measurement stability at the same time while bio-signal-based detection does. Image-based detections identify the features of drivers' head motion and eye blinking, etc. with the use of image processing. But image-based approaches are usually unstable and low stability in practical situations [44]. Behavior-based approaches compare driving behavior under normal conditions and abnormal conditions. The driving behavior reflects on the vehicle moving path such as lateral position, change in velocity, turning angle, etc. In other words, these approaches cannot provide pre-warning to abnormal drivers before they drive. Recently, wireless and wearable healthcare sensors have been raised in the market. The wearable healthcare sensors measuring bio-signals are inspired by bio-signal-based detection schemes [42, 43]. Among all bio-signals, a survey on nonintrusive driver assistance systems reported that electrocardiogram (ECG) has the highest stability on real-time measurement [44]. Therefore, in this chapter, a proposed smart scheme for studying and analyzing

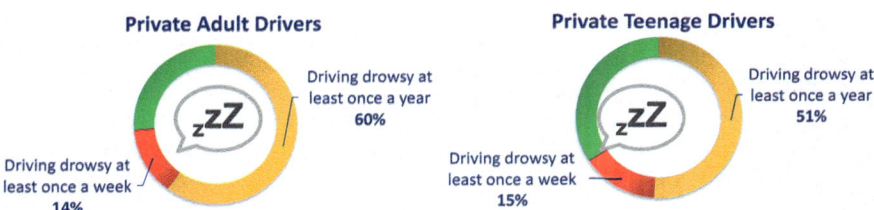

Fig. 6.3 Statistics of drowsy driving

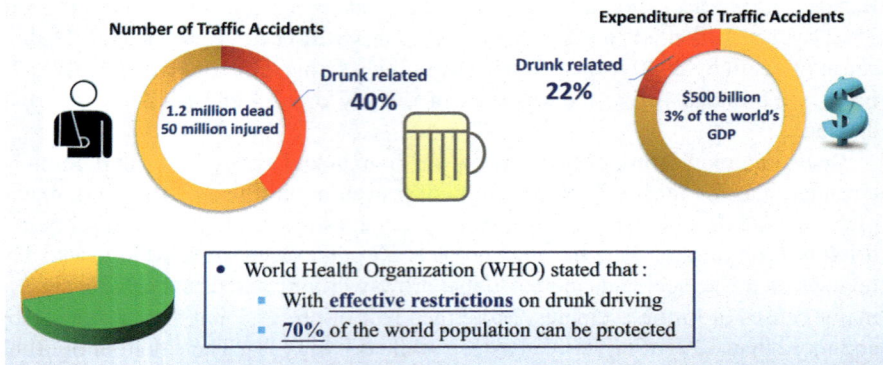

Fig. 6.4 Statistics of drunk driving

Fig. 6.5 Traditional drunk driving test

ECG signals for the detection is shown in Fig. 6.6. Figure 6.7 shows the impacts of the smart detection scheme.

The proposed ECG-based status of the human detection (ECG-HSD) consists of four stages including (1) signal pre-processing for ECG data, (2) feature extraction and building classifier, (3) multiple criteria decision making (MCDM), and (4) *K*-fold validation. The similarities of ECG samples of different statuses of the human are extracted as the feature vector. Then, the feature vector is weighted with respect to the importance of data points. After that, since the dimensionality of the feature vector affects detection performance, MCDM is applied to select the best classifier by creating a number of scenarios with various feature dimensions. The results demonstrated that the accuracy of the ECG-HSD scheme acquires satisfying accuracy of ~90% and a short testing time of ~5 s. Figure 6.8 shows some devices for electrocardiograms.

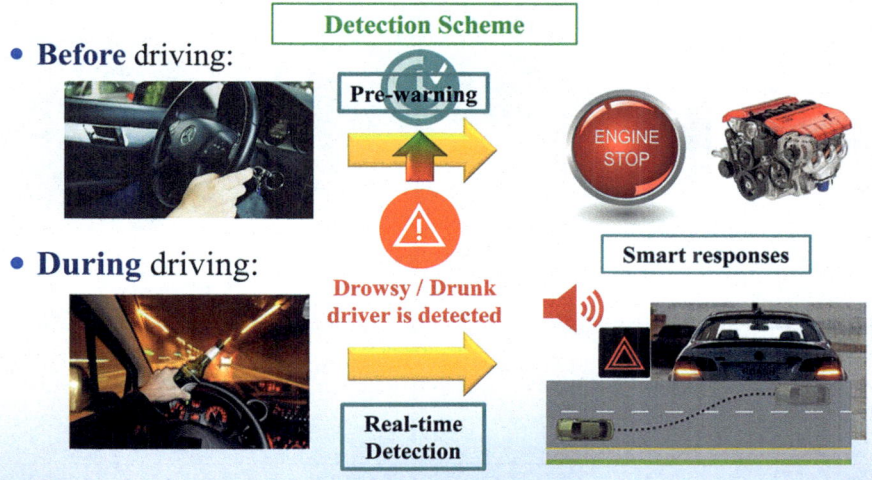

Fig. 6.6   Smart detection scheme on road safety

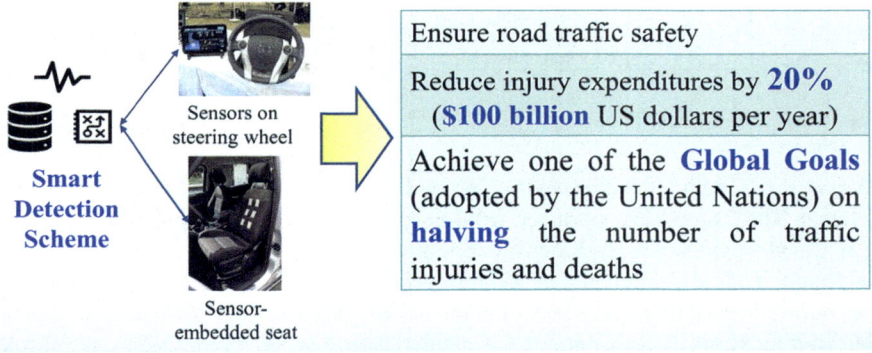

Fig. 6.7   Impacts of the smart detection scheme

6.8   ECG-Based Drunk Driver Detection Design

Figure 6.9 shows the development flow of the proposed ECG-based human status detection (ECG-HSD).

Three human conditions are considered in this section, namely Class 0: Normal, Class 1: Drowsy, and Class 2: Drunk. Four stages are categorized in the development of the ECG-HSD scheme. The first stage is signal pre-processing for ECG samples. At this stage, raw ECG signals used in training consist of noise, interference, and offset and so they cannot be directly used in the training. Signal pre-processing is carried out for noise suppression and ECG sample segmentation. The

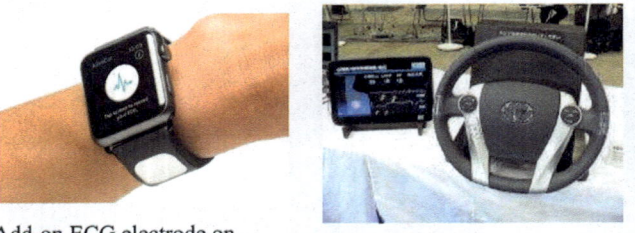

Add-on ECG electrode on
Smart Watch

ECG-embedded steering wheel

ECG-embedded seat

Fig. 6.8 Devices for electrocardiogram (ECG) measurement

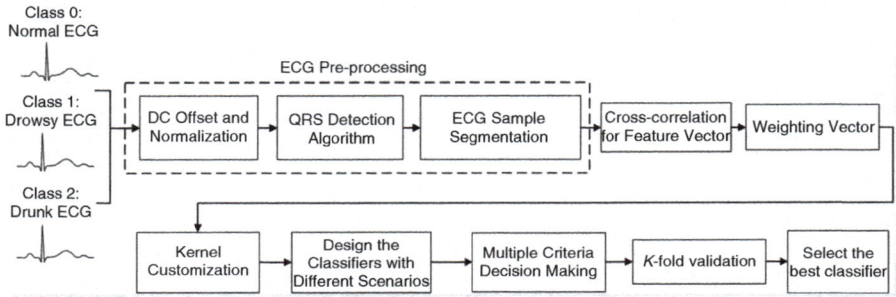

Fig. 6.9 Flowchart of proposed ECG-based drowsy driver detection scheme

second stage includes feature extraction and building classifiers for the ECG-HSD scheme. The features are extracted from the segmented ECG samples at stage 1 and then transformed to a high-dimensional feature vector for classification. To improve detection accuracy, independent weighting factors are assigned to all features during building classifiers. The assignment of features is based on multiple criteria decision making (MCDM) and it is considered as stage 3. At the last stage, K-fold cross-validations are made and evaluate the performance of the classifiers. The classifier with the best overall performance is selected for the ECG-HSD scheme.

6.8.1   Stage 1: Signal Pre-processing for ECG Data

The training ECG signals collected from body sensors are usually interrupted with noises and interferences. Therefore, signal pre-processing is necessary to suppress the noises and interferences and divides the whole ECG signals into multiple ECG samples in one heartbeat duration. As a result, feature extraction can be extracted from ECG samples in stage 2. It is worth pointing out that ECG contains five main peaks generated by P, Q, R, S, and T waves. Each wave represents an electrical signal transmitted to various heart muscles. In most cases, heart-related conditions and diseases can be revealed by examining those ECG peaks. So, they are regarded as peak-of-interest. There are three steps of ECG pre-processing as explained below:

6.8.1.1   Step 1: Bandpass Filter

Normally, the largest peak of the ECG signal is located within the QRS complex. Locating the QRS complex facilitates ECG segmentation. Theoretically, QRS complex is usually found from 5 to 15 Hz of the ECG frequency spectrum. Based on this ECG characteristic, a low pass filter with a high cut-off frequency of 11 Hz and a high pass filter with a low cut-off frequency of 5 Hz are cascaded to form the bandpass filter. The frequency components outside the range-of-interest such as muscle noise, cable noise, and wave interferences could be filtered.

6.8.1.2   Step 2: Derivative Filter

An effective way to determine the peaks of interest is to search the turning points and this can be achieved by using differentiation. As such, a five-point derivative filter is used to determine the change of slope on ECG signal with a short time interval. The change of slope represents turning points in the QRS complex and so the locations of Q, R, and S peaks could be determined.

6.8.1.3   Step 3: Squaring and Moving Window Integration

Change of slope is not sufficient to determine the QRS complex as the slope can be varied due to several factors such as heart condition and signal noises. To improve the determination of QRS complex, squaring and moving window integration is used. First, all data points are turned into positive values by squaring. Then, moving window integration is carried out to sort out more parameters, such as the interval between two waves, for determining the QRS complex. Finally, multiple QRS complexes can be determined using sorted parameters and the change of slopes.

ECG samples with one heartbeat duration are sorted out using a bandpass filter, derivative filter, and moving window integration. By applying different thresholds with respect to the typical values of ECG waves, the amplitudes, durations, and intervals of P, Q, R, S, and T waves can be found. Figures 6.10 and 6.11 show the pre-processing of ECG signal in the time domain and frequency domain, respectively.

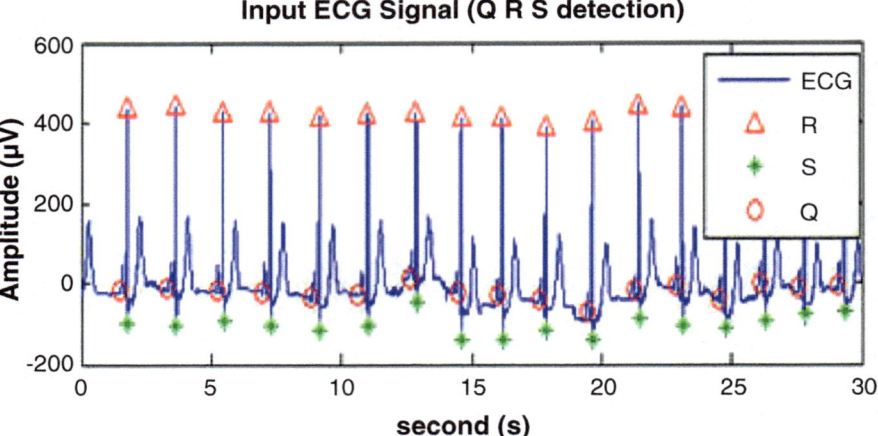

Fig. 6.10 Pre-processing of ECG signal in the time domain

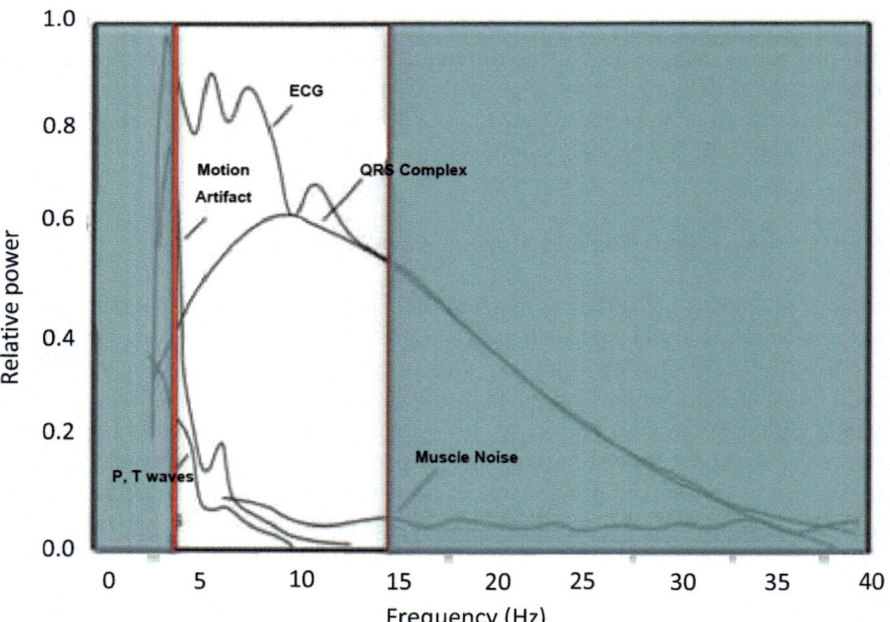

Fig. 6.11 Pre-processing of ECG signal in the frequency domain

6.8.2   *Stage 2: Feature Extraction and Building Classifiers*

The purpose of feature extraction is to sort out the characteristics from a signal which can be found in most situations and demonstrate differently in various situations. For example, the heartbeat rate varies for a person if s/he is under rest or exercise. As such, heartbeat rate could be one of the features in determining if a person is under rest or exercise. Similarly, the detection scheme could consider the status of the human as normal, drowsy, or drunk. As mentioned earlier, ECG consists of P, Q, R, S, and T waves. However, in most practical situations, no ideal ECG waveform can be measured using ECG sensors. It will affect the captured amplitudes of ECG waves and so the detection accuracy is directly affected by the waves as they are extracted as features.

In this section, the similarity of ECG signals is extracted to form a feature vector. As such, the similarity of ECG signals is measured by cross-correlation, which is a method of measuring symmetric levels between two signals. The resultant cross-correlation coefficients strongly reveal the similarity between the ECG signals between normal, drowsy, and drunk. The cross-correlation coefficient CC_{ij} between the ith ECG sample S_i and the jth ECG sample S_j is formulated as follows [10]:

$$CC_{ij}(k) = \begin{cases} \sum_{n=k}^{T-1} S_i(n) S_j(n-k), & k \geq 0 \\ \sum_{n=0}^{T-|k|-1} S_i(n) S_j(n-k), & k < 0 \end{cases} \qquad (6.5)$$

where T is the total length (number of sampling points) of an ECG sample.

In most of the problems, the data with different classes are not linearly separable. Non-linear threshold plane separating various classes increases complexity and decreases classification accuracy. Kernel trick is a widely-adopted method in transforming low-dimensional input data into high dimensional feature space. Linear hyperplanes separating different classes exist at a certain number of dimensions.

Besides, it is found that some data points on ECG signals experience larger changes at different conditions. Weighting vector $W_{ij} = [w_{ij,1} \ w_{ij,2} \ ... w_{ij,2T-1}]$ is assigned into the cross-correlation coefficient during kernel development in order to improve detection accuracy by highlighting those important data points. The weighted kernel coefficient $KC_{i,j}$ is obtained from the equation as shown below:

$$KC_{i,j} = \sum_{k=0}^{2T-1} w_{ij,k} CC_{ij}(k) \qquad (6.6)$$

After that, kernel matrix is built with weighting kernel coefficient and is expressed as:

$$K_{xcorr} = \begin{bmatrix} KC_{1,1} & \cdots & KC_{1,N_t} \\ \vdots & \ddots & \vdots \\ KC_{N_t,1} & \cdots & KC_{N_t,N_t} \end{bmatrix} \qquad (6.7)$$

The hyperplanes are the planes linearly separating the data with different classes. The data points which are closest to the hyperplanes are defined as supporting vector. The separation between the supporting vectors and the hyperplanes is defined as margin. To achieve the highest classification accuracy, the margin is expected to be maximized and this deduces an optimizing maximum margin problem. The customized equation is expressed as follows:

$$\tilde{M}(\alpha,W) = \arg\max \sum_{i=1}^{N} \alpha_i - \frac{1}{2} \sum_{i=1}^{N} \sum_{j=1}^{N} \alpha_i \alpha_j b_i b_j (K_{HSD}) \qquad (6.8)$$

s.t.

$$\text{s.t.} \begin{cases} \alpha_i \geq 0, \sum_{i=1}^{N} \alpha_i b_i = 0, \quad \text{with} \quad i = 1,\ldots,N \\[2em] \sum_{n=1}^{2T-1} w_{ij,n} = 1 \end{cases}$$

where α is the Lagrange multiplier and b is a class label. K_{HSD} is the kernel function obtained in Eqs. (6.6) and (6.7). As shown from the equation, the weighting vector W will be optimized while solving the maximum margin problem. Figures 6.12, 6.13, and 6.14 show data sampling, design for cross-correlation coefficient, and design for kernel coefficient respectively.

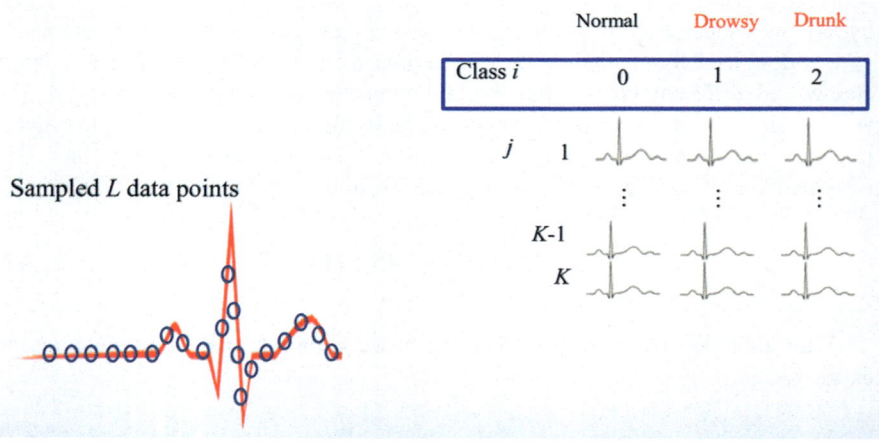

Fig. 6.12 Data sampling

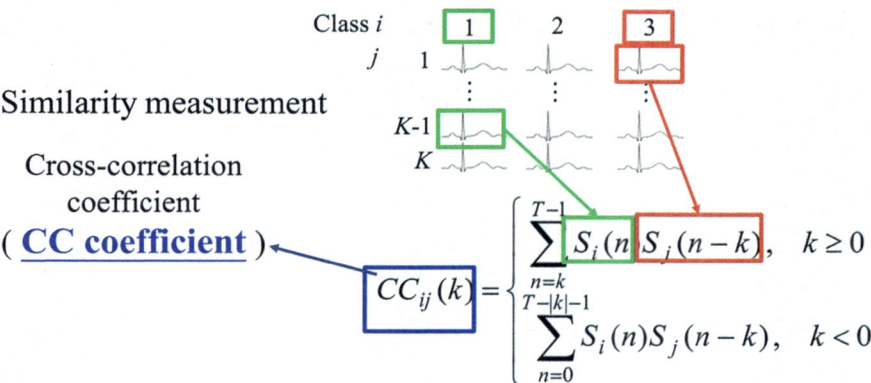

Fig. 6.13 Design for cross-correlation coefficient

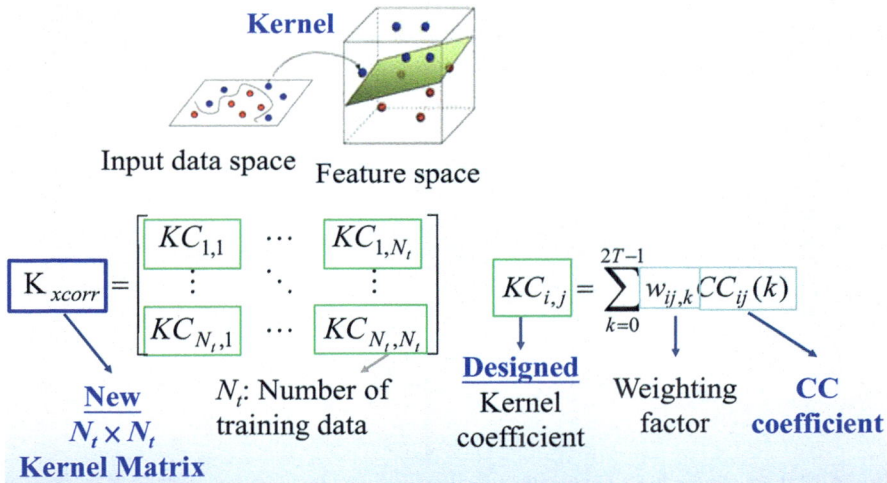

Fig. 6.14 Design for kernel coefficient

6.8.3   Stage 3: Multiple Criteria Decision Making (MCDM)

The dimensionality of the feature vector affects computational cost. In this case, taking more data points during cross-correlation will result in longer training time and testing time. Testing time could be one of the critical considerations for some applications. Therefore, the feature dimensionality is analyzed by adjusting the sampling rate on ECG samples. The trade-off between accuracy and testing time is solved by using multiple criteria decision making MCDM. Figures 6.15, 6.16, and 6.17 show

$\widetilde{M}(\alpha, W) = \arg\max \sum_{i=1}^{N} \alpha_i - \frac{1}{2} \sum_{i=1}^{N} \sum_{j=1}^{N} \alpha_i \alpha_j b_i b_j (K_{HSD})$
s.t. $\alpha_i \geq 0, \sum_{i=1}^{N} \alpha_i b_i = 0, with \quad i = 1, \ldots, N$
$\sum_{n=1}^{2T-1} w_{ij,n} = 1$
where α is the Lagrange multiplier
margin
hyperplane

Fig. 6.15 Classifier design

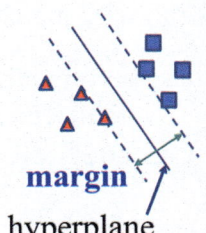

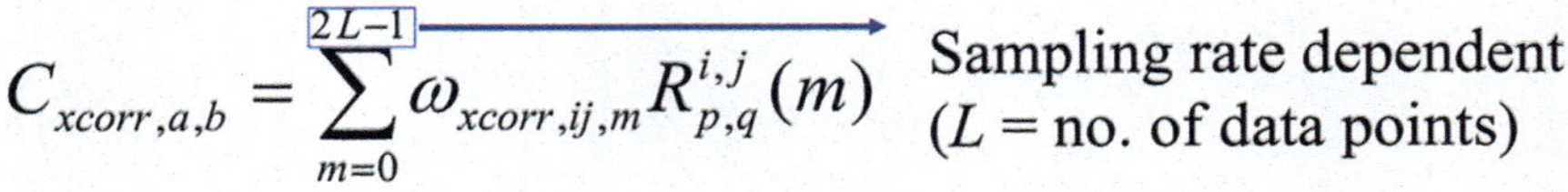

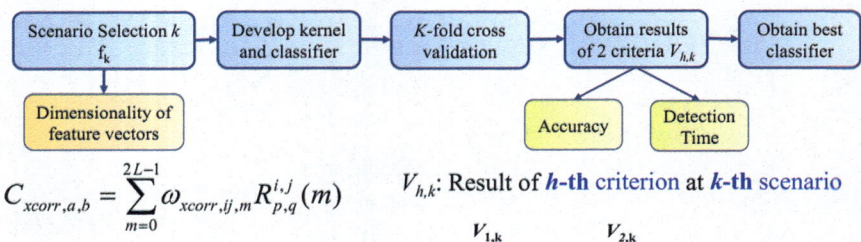

$C_{xcorr,a,b} = \sum_{m=0}^{2L-1} \omega_{xcorr,ij,m} R_{p,q}^{i,j}(m)$
Sampling rate dependent
(L = no. of data points)

Fig. 6.16 Trade-off of classifier

Scenario Selection k f_k
Develop kernel and classifier
K-fold cross validation
Obtain results of 2 criteria $V_{h,k}$
Obtain best classifier
Dimensionality of feature vectors
Accuracy
Detection Time
$C_{xcorr,a,b} = \sum_{m=0}^{2L-1} \omega_{xcorr,ij,m} R_{p,q}^{i,j}(m)$
$V_{h,k}$: Result of h-th criterion at k-th scenario
$V_{1,k}$
$V_{2,k}$

Fig. 6.17 Logic for multiple criteria decision making

the classifier design, trade-off of the classifier, and logic for multiple criteria decision-making, respectively.

First, by adjusting the sampling rate, it results in a number of feature vectors with different dimensionalities ranged from 200 to 5000. Then, a number of weighting kernels and classifiers are built by using the feature vectors with different dimensionalities. Each classifier is considered as a scenario with respect to the feature dimensionality.

The classifier performance is evaluated using K-fold cross-validation. Two criteria, namely accuracy and testing time are considered. The overall performance scoring (OPS) of the pth classifier is computed as:

$$\text{OPS}_p = \sum_q V_q \frac{S_{p,q}}{\sum_l S_{p,l}} \tag{6.9}$$

where $S_{p,q}$ is the numerical value for the pth classifier in the qth criterion. V_q is the weighting value for the qth criterion.

6.8.4   Stage 4: K-Fold Cross-Validation

K-fold cross-validation is a widely adopted method for training and evaluating the performance of a classifier [45]. At the initialization phase, all available data are divided into K sets randomly. At the first fold of validation, $K - 1$ data sets, considered as training sets, are picked up for training classifiers, and the remaining data set, considered as a validation set, is used to validate a classifier. At the next fold, the prior validation set will become a training set and it will not be selected for validating classifiers for the rest of the folds. One of the previous training sets is selected to be the validation set. The procedure repeats until all K folds are completed. The resultant performance of a classifier is an average result of all folds.

6.9   Performance Evaluation of ECG-HSD Scheme

There are two key factors for the detection schemes embedded in wearable, mobile, and light-weight devices. The first one is the detection accuracy and the second one is testing time. Early warning can be provided to users once an abnormal condition is identified in a short period. Therefore, MCDM was applied to determine the relative best classifier with respect to accuracy and testing time. Note that similar approaches can be applied for considering more criteria such as computational cost, complexity, sensitivity, and reliability.

The proposed ECG-HSD scheme is compared to other existing works and the evaluation is summarized in Figs. 6.18 and 6.19. As shown in the figures, the proposed ECG-HSD scheme achieved satisfying accuracy in both drowsy and drunk detections as compared to those individual detection schemes. Since most detection schemes have not mentioned about the testing time, so no comparison is given for this. For the proposed ECG-HSD scheme, the testing time is less than 7 s including the measurement time of ECG signals and algorithm processing time. It is sufficient to provide real-time protection to drivers.

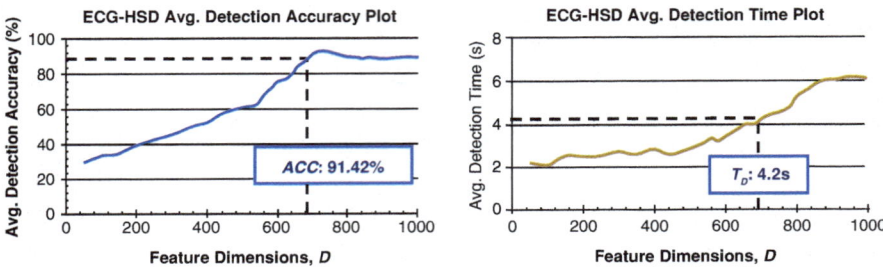

Fig. 6.18 Classification performance evaluation

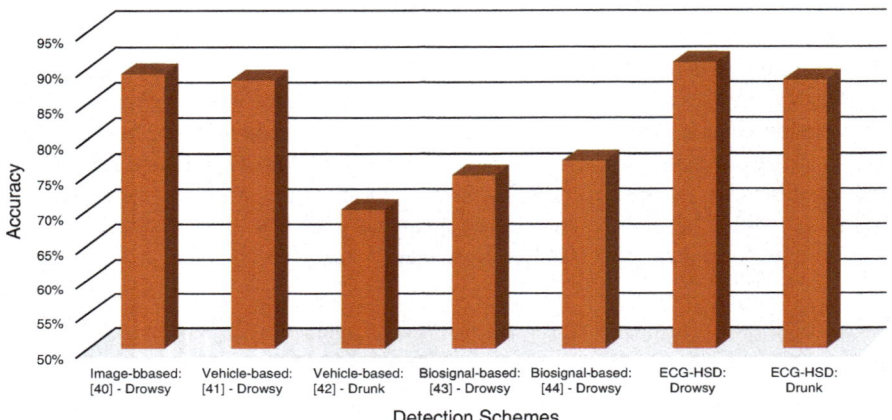

Fig. 6.19 Comparisons among different schemes

6.10   Conclusions

An optimal cardiovascular disease classifier (CDC) has been implemented by using an analytic hierarchy process (AHP) to facilitate multiple criteria decision analysis (MCDA). The four most common types of cardiovascular diseases, namely bundle branch block, myocardial infarction, heart failure, and dysrhythmia are considered. Analysis and comparison with previous works show that the speed of detecting cardiovascular diseases has been increased by 30–40% while the accuracy is retained at ~99–99.5% of traditional classifiers. In conclusion, the AHP based MCDA CDC is a reliable and speedy detection scheme for cardiovascular diseases.

A drunk driving detection scheme has been designed with ECG wearable sensor front-end developed for converting the raw ECG data into meaningful representation and attenuates the noises and interferences. The collected data will be utilized in the classifier development. The testing results demonstrate good accuracy compared to the existing method and it can meet the requirements of early and fully automated detection.

Human condition detection using electrocardiogram (ECG) signal in ECG-HSD scheme is proposed to classify normal, drowsy, and drunk status. During the development of the ECG-HSD classifier, similarities of ECG signals at normal, drowsy, and drunk conditions were extracted to construct feature vector. Then, the important data points on ECG signals were weighted to improve accuracy. The classifier performance related to the feature dimensionality and various scenarios were created by adjusting the number of feature dimension. After that, multiple criteria decision making (MCDM) was carried out to select the best classifiers with respect to accuracy and testing time. The results revealed that the proposed ECG-HSD scheme achieved satisfying accuracy compared to other related works and is suitable for real-time condition monitoring.

Acknowledgments The permission granted in using materials from the following papers is very much appreciated.

A. Wah Ching Lee, Faan Hei Hung, Kim Fung Tsang, Hoi Ching Tung, Wing Hong Lau, Veselin Rakocevic, Loi Lei Lai, A speedy cardiovascular diseases classifier using multiple criteria decision analysis. Sensors, MDPI **15**, 1312–1320 (2015)
B. Cheon Hoi Koo, Hongxu Zhu, Yee Ting Tsang, Tsz Tat Yu, Kim Fung Tsang, Loi Lei Lai, A humans' status detection scheme for industrial safety, in *2018 IEEE 27th International Symposium on Industrial Electronics (ISIE2018)*, 13–15 June 2018, Australia

References

1. World Health Organization, *Global Status Report on Road Safety 2013: Supporting a Decade of Action* (World Health Organization, Geneva, 2013)
2. World Health Organization, *World Report on Road Traffic Injury Prevention* (World Health Organization, Geneva, 2004)
3. K. Dalal, Z. Lin, M. Gifford, L. Svanström, Economics of global burden of road traffic injuries and their relationship with health system variables. Int. J. Prev. Med. **4**, 1442–1450 (2013)
4. Global Road Safety Partnership, *Drinking and Driving: A Road Safety Manual for Decision-Makers and Practitioners* (Global Road Safety Partnership, Geneva, 2007)
5. Z. Li, X. Jin, X. Zhao, Drunk driving detection based on classification of multivariate time series. J. Saf. Res. **54**, 61–67 (2015)
6. J. Dai, J. Teng, X. Bai, Z. Shen, D. Xuan, Mobile phone based drunk driving detection, in *2010 IEEE 4th International Conference on Pervasive Computing Technologies for Healthcare (PervasiveHealth)* (2010), pp. 1–8
7. K. Murata, E. Fujita, S. Kojima, S. Maeda, Y. Ogura, T. Kamei, T. Tsuji, S. Kaneko, M. Yoshizumi, N. Suzuki, Noninvasive biological sensor system for detection of drunk driving. IEEE Trans. Inf. Technol. Biomed. **15**, 19–25 (2011)
8. Y.S. Lee, W.Y. Chung, Visual sensor based abnormal event detection with moving shadow removal in home healthcare applications. Sensors **12**, 573–584 (2012)
9. Y.H. Noh, D.U. Jeong, Implementation of a data packet generator using pattern matching for wearable ECG monitoring systems. Sensors **14**, 12623–12639 (2014)
10. W. Liang, Y. Zhang, J. Tan, Y. Li, A novel approach to ECG classification based upon two layered HMMs in body sensor networks. Sensors **14**, 5994–6011 (2014)

11. T.-P.V. Staa, M. Gulliford, E.S.-W. Ng, B. Goldacre, L. Smeeth, Prediction of cardiovascular risk using Framingham, ASSIGN and QRISK2: how well do they predict individual rather than population risk? PLoS One **9**, 1–10 (2014)
12. J.A. Sanz, M. Galar, A. Jurio, A. Brugos, M. Pagola, H. Bustince, Medical diagnosis of cardiovascular diseases using an interval-valued fuzzy rule-based classification system. Appl. Soft Comput. **20**, 103–111 (2014)
13. K.-K. Tseng, X. He, W.-M. Kung, S.-T. Chen, M. Liao, H.-N. Huang, Wavelet-based watermarking and compression for ECG signals with verification evaluation. Sensors **14**, 3721–3736 (2014)
14. S. Banerjee, M. Mitra, Application of cross wavelet transform for ECG pattern analysis and classification. IEEE Trans. Instrum. Meas. **63**, 326–333 (2014)
15. Q. Li, C. Rajagopalan, G.D. Clifford, Ventricular fibrillation and tachycardia classification using machine learning approach. IEEE Trans. Biomed. Eng. **61**, 1607–1613 (2014)
16. L. Sun, Y. Lu, K. Yang, S. Li, ECG analysis using multiple instance learning for myocardial infarction detection. IEEE Trans. Biomed. Eng. **59**, 3348–3356 (2012)
17. B. Xie, H. Minn, Real-time sleep apnea detection by classifier combination. IEEE Trans. Inf. Technol. Biomed. **16**, 469–477 (2012)
18. The PTB Diagnostic ECG Database, Physionet, http://www.physionet.org/physiobank/database/ptbdb/. Accessed 10 Sept 2014
19. A.L. Goldberger, L.A.N. Amaral, L. Class, J.M. Hausdorff, P.C.H. Ivanov, R.G. Mark, J.E. Mietus, G.B. Moody, C.-K. Peng, H.E. Stanley, PhysioBank, PhysioToolkit, and PhysioNet: components of a new research resource for complex physiologic signals. Circulation **101**, e215–e220 (2000)
20. V.N. Vapnik, *The Nature of Statistical Learning* (Springer, Berlin, 1995)
21. D.L. Kuchar, C.W. Thorburn, N.L. Sammel, Prediction of serious arrhythmic events after myocardial infarction: signal-averaged electrocardiogram, Holter monitoring and radionuclide ventriculography. J. Am. Coll. Cardiol. **9**, 531–538 (1987)
22. M. Rotman, J.H. Triebwasser, A clinical and follow-up study of right and left bundle branch block. Circulation **51**, 477–484 (1975)
23. M.J. Krowka, P.C. Pairolero, V.F. Trastek, W.S. Payne, P.E. Bernatz, Cardiac dysrhythmia following pneumonectomy. Clinical correlates and prognostic significance. Chest **91**, 490–495 (1987)
24. J.S. Gottdiener, A.M. Arnold, G.P. Aurigemma, J.F. Polak, R.P. Tracy, D.W. Kitzman, J.N. Gardin, J.E. Rutledge, R.C. Boineau, Predictors of congestive heart failure in the elderly: the cardiovascular health study. J. Am. Coll. Cardiol. **35**, 1628–1637 (2000)
25. W.J. Tompkins, *Biomedical Digital Signal Processing C-Language Examples and Laboratory Experiments for the IBMPC* (Prentice Hall, Upper Saddle River, 2000), pp. 236–264
26. C.M. Bishop, *Pattern Recognition and Machine Learning* (Springer, Singapore, 2006), pp. 325–343
27. C. Cortes, V. Vapnik, Support-vector networks. Mach. Learn. **20**, 273–297 (1995)
28. G.J. McLachlan, K.A. Do, C. Ambroise, *Analyzing Microarray Gene Expression Data. Supervised Classification of Tissue Samples* (Wiley, New York, 2004), pp. 221–251
29. M. Köksalan, J. Wallenius, S. Zionts, *Multiple Criteria Decision Making: From Early History to the 21st Century* (World Scientific Publishing, Singapore, 2001), pp. 43–62
30. M.S. Ozdemir, Validity and inconsistency in the analytic hierarchy process. Appl. Math. Comput. **161**, 707–720 (2005)
31. W.C. Lee, F.H. Hung, K.F. Tsang, H.C. Tung, W.H. Lau, V. Rakocevic, L.L. Lai, A speedy cardiovascular diseases classifier using multiple criteria decision analysis. Sensors **15**, 1312–1320 (2015)
32. W.J. Tompkins, *Biomedical Digital Signal Processing C-Language Examples and Laboratory Experiments for the IBM®PC* (Prentice Hall, New Jersey, 2000), pp. 245–264
33. W. Zhu, N. Zeng, N. Wang, Sensitivity, specificity, accuracy, associated confidence interval and ROC analysis with practical SAS® implementations, in *NESUG Proceedings: Health Care and Life Sciences* (2010), pp. 1–9

34. C.K. Wu, K.F. Tsang, H.R. Chi, A wearable drunk detection scheme for healthcare applications, in *2016 IEEE 14th International Conference on Industrial Informatics (INDIN)* (2016), pp. 19–21

35. UN General Assembly, Transforming Our World: The 2030 Agenda for Sustainable Development, 21 October 2015

36. World Health Organization, *Global Status Report on Road Safety 2015* (World Health Organization, Geneva, 2015)

37. World Health Organization, Road Traffic Injuries—Fact Sheet, updated May 2017

38. NHTSA, *Drowsy Driving and Automobile Crashes* (Nat. Highway Traffic Safety Admin., Washington, DC, USA, Rep., 2012)

39. Department of Transportation (US), National Highway Traffic Safety Administration (NHTSA), Traffic Safety Facts 2014 Data: Alcohol-Impaired Driving (Washington, 2015)

40. R. Ahmad, J.N. Borole, Drowsy driver identification using eye blink detection. Int. J. Comput. Sci. Inf. Technol. (IJISET) **6**, 270–274 (2015)

41. J.W. Lee, S.K. Lee, C.H. Kim, K.H. Kim, O.C. Kwon, Detection of drowsy driving based on driving information, in *2014 International Conference on Information and Communication Technology Convergence (ICTC)* (2014), pp. 607–608

42. S. Hu, G. Zheng, B. Peters, Driver fatigue detection from electroencephalogram spectrum after electrooculography artefact removal. IET Intell. Transp. Syst. **7**, 105–113 (2013)

43. K.T. Chui, K.F. Tsang, H.R. Chi, C.K. Wu, B.W.K. Ling, Electrocardiogram based classifier for driver drowsiness detection, in *2015 IEEE 13th International Conference on Industrial Informatics (INDIN)* (2015), pp. 600–603

44. Y. Sun, X.B. Yu, An innovative nonintrusive driver assistance system for vital signal monitoring. IEEE J. Biomed. Health Inf **18**, 1932–1939 (2014)

45. C.K. Wu, K.F. Tsang, H.R. Chi, F.H. Hung, A precise drunk driving detection using weighted kernel based on electrocardiogram. Sensors **16**, 659–667 (2016)

Index

A

Absorbent glass mat (AGM), 353
Active network management (ANM), 266
Adaptive Synthetic (ADASYN), 228, 229
AD Biogas power plant, 238
Admission mechanism, 278
Advanced metering infrastructure (AMI), 266,
 270, 307, 308, 310
Ambient temperature (AT), 356
Analytic hierarchy process (AHP), 366,
 370, 388
Annual sizing case, 247
Apache Hadoop, 132
Application layer, 274
Artificial intelligence (AI), 265
Artificial neural network (ANN), 98
Asthma, 145
Augmented Lagrangian relaxation, 31
Automatic meter reading (AMR), 270
Automatic voltage control (AVC), 266

B

Baltimore Gas & Electric (BGE), 138
Bandpass filter, 381
Battery energy storage systems, 176, 177
Behavior-based approaches, 377
Beneficial correlated regularization (BCR), 98,
 99, 102–106, 109, 110
Beta distribution, 187
Big data
 application, 131
 challenges, 130
 cloud computing, 132

clustering methods, 131
data analysis and visualization, 135
and data analytics, 130
data center, 133
data explosion, 135
data mining, 131
data origin, 135
data volumes, 130
EDF, 137
E.ON Metering, 137
Exelon, 138
goals, 136
GrC, 133
IoT, 132
IT tools, 131
KEPCO, 139
optimization methods, 131
PG&E, 138
potential application, 136, 137
quantum computing, 133
security, 134
shortage of talent, 136
in smart grid, 130
standards, 136
technologies and
 architectures, 130
tools, 132
uncertainty, 134
utilities, 134
variety, 130
velocity, 130
Bio-signal based approaches, 365
Bitcoin system, 274
Bit error rate (BER), 322

Printed by Printforce, the Netherlands